As one of the major constituents of plants, lipids have received considerable attention from plant biochemists. Interest has recently intensified and broadened following exciting new discoveries of particular relevance to industry, ranging from the elucidation of new biosynthetic pathways sensitive to herbicide action to the successful genetic manipulation of the quality of food oils. This volume brings together contributions from those working on plant lipid biosynthesis, both with respect to the fundamentals of the process and also on more specifically applied aspects. Together, these contributions provide an up-to-date review of the most exciting recent advances which illustrate that the fundamental and applied approaches are complementary and synergistic.

SOCIETY FOR EXPERIMENTAL BIOLOGY
SEMINAR SERIES: 67

PLANT LIPID BIOSYNTHESIS
FUNDAMENTALS AND AGRICULTURAL APPLICATIONS

SOCIETY FOR EXPERIMENTAL BIOLOGY SEMINAR SERIES

A series of multi-author volumes developed from seminars held by the Society for Experimental Biology. Each volume serves not only as an introductory review of a specific topic, but also introduces the reader to experimental evidence to support the theories and principles discussed, and points the way to new research.

6. Neurones without impulses: their significance for vertebrate and invertebrate systems. *Edited by A. Roberts and B.M.H. Bush*
8. Stomatal physiology. *Edited by P.G. Jarvis and T.A. Mansfield*
16. Gills. *Edited by D.F. Houlihan, J.C. Rankin and T.J. Shuttleworth*
31. Plant canopies: their growth, form and function. *Edited by G. Russell, B. Marshall and P.G. Jarvis*
33. Neurohormones in invertebrates. *Edited by M. Thorndyke and G. Goldsworthy*
34. Acid toxicity and aquatic animals. *Edited by R. Morris, E.W. Taylor, D.J.A. Brown and J.A. Brown*
35. Division and segregation of organelles. *Edited by S.A. Boffey and D. Lloyd*
36. Biomechanics in evolution. *Edited by J.M.V. Rayner and R.J. Wootton*
37. Techniques in comparative respiratory physiology: an experimental approach. *Edited by C.R. Bridges and P.J. Butler*
38. Herbicides and plant metabolism. *Edited by A.D. Dodge*
39. Plants under stress. *Edited by H.C. Jones, T.J. Flowers and M.B. Jones*
40. *In situ* hybridisation: application to developmental biology and medicine. *Edited by N. Harris and D.G. Wilkinson*
41. Physiological strategies for gas exchange and metabolism. *Edited by A.J. Woakes, M.K. Grieshaber and C.R. Bridges*
42. Compartmentation of plant metabolism in non-photosynthesis tissues. *Edited by M.J. Emes*
43. Plant growth: interactions with nutrition and environment. *Edited by J.R. Porter and D.W. Lawlor*
44. Feeding and texture of foods. *Edited by J.F.V. Vincent and P.J. Lillford*
45. Endocytosis, exocytosis and vesicle traffic in plants. *Edited by G.R. Hawes, J.O.D. Coleman and D.E. Evans*
46. Calcium, oxygen radicals and cellular damage. *Edited by C.J. Duncan*
47. Fruit and seed production: aspects of development, environmental physiology and ecology. *Edited by C. Marshall and J. Grace*
48. Perspectives in plant cell recognition. *Edited by J.A. Callow and J.R. Green*
49. Inducible plant proteins: their biochemistry and molecular biology. *Edited by J.L. Wray*
50. Plant organelles: compartmentation of metabolism in photosynthetic cells. *Edited by A.K. Tobin*
51. Oxygen transport in biological systems: modelling of pathways from environment to cell. *Edited by S. Eggington and H.F. Ross*
52. New insights in vertebrate kidney function. *Edited by J.A. Brown, R.J. Balment and J.C. Rankin*
53. Post-translational modifications in plants. *Edited by N.H. Battey, H.G. Dickinson and A.M. Hetherington*
54. Biomechanics and cells. *Edited by F.J. Lyall and A.J. El Haj*
55. Molecular and cellular aspects of plant reproduction. *Edited by R. Scott and A.D. Stead*
56. Amino acids and 'their derivatives in plants. *Edited by R.M. Wallsgrove*
57. Toxicology of aquatic pollution: physiological, cellular and molecular approaches. *Edited by E. W. Taylor*
58. Gene expression and adaptation in aquatic organisms. *Edited by S. Ennion and G. Goldspink*
59. Animals and temperature: phenotypic and evolutionary adaptation. *Edited by I.A. Johnston and A.F. Bennett*
60. Molecular physiology of growth. *Edited by P.T. Loughra and J.M. Pell*
61. Global warming: implications for freshwater and marine fish. *Edited by C. Wood and G. McDonald*
62. Fish stress and health in aquaculture. *Edited by G.K. Iwama, A.D. Pickering, J. Sumpter and C. Schreck*
63. Scaling-up. *Edited by P. van Gardingen, P. Curran and G. Foody*
64. Plants and UV-B: responses to environmental change. *Edited by P. Lumsden*
65. Recent advances in anthropod endocrinology. *Edited by G.M. Coast and S.G. Webster*
66. Cold ocean physiology. *Edited by H-O. Pörtner and R. Playle*

PLANT LIPID BIOSYNTHESIS

FUNDAMENTALS AND AGRICULTURAL APPLICATIONS

Edited by

John L. Harwood

Professor, School of Molecular and Medical Biosciences, University of Wales, Cardiff

PUBLISHED BY THE PRESS SYNDICATE OF THE UNIVERSITY OF CAMBRIDGE
The Pitt Building, Trumpington Street, Cambridge CB2 1RP, United Kingdom

CAMBRIDGE UNIVERSITY PRESS
The Edinburgh Building, Cambridge CB2 2RU, UK http://www.cup.cam.ac.uk
40 West 20th Street, New York, NY 10011-4211, USA http://www.cup.org
10 Stamford Road, Oakleigh, Melbourne 3166, Australia

First published 1998

Printed in the United Kingdom at the University Press, Cambridge

Typeset in Times 10/12 pt [wv]

A catalogue record for this book is available from the British Library

Library of Congress Cataloguing in Publication data

Plant lipid biosynthesis: fundamentals and agricultural applications / edited by John L. Harwood
p. cm. – (Society for Experimental Biology seminar series; 67)
Based on contributions made at the S.E.B. meeting held at the University of Kent in April 1997.
Includes index.
ISBN 0 521 62074 0 (hb)
1. Plant lipids. – 2. Plant lipids – Synthesis – Congresses. I. Harwood, John L. II. Series: Seminar series
(Society for Experimental Biology seminar series (Great Britain)); 67.
QK898.L56P58 1998 572′.572–dc21 98–21343 CIP

ISBN 0 521 62074 0 hardback

Contents

Contributors

ALBAN, C.
Laboratoire Mixte CNRS/Rhône-Poulenc (UM41 associée au Centre National de la Recherche Scientifique), Rhône-Poulenc Agrochimie, 14–20 rue Pierre Baizet, 69263 Lyon Cédex 9, France

BALDOCK, C.
Krebs Institute for Biomolecular Research, Department of Molecular Biology and Biotechnology, The University of Sheffield, Sheffield S10 2TN, UK

BESSOULE, J-J.
Laboratoire de Biogenèse membranaire, CNRS and Université Victor Ségalen Bordeaux, UMR 5544, 146 Rue Léo Saignat, 33076 Bordeaux Cédex, France

BRIGHT, S.W.J.
Zeneca Agrochemicals and Plant Science, Jealott's Hill Research Station, Bracknell, Berks RG42 6ET, UK

BROWSE, J.
Institute of Biological Chemistry, Washington State University, Pullman, WA 99164-6340, USA

CASSAGNE, C.
Laboratoire de Biogenèse membranaire, CNRS and Université Victor Ségalen Bordeaux, UMR 5544, 146 Rue Léo Saignat, 33076 Bordeaux Cédex, France

DOMERGUE, F.
Laboratoire de Biogenèse membranaire, CNRS and Université Victor Ségalen Bordeaux, UMR 5544, 146 Rue Léo Saignat, 33076 Bordeaux Cédex, France

DOUCE, R.
Laboratoire Mixte CNRS/Rhône-Poulenc (UM41 associée au Centre National de la Recherche Scientifique), Rhône-Poulenc Agrochimie, 14–20 rue Pierre Baizet, 69263 Lyon Cédex 9, France

ELBOROUGH, K.M.
Lipid Molecular Biology Group, Department of Biological Sciences, University of Durham, Durham DH1 3LE, UK

FRENTZEN, M.
Universität Hamburg, Institut für Allgemeine Botanik, Ohnhorststr 18, D-22609, Hamburg, Germany

HANLEY, S.Z.
Lipid Molecular Biology Group, Department of Biological Sciences, University of Durham, Durham DH1 3LE, UK

HARWOOD, J.L.
School of Molecular and Medical Biosciences, University of Wales, Cardiff, Cardiff CF1 3US, UK

HAWKES, T.R.
Zeneca Agrochemicals and Plant Science, Jealott's Hill Research Station, Bracknell, Berks RG42 6ET, UK

KINNEY, A.J.
DuPont Experimental Station, PO Box 80402, Wilmington, DE 19880-0402, USA

KNAUF, V.C.
Calgene Inc., 1920 Fifth Street, Daris, CA 95616, USA

KNUTZON, D.S.
Calgene Inc., 1920 Fifth Street, Daris, CA 95616, USA

LESSIRE, R.
Laboratoire de Biogenèse membranaire, CNRS and Université Victor Ségalen Bordeaux, UMR 5544, 146 Rue Léo Saignat, 33076 Bordeaux Cédex, France

LIGHTNER, J.
Institute of Biological Chemistry, Washington State University, Pullman, WA 99164-6340, USA

MOREAU, P.
Laboratoire de Biogenèse membranaire, CNRS and Université Victor Ségalen Bordeaux, UMR 5544, 146 Rue Léo Saignat, 33076 Bordeaux Cédex, France

MURPHY, D.J.
Department of Brassica and Oilseeds Reseach, John Innes Centre, Norwich Research Park, Colney Lane, Norwich NR4 7UH, UK

NAPIER, J.
Institute of Arable Crops Research, Long Ashton Research Station, University of Bristol, Long Ashton, Bristol BS18 9AF, UK

OKULEY, J.
Institute of Biological Chemistry, Washington State University, Pullman, WA 99164-6340, USA

PIFFANELLI, P.
Department of Brassica and Oilseeds Research, John Innes Centre, Norwich Research Park, Colney Lane, Norwich NR4 7UH, UK

QUINN, P.J.
Division of Life Sciences, King's College London, Campden Hill, London W8 7AH, UK

RAFFERTY, J.B.
Krebs Institute for Biomolecular Research, Department of Molecular Biology and Biotechnology, The University of Sheffield, Sheffield S10 2TN, UK

RICE, D.W.
Krebs Institute for Biomolecular Research, Department of Molecular Biology and Biotechnology, The University of Sheffield, Sheffield S10 2TN, UK

SHEWRY, P.R.
Institute of Arable Crops Research, Long Ashton Research Station, University of Bristol, Long Ashton, Bristol BS18 9AF, UK

SLABAS, A.R.
Lipid Molecular Biology Group, Department of Biological Sciences, University of Durham, Durham DH1 3LE, UK

SPYCHALLA, J.
Institute of Biological Chemistry, Washington State University, Pullman, WA 99164-6340, USA

STOBART, A.K.
School of Biology, University of Bristol, Bristol BS8 1UG, UK

STUITJE, A.R.
Department of Genetics, Institute of Molecular Biological Studies (IMBW), Vrije Universiteit, Biocenter Amsterdan, De Boelelaan 1087, 1081 HV Amsterdam, The Netherlands

STYMNE, S.
Department of Plant Breeding Research, Swedish University of Agricultural Research, S-268 31 Svalov, Sweden

WHITE, A.J.
Lipid Molecular Biology Group, Department of Biological Sciences, University of Durham, Durham DH1 3LE, UK

WOLTER, F.P.
Universität Hamburg, Institut für Allgemeine Botanik, Ohnhorststr 18, D-22609, Hamburg, Germany

Preface

This book is based on contributions made at the SEB meeting held at the University of Kent in April 1997. I was encouraged to organise both the meeting and, later, the book from a multitude of sources. It seemed that there was a widespread interest in the topic of plant lipids not just from specialists working in the field but from within the biological community as a whole.

Why should this be? Well, of course, lipids are one of the major constituents of plants so there has always been a base level of interest there. However, in recent years there have been a number of exciting developments. These range from the elucidation of new pathways and the recognition of lipid signalling molecules to the successful genetic manipulation of lipid quality. Several aspects of plant lipid biochemistry have very important implications for the agricultural industry. For example, some of the most successful recent herbicides have targeted fatty acid biosynthesis, while the production of genetically modified oil crops is an excellent example for illustrating the potential of such manipulations not only for food production but also to act as renewable sources of industrial chemicals.

The chapters in this book reflect the very real application of basic biochemistry to the agricultural (and other) industries. Even when probing the fundamentals of a process, it is possible to see a future application. Moreover, in some cases it is the agricultural application which has led to fundamental discoveries – such as the stimulus of trying to find out how graminicide herbicides work, leading to new, unexpected, findings about acetyl-CoA carboxylase.

These are exciting times in which to be a lipidologist and I hope that readers will not only be able to learn facts from this book but also be stimulated. Perhaps, if the writers' enthusiasms are conveyed well enough, we may even recruit new scientists to the area! Certainly, no one who reads Simon Bright's and Tim Hawkes' closing chapter will be in any doubt about the importance of plant lipids for the agricultural industry of the future.

John L. Harwood
Cardiff, 1998

J.L. HARWOOD

1 What's so special about plant lipids?

In plants, lipids play several roles (Table 1) – most of which have echoes in their functions in animals or microorganisms. Thus, lipids act as an efficient store of reserve energy, in a form which is both compact and stable as well as being easily metabolisable. A second, ubiquitous, role is the contribution that lipids play in membrane structure. Thus, the basic bimolecular layer of membranes is composed of a mixture of amphipathic lipid molecules and certain lipids, which adopt non-bilayer structures may also have special roles such as in membrane fusion (Chapter 7). Thirdly, there are many examples of what could be termed biologically active lipids. These are molecules which are usually present in small amounts but whose presence or metabolism is associated with profound effects in terms of cellular and tissue regulation. Fourthly, the surfaces of plants are covered in lipid-rich layers (wax, cutin, suberin) which function both to reduce water loss as well as to protect plants from the entry by (and hence damage from) microorganisms or noxious chemicals. Finally, there are a number of examples of lipids which have specialised functions not categorisable under the above headings. The photosynthetic roles of chlorophylls and carotenoids are the most important of these functions.

Storage lipids

Triacylglycerols are a compact, easily metabolised and non-hydrated energy store. They can even be accumulated (such as in adipose tissue or oil fruits) without the need for directed biosynthesis into an organelle. Small wonder therefore, that they are a favourite form for storage in many organisms. In plants, also, the vast majority of species use triacylglycerols as their fat store. The only important agricultural exception to this is the desert plant jojoba (*Simmondsia californica*) where wax esters are accumulated.

For most commercially important oil plants, lipid accumulates in the seed where it serves as an energy reserve during germination. The oil is biosynthesised during the second stage of seed maturation (Harwood & Page, 1994) at which time the relevant biosynthetic

Table 1. *Summary of the basic functions of plant lipids*

Function	Lipid characteristic	Special features of plant lipids
1. Storage	Usually triacylglycerols	Huge variety of acyl constituents possible; wax esters known; plants engineered to accumulate novel products
2. Membrane constituent	Amphipathic lipids	Common 'animal' phosphoglycerides but phosphatidylglycerol also prominent; chloroplasts rich in glycosylglycerides; sitosterol and stigmasterol are 'plant' sterols
3. Surface coverings	Very hydrophobic Often form stable polymers	Waxes, cutin and suberin all have special 'plant' characteristics
4. Biological activity	Diverse structures (inositol lipids, sphingolipids, oxidation products, etc.)	Lipoxygenase-generated products especially important
5. Other	Diverse structures related to their functions	Chlorophyll *a* as reaction centre pigment; chlorophylls *a* and *b* and carotenoids for light-harvesting; β-carotene to protect against photooxidation; tocopherols as antioxidants, etc.

enzymes are expressed at high activity. In the subsequent (desiccation) phase, most of these enzyme activities are reduced to very low levels. The nature of the acyl composition of the triacylglycerol is dependent on the availability of fatty acids from the acyl-CoA substrate pool as well as the selectivity of the acyltransferases of the Kennedy pathway. In addition, certain modification reactions may allow alteration of the basic triacylglycerol quality after oil production (see Stymne & Stobart, 1987; Harwood, 1996). The triacylglycerol stores in plants exist (almost without exception) as oils because ambient temperatures are above their Tc (gel–liquid phase transition).

When the acyl compositions of plant triacylglycerols were examined, it became clear early on that, unlike membrane lipids, the storage lipids could have all manner of fatty acid compositions (Hilditch & Williams, 1964). The lack of constraint in this latter parameter is presumably dictated by the fact that only general physical properties have to be achieved rather than the precise functions needed in membranes. Clearly, the lipid stores have to be capable of being catabolised but the degradative enzymes involved (e.g. lipases) are often rather non-selective. This non-discriminatory compositional requirement for storage oils has important implications for the agricultural industry. It means that it should be possible to breed or, more likely, genetically engineer crops which have strongly modified storage lipids. In extreme cases, quite unusual fatty acids could form the bulk of the acyl groups and many of these could be used by the chemical industry (or should one call it the biochemical industry?). Some remarks on the possible production of what have been termed 'designer oils' are made later (see Murphy, 1994*a* and Chapters 11 and 12).

Having pointed to the wide range of triacylglycerol species which can be found in Nature, the vast majority of important food crops have a rather limited range of fatty acids. Palmitate, oleate and linoleate represent about 80% of the total fatty acids with laurate, myristate, stearate and α-linolenate a further 15%. It is unlikely that this situation will change for edible oils, not least because of problems with regulatory agencies and inate consumer resistance. Even in cases where an oil has formed a staple component in a developed country's diet for many years (apparently without obvious harmful effects), it is possible for adverse findings, following animal experiments, to lead to its prohibition. Thus, rape seed oil which originally had about 3–15% erucate was used as a spread in Sweden for many years (without any detectable ill effects) but FDA (and then EC) derivatives have put strict limits on the percentage of erucate allowed and led to the breeding of LEAR (low erucic acid rape) or Canola varieties.

A sample of important food oils are shown in Table 2, with each example being enriched in a particular fatty acid. In one unique case, the oil palm, two important products are produced. The fleshy mesocarp produces about 90% of the total oil and this is enriched in palmitic acid (hence the name!). The kernel, on the other hand, contains an oil rather similar to coconut in being enriched in medium-chain products. Cocoa butter is rather unusual in that the dominant triacylglycerols are of the saturated/oleate/saturated type with equal amounts of the saturated com-

Table 2. *Fatty acid compositions of some commercially-important fats and oils*

Plant lipid	Fatty acid composition (%)						
	8 : 0– 12 : 0	16 : 0	18 : 0	18 : 1	18 : 2	18 : 3	Other[a]
Coconut oil	62	9	2	7	2	–	18
Palm kernel oil	57	8	2	15	2	trace	16
Palm oil	1	45	4	40	9	trace	1
Cocoa butter	trace	26	34	35	3	trace	2
Olive oil	trace	10	2	78	7	1	2
Rapeseed oil[b]	trace	4	2	56	26	10	2
Sunflower oil	trace	6	5	18	69	trace	2
Linseed oil	trace	6	3	17	14	60	trace
Castor bean oil	trace	1	1	3	4	trace	90

[a]In castor bean, ricinoleic acid is the main constituent. In coconut, 14 : 0 is 17%. In palm kernel, 14 : 0 is 15%.
[b]For modern, low-erucic acid rape (LEAR) varieties.
Fatty acids are abbreviated with first number indicating carbon number and number after colon indicating number of double bonds. In this table, 18 : 1 corresponds to oleic acid, 18 : 2 to linoleic acid and 18 : 3 to α-linolenic acid.

ponents palmitate and stearate. Together with the flavour components dissolved in the fat, the unique composition, and hence properties of cocoa butter, make it a high-value commodity (Harwood, 1993). Only shea butter, of the natural alternatives, comes near to its properties. Indeed, when market prices are compared (Table 3) only olive oil is close to that of cocoa butter. Olive oil's position as another expensive oil comes from the consumer's appreciation of its unique flavour and aroma characteristics.

Also listed in Table 3, are some comparisons of the productivity and total production of various commercial oils. Of these, palm oil is the world's main crop which is produced mainly for its oil and it is also the most productive on a per hectare basis. Malaysia alone currently produces over half of the world's palm oil but this is likely to be reduced soon as Indonesia plants huge acreages. Although oilseed crops produce only about 20–30% of the yield of oil palm, their return is quite sufficient for most agricultural areas. Additionally, certain crops such as, say, jojoba or olive, may occupy particular niches in allowing marginal or difficult terrain to be utilised.

Table 3. *Market prices, productivity and total production of some oils and fats*

	Price (£/tn)	Oil yield (Tn/ha)	Total production (% world edible oils)
Soya bean	335	0.7	26
Palm	380	6.0	19
Rapeseed	340	1.8	15
Sunflower	390	1.6	12
Coconut	560	2.2	4
Olive (extra virgin)	2800	–	4
Palm kernel	530	0.8	2
Cocoa butter	4920	–	1

Prices for crude oils are based on average values in April 1997.
Prices are increased with refining and hardening, where appropriate.
Information on yields and production are taken from Luhs and Friedt (1994).

For an oil to be modified successfully a number of criteria have to be met and manipulations accomplished successfully (Dale & Irwin, 1994; Murphy, 1994*b*). The necessary changes to fatty acid synthesizing enzymes have to be made (anti-sense, up-regulation, new genes introduced) with the alterations occurring in a temporal and tissue-specific manner. Furthermore, the ability of three acylation enzymes of the Kennedy pathway to deal with the unusual substrates has also to be considered (Harwood & Page, 1994). This question of specificity is also relevant to efforts at increasing the proportion of naturally occurring acyl chains in a particular seed's oil. A particular example might be with the erucate content of high erucic acid rape (HEAR) varieties of rape. Because of the tight specificity of acyl-CoA : 1-acylglycerol 3-phosphate acyltransferase (the second enzyme in the Kennedy pathway), HEAR will normally only accumulate triacylglycerol with erucate at the 1- and 3-positions. Since trierucin is the desirable structure for a high temperature industrial lubricant, the solution has to involve introduction of a gene coding an acyltransferase able to use erucate. (Some examples of these problems and efforts to overcome them will be found later in the book in Chapters 10 and 12.)

There is extensive work currently on efforts to produce modified oils for a whole variety of purposes (Table 4). Much of this work was

Table 4. *Examples of unusual fatty acids in plant oils which can be used for specific purposes*

Acid	Application
Cyclic fatty acid	
Chaulmoogric	Used for leprosy treatment
Epoxy fatty acid	
Vernolic	Polymers, plasticisers
Hydroxy fatty acids	
Ricinoleic	Triricinolein as a lubricant
Dimorphecolic	Coatings, pharmaceuticals, flavours, lubricants, polymers
Unsaturated fatty acids	
Calendic	Paints, coatings
Erucic	Trierucin as lubricant
Petroselinic	Detergents, polymers

discussed in the book by Murphy (1994*a*). Some more recent developments have been covered in two symposia volumes (Kader & Mazliak, 1995; Williams *et al.*, 1997) as well as in Chapters 11 and 12.

Two special aspects of storage lipids can be mentioned. First, as noted above, jojoba produces a wax ester oil. The oil consists mainly (about 80%) of C_{40} and C_{42} monoesters rich in 20 : 1 and 22 : 1 acids and alcohols (Gunstone *et al.*, 1994). Wax esters have a niche market status in that they can substitute for the use of sperm (and other) whale oil as well as being used for particular purposes like skin or hair-care products.

Secondly, there has been some interest in polyhydroxybutyrate (PHB). This polymer has application as a biodegradable plastic and has been produced by large-scale microbial culture (Anderson & Dawes, 1990). Initial experiments where two of the bacterial (from *Alcaligenes eutrophus*) genes for PHB synthesis were placed under the control of a constitutive plant promoter and introduced into *Arabidopsis* were successful in that PHB was found in the transgenic plants (Poirier *et al.*, 1992). However, the transformed plants showed stunted growth and poor seed set. More recent experiments, where all three genes from *Alcaligenes eutrophus* needed for PHB synthesis were targeted to the plastid, were successful. Transformed *Arabidopsis* plants accumulated up to 14% PHB as dry weight and there was no obvious effect on growth or fertility (Nawrath, Poirier & Somerville, 1994). The experiments showed that it is possible to engineer plants to produce high amounts of novel compounds without simultaneous yield penalties being incurred.

Membrane lipids

The various subcellular membranes of plant tissues have distinct and rather well-conserved lipid compositions (Table 5). For the extra-chloroplast membranes, phosphoglycerides predominate, with phosphatidylcholine and phosphatidylethanolamine the major constituents (Table 5). Some other particular points to note are that sphingolipids (primarily ceramides) as well as polyphosphoinositides (phosphatidylinositol-4-phosphate, phosphatidylinositol-4,5-*bis*phosphate) are concentrated in the plasma membrane, while diphosphatidylglycerol (cardiolipin) is uniquely located in the inner mitochondrial membrane. Plant membrane compositions have been reviewed generally (Mazliak, 1977; Harwood, 1980).

The thylakoid membranes of plants, algae and cyanobacteria are virtually unique in nature because of their preponderance of glycosylglycerides (Fig. 1). Monogalactosyldiacylglycerol (MGDG) accounts for 40–50%, digalactosyldiacylglycerol (DGDG) for 20–40%, sulphoquinovosyldiacylglycerol (SQDG) for 7–10% and the only significant, phospholipid, phosphatidylglycerol (PtdGly) for 9–17% of thylakoid membrane lipids (Hitchcock & Nichols, 1971; Harwood, 1980). Plants differ slightly in the acyl composition of these four thylakoid lipids. Thus, species can be categorised as '16 : 3' or '18 : 3' plants which differ in whether or not hexadecatrienoic acid is present in their MGDG (Table 6). The enzymatic reasons for this distinguishing feature have been well discussed (see Heinz & Roughan, 1983; Browse & Somerville, 1991). Moreover, although otherwise the acyl compositions of MGDG and DGDG are rather constant, SQDG can differ considerably in its palmitate content (Heinz, 1977) and PtdGly in its palmitate and *trans*-Δ3-hexadecenoate levels. The latter property has implications for chilling sensitivity (Murata, 1983; Chapter 7).

It is, perhaps, pertinent to consider why thylakoid lipid compositions have been so well preserved through evolution. Also, why is there such a high degree of unsaturation in typical plant membranes? For an answer to the latter, one cannot look to considerations about transition temperatures and membrane fluidity because there is, for example, no significant difference between linoleoyl or linolenoyl species in that regard. It has been suggested that (α-linolenate could be important for the packing characteristics of lipids in thylakoids, enriched as they are in bulky photosynthetic complexes (c.f. Quinn & Williams, 1983; also

Table 5. *The lipid compositions of some plant subcellular membranes*

	Lipid composition (%)							
	Glycosylglycerides			Phosphoglycerides				
	MGDG	DGDG	SQDG	PtdCho	PtdEtn	PtdIns	PtdGly	DiPtdGly
Plasma[a]	–	–	–	32	46	19	–	3
Chloroplast envelope[b]	22	32	5	27	–	1	8	–
Chloroplast thylakoid[c]	51	26	7	3	–	1	9	–
Mitochondrial outer[d]	–	–	–	54	30	11	5	0
Mitochondrial inner[e]	–	–	–	41	37	5	3	15
Glyoxysomal[f]	–	–	–	47	27	8	2	tr.
Endoplasmic reticulum[g]	–	–	–	45	29	13	4	3

Sources: [a] = potato tuber; [b,c] = spinach; [d,e] = potato; [f,g] = castor bean.
Abbreviations: MGDG, monogalactosyldiacylglycerol; DGDG, digalactosyldiacylglycerol; SQDG, sulphoquinovosyldiacylglycerol; PtdCho, phosphatidylcholine; PtdEtn, phosphatidylethanolamine; PtdIns, phosphatidylinositol; PtdGly, phosphatidylglycerol; DiPtdGly, diphosphatidylglycerol (cardiolipin).
Information taken from Harwood (1980) and Packer and Douce (1987) where further details will be found.

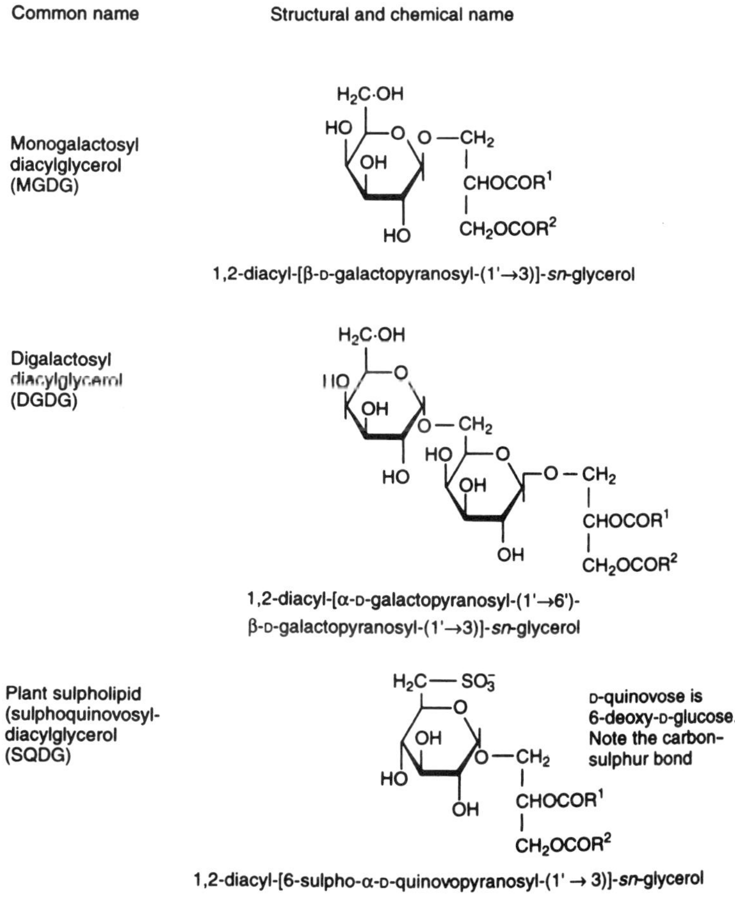

Common name
Structural and chemical name
Monogalactosyl diacylglycerol (MGDG)
1,2-diacyl-[β-D-galactopyranosyl-(1'→3)]-sn-glycerol
Digalactosyl diacylglycerol (DGDG)
1,2-diacyl-[α-D-galactopyranosyl-(1'→6')-β-D-galactopyranosyl-(1'→3)]-sn-glycerol
Plant sulpholipid (sulphoquinovosyl-diacylglycerol (SQDG)
D-quinovose is 6-deoxy-D-glucose. Note the carbon-sulphur bond
1,2-diacyl-[6-sulpho-α-D-quinovopyranosyl-(1' → 3)]-sn-glycerol

Fig. 1. The structure of plant glycosylglycerides.

Chapter 7). As to why glycosylglycerides rather than phosphoglycerides are characteristic of thylakoids, we have no really convincing theories at present. For marine organisms, there are obvious advantages to not having a significant phosphate requirement for membrane synthesis since this anion is in short supply. However, the same argument cannot be applied to terrestrial organisms. It has also been suggested that the galactose head groups of galactosylglycerides may participate in hydrogen bonding (see Sen *et al.*, 1983) as proposed for other glycolipids. Moreover, the ready conversion of MGDG to DGDG not only alters melting characteristics (Harwood *et al.*, 1994) but also alters the

Table 6. *Chloroplast thylakoid lipids from a '16 : 3' and an '18 : 3' plant*

Plant		% total lipids	Fatty acid composition (% total)					
			16 : 0	16 : 1[a]	16 : 3	18 : 1	18 : 2	18 : 3
Spinach	MGDG	41	tr.	tr.	25	1	2	72
('16 : 3')	DGDG	29	3	tr.	5	2	2	87
	SQDG	16	39	tr.	–	1	7	53
	PtdGly	12	11	34	–	2	4	47
Pea	MGDG	42	4	1	–	1	3	90
('18 : 3')	DGDG	31	9	1	–	3	7	78
	SQDG	11	32	2	–	2	5	58
	PtdGly	12	18	29	–	2	11	38

[a]16 : 1 = palmitoleate for all lipids except PtdGly where it is mainly *trans*-Δ3-hexadecenoate. For lipid abbreviations, see Table 5. For fatty acid abbreviations, see Table 2 and 16 : 3 is all *cis*-Δ7,10,13-hexadecatrienoate.

preferred molecular conformation from hexagonal II to bilayer (Gurr & Harwood, 1991 and Chapter 7). Murphy (1986) has speculated about the relevance of the latter to the high radius of curvature at the ends of the granal stacks but sided-distribution studies (see Gounaris, Barber & Harwood, 1986) do not give the right distribution for this to be a major function. Some general speculations about the roles of thylakoid lipids and evidence for possible functions have been made (Gounaris *et al.*, 1986; Murphy, 1986; Williams, 1994). Recently, the availability of a few lipid-deficient mutants, if not in plants at least in simpler photosynthetic organisms, has enabled physiological tests to be performed (Guler *et al.*, 1996; Dormann *et al.*, 1996). As with many deficient mutants, the results are not as startling as one might have predicted. SQDG lack, for example, only seems to become a problem when cells are stressed (Benning *et al.*, 1993). For thylakoid composition to be so conserved, there must be clear advantages. However, in evolutionary terms a 10% advantage may be at least enough since organisms are in direct competition with each other *in vivo*. A 10% change in a biochemical or physiological measure may, however, be too subtle to demonstrate experimentally.

PtdGly is the only phospholipid (of any importance) in thylakoids. Again, we have little concept of the functional significance of this.

It is, perhaps, worth noting that in photosynthetic bacteria the same phospholipid is also enriched in the photosynthetic membranes (Russell & Harwood, 1979) although, in this case (*Rhodobacter sphaeroides*), photosynthesis is non-oxygenic. The unique *trans*-Δ3-hexadecenoic acid, found in PtdGly from a wide range of species has been a fertile subject of speculation for some time. The first suggestion as to a specific function for PtdGly containing *trans*-Δ3-hexadecenoic acid was that it was involved in granal stacking (Tuquet *et al.*, 1977). However, examination of chlorophyll *b*-less mutants (with hardly any stacking) (Bolton, Wharfe & Harwood, 1978) and other chloroplasts where granal stacking ranged considerably showed very little agreement with the theory which can now be discounted (Gounaris *et al.*, 1986).

Since that time there have been various other ideas for *trans*-Δ3-hexadecenoate function, most of which relate to interactions with light-harvesting pigment proteins. This developed from ideas that the acid could influence the formation of the oligomeric form of light-harvesting chlorophyll protein$_2$ (LHC_2). In an early paper, Remy, Trémolières and Ambard-Bretteville (1984) showed that the *trans*-Δ3-hexadecenoate content of thylakoid phosphatidylglycerol was correlated with the stabilisation of LHC_2. Additional evidence for this association including the restoration of oligomeric LHC_2 in deficient mutants of *Chlamydomonas* (Garnier *et al.*, 1990) has been obtained by several groups (see Krupa *et al.*, 1987). In a series of experiments Huner *et al.* (1987) showed that exposure of rye to cold-hardening growth temperatures resulted in a specific reduction in *trans*-Δ3-hexadecenoate content. This decrease was associated with increased levels of monomeric LHC_2. Moreover, phosphatidylglycerol was associated with LHC_2 during purification (Krol *et al.*, 1988) and only its *trans*-Δ3-hexadecenoate molecular species were associated with oligomeric LHC_2. Digestion of phosphatidylglycerol with phospholipases and reconstitution experiments confirmed the idea of a role of *trans*-Δ3-hexadecenoate in stabilising oligomeric LHC_2 (Krupa *et al.*, 1987). An alternative idea, which is not necessarily unconnected, is that *trans*-Δ3-hexadecenoate could be involved in State1/State 2 transitions – those rapid adaptations to ambient light quality which allow light energy to be re-distributed between the two photosystems (Allen *et al.*, 1981). Harwood (1984) suggested a role in State 1/State 2 transitions, his idea being developed originally from the observation that palmitate and *trans*-Δ3-hexadecenoate appeared to undergo an oxidation/reduction cycle under some conditions (Harwood & James, 1974). Confirmatory evidence for the theory came from experiments (H. Jones & J.L. Harwood, 1984, unpublished results) where chloroplasts could be manipulated into either of the two

States *in vitro* (Telfer, Hodges & Barber, 1983). This hypothesis has been followed up more recently by Trémolières, Garnier and Dubertrot (1992), who have discussed the possible role of *trans*-Δ3-hexadecenoate in influencing lipid–protein interactions during light energy distribution changes.

Lipids as bioregulators in cell signalling

Over the last dozen years or so there has been an explosion of interest in animals about the role that various lipids can play in cellular regulation, development and differentiation. The Plant Kingdom also has many similar systems as well as some distinct ones. In the latter category, some growth regulators (plant hormones) could be included, which, by virtue of their solubility properties can count as lipids, but here only acyl lipids will be considered.

The inositide cycle is well known as a provider of two second messengers (inositol 1,4,5-*tris*phosphate, diacylglycerol) in animals. In plants too, plasma membrane-located kinases ensure that the higher inositides accumulate there (Drobak, 1992) and turnover of phosphatidylinositol-4,5-*bis*phosphate (PIP_2) accompanies many changing environmental, agonist-driven or stress situations (Hetherington & Drobak, 1992) (Fig. 2). Most of the enzymes identified in the inositide cycle (Berridge & Irvine, 1984) have been defined by molecular cloning in animals and some in plants as well (Mazliak, 1997). However, no enzyme equivalent to a protein kinase C has been identified, although both heterotrimeric and small G proteins have been demonstrated (Mazliak, 1997).

More recently in animals there has been interest in the phospholipase D-mediated breakdown of other phospholipids (mainly phosphatidylcholine). In conjunction with phosphatidate phosphohydrolase this is capable of providing a sustained source of diacylglycerol for second messenger functions (Exton, 1994). There is some evidence for a role of phospholipase D in cell signalling in plants (Munnik *et al.*, 1995; Wang, 1993) and a gene expressing the enzyme has been cloned from castor bean (Wang, Xu & Zheng, 1994). In addition, transgenic *Arabidopsis* and tobacco plants were produced with overexpressed or antisensed phospholipase D genes, in order to study the physiological role of this enzyme's action (Wang *et al.*, 1997).

More recently, there has been a realisation that sphingolipids may be involved in another cycle – one which also has ramifications for cell regulation, differentiation and death (Fig. 3). Indeed, sphingolipids have been considered in cellular signalling for about a decade (see Kolesnick, 1991). However, sphingolipids have never been considered much in plants so, of

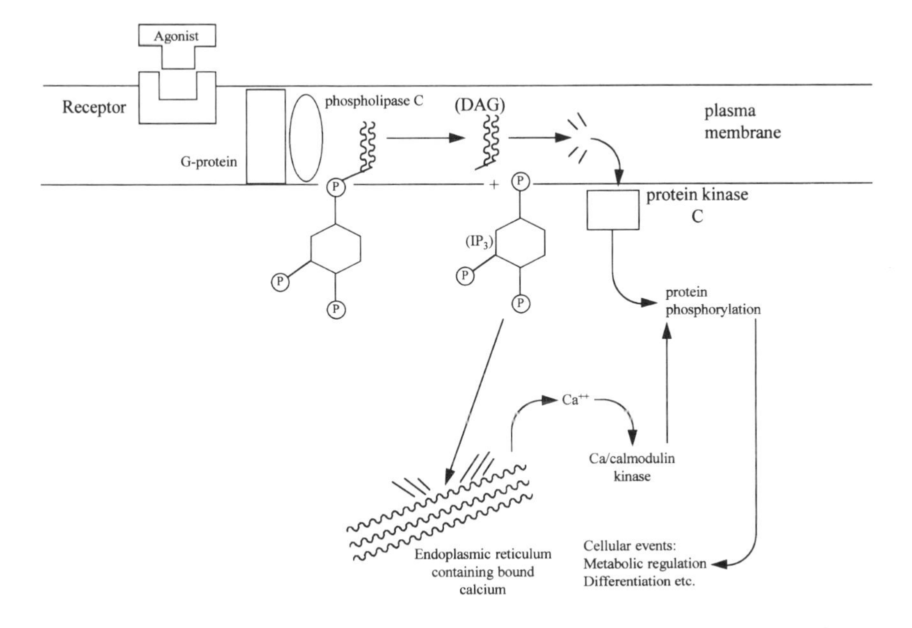

Fig. 2. Phosphatidylinositol-4,5-*bis*phosphate hydrolysis and the generation of second messengers.

course, it is difficult to know what, if any, similar function they may have in signalling there. Nevertheless, cerebrosides (glycosylceramides), in particular, are significant components (6–30%) of those plant plasma membranes which have been analysed (Lynch & Phinney, 1995) so that it would seem worthwhile to look for evidence of their participation in regulatory phenomena similar to animals. Indeed, changes in the glucosylceramide content of plasma membranes has been observed following acclimation to environmental stresses (e.g. Uemura & Steponkus, 1994; Norberg & Liljenberg, 1991). Glucosylceramides, the predominant sphingolipids in plant tissues, are also significant components of plant tonoplast membranes (Lynch & Phinney, 1995).

Lipoxygenases were first recognised in plant tissues, where they have generally high activities. They are particularly important in food plants where they destroy the polyunsaturated fatty acids to produce derivatives with characteristic tastes, flavours and aromas. Classically, they are divided into Type I (alkaline) and Type II lipoxygenases (Galliard & Chan, 1980). In animals, lipoxygenases play a notable role in one branch of the eicosanoid production pathways (Gurr & Harwood,

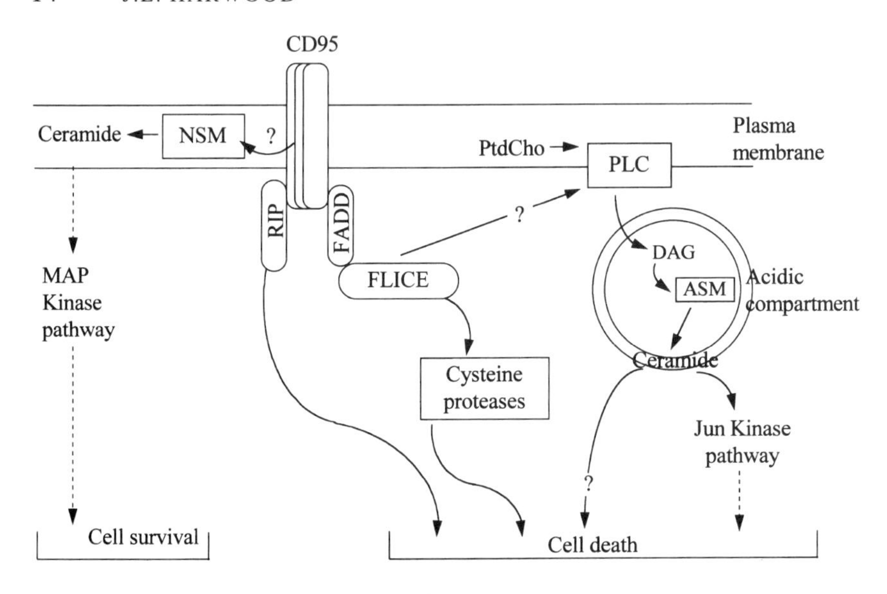

Fig. 3. Ceramides and cell signalling in mammals. This diagram is based on part of an illustration in Testi (1996) where ceramide-dependent pathways initiated at the tumour necrosis factor or the interleukin-1β(CD95) receptors are shown. By analogy with most control pathways, it seems likely that plants will show at least some of these regulatory activities. The acidic compartment could represent caveolae, endosomes or lysosomes in animals and the vacuole in plants (ceramides are found in tonoplast membranes). CD95 directly interacts with at least two proteins, a 23 kDa FADD (*F*as-*a*ssociated through *d*eath *d*omain) protein and a 74 kDa *r*eceptor-*i*nteracting *p*rotein (RIP). FLICE is a protease which can initiate the cysteine protease cascade. NSM = neutral sphingomyelinase; ASM = acidic sphingomyelinase – both of which lead to changes in ceramide levels and hence cell signalling phenomena; PtdCho = phosphatidylcholine and this can be acted upon by a phospholipase C (PLC) to generate the second messenger, diacylglycerol (DAG).

1991). In plants, lipoxygenase attack, for example on α-linolenic acid, initiates a series of enzymatic reactions which can lead to the generation of short chain aldehydes and alcohols as well as ω-oxoacids (Fig. 4). The C_6 aldehydes and alcohols give the flavours and aromas so characteristic of many plants and plant foods and their production relies on the presence of hydroperoxide lyase, an isomerase (which converts the

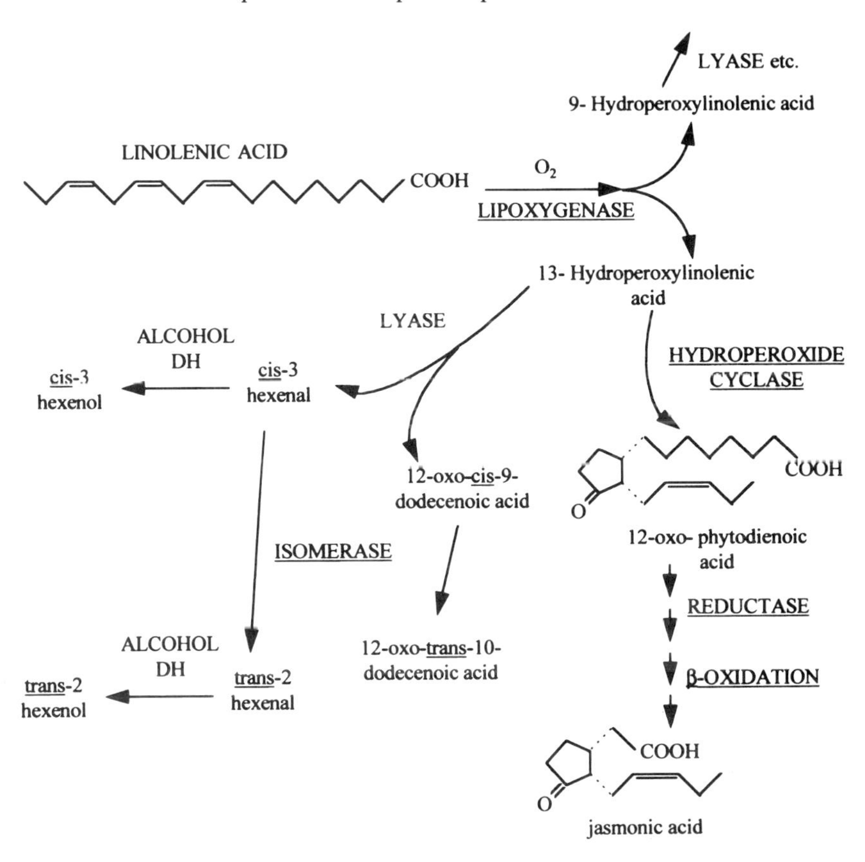

Fig. 4. Simplified lipoxygenase pathway.

cis-3-enal structure to the *trans*-2-enal form) and alcohol dehydrogenase (Vick & Zimmerman, 1987).

Instead of the 13-hydroperoxide product of lipoxygenase attack on α-linolenic acid being cleaved by the lyase, it can be acted upon by hydroperoxide cyclase. The product of the latter enzyme is 12-oxo-phytodienoic acid and this can be converted in a series of reactions to jasmonic acid (Vick & Zimmerman, 1987). The latter compound plus other intermediates of the lipoxygenase pathway can have a variety of important physiological roles.

For a more comprehensive review of plant lipoxygenases, the reader is referred to Siedow (1991) and to Gardner (1991).

Plant surface layers

Aerial plant tissues are covered in a layer of cutin, while roots have suberin on their surface. Suberin is also formed to cover wounds regardless of whether these are above or below ground. Overlaid on the cutin is a layer of wax. The function of all these surface components is protective. On the one hand, they provide a strong structure capable of effectively preventing water loss. They also stop the easy penetration of plant tissues by bacteria, fungi and other toxic organisms. For the agricultural industry, knowledge of the surface architecture of plants is very important because this layer controls how agrochemicals penetrate and, hence, their efficacy.

Typical components of waxes, cutin and suberin are shown in Table 7. Hydrocarbons (alkanes) are ubiquitous in plant waxes and usually constitute a high percentage of the total lipids. Because of the way in which they are formed (by decarbonylation of aldehydes: Dennis & Kolattukudy, 1992), most alkanes are odd-chain. Monoesters of long-chain alcohols and acids are also found in all plant waxes examined. In addition, very long chain (usually C_{22}–C_{34}) alcohols and very long chain fatty acids are frequent components, but in very variable amounts (Kolattakudy, 1980; Bianchi, 1995). In some waxes, ketones and β-diketones may be major constituents.

For cutin, monomers are chiefly provided by C_{16} or $C_{18:1}$ chains. The polyester structure (Kolattukudy, 1980) is then built up by ester links between hydroxyl groups on the monomers, e.g. 16-hydroxypalmitic or 10,16-dihydroxypalmitic acids. Although ester links provide the main polymer architecture, other structures may be present as evidenced by the significant residue (20–30%) left after depolymerisation with

Table 7. *Major features of waxes, cutins and suberins*

Waxes
Alkanes major component; *ketones* usually minor; β-*diketones* sometimes major; *esters* minor but common; *very long chain acids* and *alcohols* common
Cutins
Very long chain acids and *alcohols* rare and minor, d*icarboxylic acids* minor, *in-chain substituted acids* major, *phenolics* low.
Suberin
Very long chain acids and *alcohols* common and major, *dicarboxylic acids* major; *in-chain substituted acids* minor; *phenolics* high.

reagents that cleave ester bonds (Walton, 1990). A lipoxygenase/peroxygenase pathway appears to be involved in cutin monomer formation (Blée & Schuber, 1993).

Suberin is also a polymeric structure but its details are largely unknown. Cross-linking of condensed aromatic domains via ether and carbon–carbon linkages in a polymer structure similar to, but distinct from, lignin have been suggested (Walton, 1990). Additionally, the aromatic domain is probably linked via ester bonds to aliphatic regions (Kolattukudy, 1980). Of the latter, which represent 20–50% of suberin-rich samples, major components are ω-hydroxy acids, the corresponding dicarboxylic acids, very long chain ($>C_{18}$) acids and alcohols. The phenolic core has ρ-coumaric acid as a major constituent. von Wettstein-Knowles (1993) has drawn a distinction between cutin and suberin in that cutin monomers generally have mid-chain substitutions (mainly hydroxy and epoxy), while suberin monomers generally range from C_{16}–C_{24} and have α-substitutions.

For further information about plant surface coverings, the reader is referred to Kolattukudy (1980, 1987), Walton (1990) and Hamilton (1995). For their biosynthesis, also see von Wettstein-Knowles (1993) and Lemieux (1996).

An introduction to plant lipid biosynthesis

As a scene-setter for further chapters in this book, some preliminary words about lipid biosynthesis are in order. Since this book is concerned virtually exclusively with acyl lipids then I shall start with a summary of fatty acid formation.

Fatty acid biosynthesis is summarised in Fig. 5. A combination of acetyl-CoA carboxylase (ACCase) and fatty acid synthase (FAS) allows the *de novo* formation of long-chain fatty acids from acetyl-CoA, the source of which has been a matter of some controversy (Harwood, 1988). The characteristics of the component enzymes of FAS, together with the substrate selectivity of acyl-ACP thioesterase ensures that, in most situations, the products of *de novo* synthesis are palmitate and stearate. The latter can be rapidly desaturated (as stearoyl-ACP) to oleate by a Δ9-desaturase. All these enzymes are present in the chloroplast (plastid) stroma. Shorter-chain products can be produced, in certain cases (see Chapters 10–12) by the intervention of a thioesterase capable of hydrolysing medium chain acyl-ACPs.

Once palmitate and oleate have been formed in the plastid, their further metabolism is mainly dependent on their export to outside the organelle. The plastid stromal acyl-ACP thioesterase rapidly hydrolyses

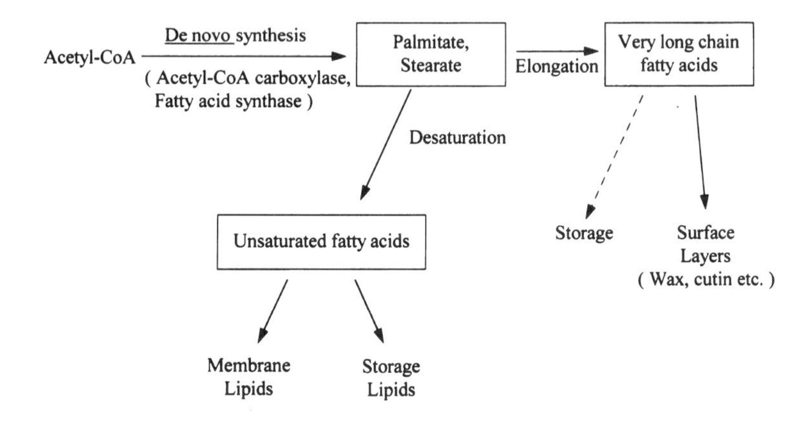

Fig. 5. Simplified depiction of fatty acid synthesis in plants.

the fatty acids which are re-esterified to acyl-CoAs on the plastid envelope (see Harwood, 1996). The acyl-CoA pool thus created provides substrates for the acyltransferases involved in the Kennedy pathway for complex lipid synthesis (Fig. 6) as well as for other acyltransferases that may be used for other reactions connected with lipid metabolism. For example, acyltransferases allow exchange of fatty acids between the acyl-CoA pool and phosphatidylcholine and thus play an important role in fatty acid desaturation (Stymne & Stobart, 1987).

Fatty acids, as CoA-esters, are substrates for elongation enzymes (Chapter 8) of which there are several chain-length specific complexes. These complexes, although using malonyl-CoA and NADPH as co-substrates, differ from fatty acid synthase in being membrane-located (Harwood, 1996). They provide very long chain fatty acids mainly for wax, cutin or suberin synthesis (Fig. 5) although some seeds (e.g. high erucic acid rape, HEAR) will use such acids for storage lipids.

An alternative, and quantitatively more important, modification of fatty acids is aerobic desaturation. These enzymes show cyanide-sensitivity, produce *cis* double bonds and require an exogenous reductant, oxygen and (usually) a complex lipid substrate. Phosphatidylcholine is particularly important as a substrate for Δ12-desaturation of oleate to linoleate and for the subsequent desaturation (Δ15) to α-linolenate in seeds. For leaves, much of the linoleate is returned to the chloroplast for desaturation on monogalactosyldiacylglycerol substrate. Alternative pathways and substrates are also possible, especially those concentrated in the plastid for 16 : 3-plants (see Harwood, 1996 for more details). The membrane-bound desaturases show considerable

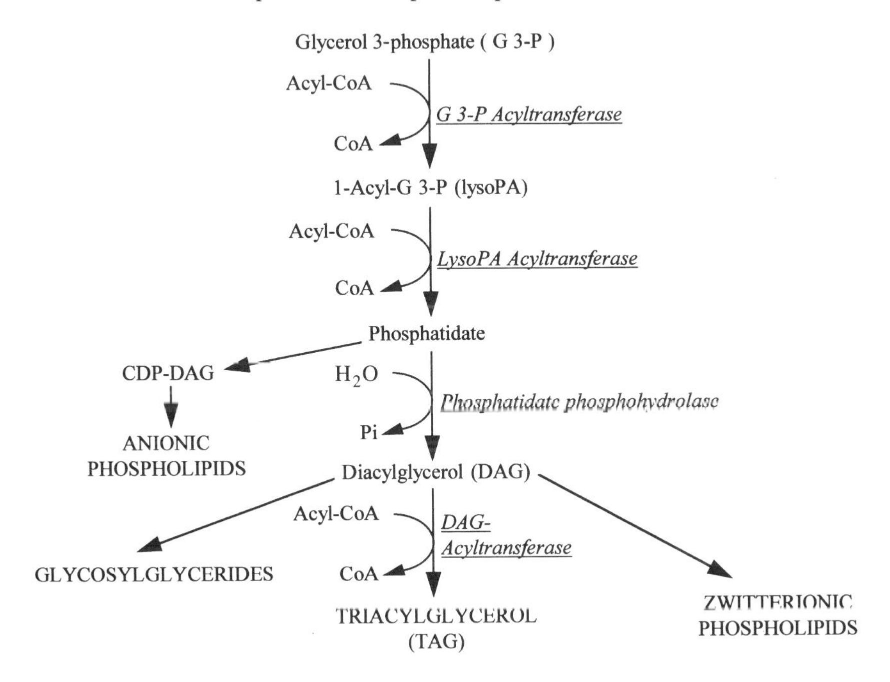

Fig. 6. The Kennedy pathway for acyl lipid formation in plants. Abbreviations: G 3-P = glycerol 3-phosphate; PA = phosphatidic acid; DAG = diacylglycerol; TAG = triacylglycerol.

sequence conservation (Shanklin, Whittle & Fox, 1995), particularly around three conserved HXXXH sequences (Schmidt, Sperling & Heinz, 1995). Unsaturated fatty acids, of course, are mainly used for membrane or storage lipid production but may have other biological roles such as substrates for lipoxygenases.

Many of the enzymes mentioned above are used by other types of organisms rather than plants. However, there are certain features of the plant enzymes which can be highlighted (Table 8). Further details will be found in Harwood (1996).

The Kennedy pathway (Fig. 6) is central for complex lipid formation. Two acyltransferases are needed to generate the important intermediate, phosphatidic acid, from glycerol 3-phosphate. A fork then appears in the pathway. Cytidylyltransferase activity can generate CDP-diacylglycerol which is used for anionic phosphoglyceride (phosphatidylglycerol, diphosphatidylglycerol, phosphatidylinositol) production. Alternatively, phosphatidate phosphohydrolase action yields diacylglycerol, which is used for membrane lipid synthesis

Table 8. *Some important features of plant fatty acid synthesis*

Enzyme system	Points of note
Acetyl-CoA carboxylase	Isoforms with chloroplast and cytoplasmic location. Multifunctional proteins except for dicotyledon chloroplast isoforms (multienzyme complex).
Fatty acid synthase	Type II (multienzyme complex). Three separate condensing enzymes. Isoforms of both reductases – differ in nucleotide specificity. Pre-mature termination by thioesterase possible.
Elongases	Membrane-bound, chain-length specific. Use acyl-CoA substrates, malonyl-CoA and NADPH.
Desaturases	Stearoyl-Δ9-desaturases uses stearoyl-ACP substrate but others use complex lipids. Aerobic, cyanide-sensitive, have diiron centres. Products usually *cis* and for polyunsaturates are methylene-interrupted.

(glycosylglycerides, zwitterionic phosphoglycerides) or triacylglycerol formation. The role of the Kennedy pathway in storage oil synthesis has been discussed in detail (Harwood & Page, 1994) while for galactosylglyceride synthesis the reader is referred to Joyard and Douce (1987). For further information on sulpholipid formation, refer to Mudd and Kleppinger-Sparace (1987) and Pugh *et al.* (1995). Phosphoglyceride synthesis has been discussed by Mudd (1980) and by Harwood (1989). While many of the enzymes involved in complex lipid formation in plants share similarities with those from animals or microorganisms, there are differences in detail. Of course, formation of galactosylglycerides and sulpholipid is almost exclusively confined to plants, algae and cyanobacteria – showing that there is something (very) special about these lipids.

References

Allen, J.F., Bennett, J., Steinback, K.E. & Arntzen, C.J. (1981). Chloroplast protein phosphorylation couples plastoquinone redox state to distribution of excitation energy between photosystems. *Nature*, **291**, 25–9.

Anderson, A.J. & Dawes, E.A. (1990). Occurrence, metabolism, metabolic role and industrial uses of bacterial polyhydroxyalkanates. *Microbiological Reviews*, **54**, 450–72.

Benning, C., Beatty, J.T., Prince, R.C. & Somerville, C. (1993). The sulfolipid, sulphoquinovosyldiacylglycerol is not required for photosynthetic electron transport in *Rhodobacter sphaeroides* but enhances growth under phosphate limitation. *Proceedings of the National Academy of Sciences, USA*, **90**, 1561–5.

Berridge, M.J. & Irvine, R.F. (1984). Inositol trisphosphate, a novel second messenger in cellular signal transduction. *Nature*, **312**, 315–21.

Bianchi, G. (1995). Plant waxes. In *Waxes: Chemistry, Molecular Biology and Functions*, ed. R.J. Hamilton, pp. 175–222. Dundee: The Oily Press.

Blée, E. & Schuber, F. (1993). Biosynthesis of cutin monomers: involvement of a lipoxygenase/peroxygenase pathway, *The Plant Journal*, **4**, 113–23.

Bolton, P., Wharfe, J. & Harwood, J.L. (1978). The lipid composition of a barley mutant lacking chlorophyll b. *The Biochemical Journal*, **174**, 67–72.

Browse, J. & Somerville, C. (1991). Glycerolipid synthesis: biochemistry and regulation. *Annual Review of Plant Physiology and Plant Molecular Biology*, **42**, 467–506.

Dale, P.J. & Irwin, J.A. (1994). Transformation of oil crops. In *Designer Oil Crops*, ed. D.J. Murphy, pp. 195–218. Weinheim: VCH.

Dennis, M. & Kolattukudy, P.E. (1992). A cobalt-porphyrin enzyme converts a fatty aldehyde to a hydrocarbon and CO. *Proceedings of the National Academy of Sciences, USA*, **89**, 5306–10.

Dormann, P., Hartel, H., Lokstein, H., Trethewey, R. & Benning, C. (1996). Altered photosynthesis in an *Arabidopsis* mutant deficient in the thylakoid lipid digalactosyldiacylglycerol. In *12th International Symposium on Plant Lipids*, Toronto 7th–12th July, Abstracts II6.

Drobak, B.K. (1992). The plant phosphoinositide system. *The Biochemical Journal*, **288**, 697–712.

Exton, J.H. (1994). Phosphatidylcholine breakdown and signal transduction. *Biochimica et Biophysica Acta*, **1212**, 26–42.

Galliard, T. & Chan, H.W-S. (1980). Lipoxygenases. In *The Biochemistry of Plants*, vol. 4, ed. P.K. Stumpf, pp. 131–61. New York: Academic Press.

Gardner, H.W. (1991). Recent investigations into the lipoxygenase pathway of plants. *Biochimica et Biophysica Acta*, **1084**, 221–39.

Garnier, J., Wu, B., Maroc, J., Guyon, D. & Trémolières, A. (1990). Restoration of both an oligomeric form of the light-harvesting antenna CPII and a fluorescence state II – state I transition by Δ^3-

trans-hexadecenoic acid-containing phosphatidylglycerol in cells of a mutant of *Chlamydomonas reinhardtii*. *Biochimica et Biophysica Acta*, **1020**, 153–62.

Gounaris, K., Barber, J. & Harwood, J.L. (1986). The thylakoid membranes of higher plant chloroplasts. *The Biochemical Journal*, **237**, 313–26.

Guler, S., Narang, R.A., Essigmann, B., Rossak, M., Hartel, H., Seeligo, A., Renger, G. & Benning, C. (1996). Genetic analysis of sulpholipid function in bacteria and plants. In *12th Int. Symposium on Plant Lipids*, Toronto 7th–12th July, Abstracts II4.

Gunstone, F.D., Harwood, J.L. & Padley, F.B. (eds.) (1994). *The Lipid Handbook*. 2nd edn. London: Chapman and Hall.

Gurr, M.I. & Harwood, J.L. (1991). *Lipid Biochemistry*, 4th edn. London: Chapman and Hall.

Hamilton, R.J. (ed.) (1995). *Waxes: Chemistry, Molecular Biology and Function*. Dundee: The Oily Press.

Harwood, J.L. (1980). Plant acyl lipids: structure, distribution and analysis. In *The Biochemistry of Plants*, vol. 4, ed. P.K. Stumpf, pp. 1–55. New York: Academic Press.

Harwood, J.L. (1984). Effects of the environment on the acyl lipids of algae and higher plants. In *Structure, Function and Metabolism of Plant Lipids*, ed. P-A. Siegenthaler & W. Eichenberger, pp. 543–50. Amsterdam: Elsevier.

Harwood, J.L. (1988). Fatty acid metabolism. *Annual Review of Plant Physiology and Plant Molecular Biology*, **39**, 101–38.

Harwood, J.L. (1989). Lipid metabolism in plants. *Critical Reviews in Plant Science*, **8**, 1–43.

Harwood, J.L. (1993). Who likes eating a lump of fat? *The Biochemist*, April 1993, pp. 6–9. London: The Biochemical Society.

Harwood, J.L. (1996). Recent advances in the biosynthesis of plant fatty acids. *Biochimica et Biophysica Acta*, **1301**, 7–56.

Harwood, J.L. & James, A.T. (1974). Metabolism of *trans*-3-hexadecenoic acid in broad bean. *European Journal of Biochemistry*, **50**, 325–34.

Harwood, J.L. & Page, R.A. (1994). Biochemistry of oil synthesis. In *Designer Oil Crops*, ed. D.J. Murphy, pp. 165–94. Weinheim: VCH.

Harwood, J.L., Jones, A.L., Perry, H.J., Rutter, A.J., Smith, K.L. & Williams, M. (1994). Changes in plant lipids during temperature adaptation. In *Temperature Adaptation of Biological Membranes*, ed. A.R. Cossins, pp. 107–18. London: Portland Press.

Heinz, E. (1977). Enzymatic reactions in galactolipid biosynthesis. In *Lipids and Lipid Polymers in Higher Plants*, ed. M. Tevini & H.K. Lichtenthaler, pp. 102–20. Berlin: Springer-Verlag.

Heinz, E. & Roughan, P.G. (1983). Similarities and differences in lipid metabolism of chloroplasts isolated from 18 : 3 and 16 : 3 plants. *Plant Physiology*, **72**, 273–9.

Hetherington, A.M. & Drobak, B.K. (1992). Inositol-containing lipids in higher plants. *Progress in Lipid Research*, **31**, 53–64.

Hilditch, R.P. & Williams, P.N. (1964). *The Chemical Constitution of Natural Fats*. London: Chapman and Hall.

Hitchcock, C. & Nichols, B.W. (1971). *Plant Lipid Biochemistry*. London: Academic Press.

Huner, N.P.A., Kool, M., Williams, J.P., Maisson, E., Low, P., Roberts, D. & Thompson, J.E. (1987). A low temperature induced decrease in 3-*trans*-hexadecenoic acid content and its influence on LHCII organisation. *Plant Physiology*, **84**, 12–18.

Joyard, J. & Douce, R. (1987). Galactolipid synthesis. In *The Biochemistry of Plants*, vol. 9, ed. P.K. Stumpf, pp. 215–74. New York: Academic Press.

Kader, J-C. & Mazliak, P. (eds.) (1995). *Plant Lipid Metabolism*. Dordrecht: Kluwer.

Kolattukudy, P.E. (1980). Cutin, suberin and waxes. In *The Biochemistry of Plants*, vol. 4, ed. P.K. Stumpf, pp. 571–646. New York: Academic Press.

Kolattukudy, P.E. (1987). Lipid-derived defensive polymers and waxes and their role in plant-microbe interaction. In *The Biochemistry of Plants*, vol. 9, ed. P.K. Stumpf, pp. 291–314. New York: Academic Press.

Kolesnick, R.N. (1991). Sphingomyelin and derivatives as cellular signals. *Progress in Lipid Research*, **30**, 1–38.

Krol, K., Huner, N.P.A., Williams, J.P. & Maissan, E. (1988). Chloroplast biogenesis at cold hardening temperatures. Kinetics of *trans*-3-hexadecenoic acid accumulation and the assembly of LHCII. *Photosynthesis Research*, **15**, 115–32.

Krupa, Z., Huner, N.P.A., Williams, J.P., Maissan, E. & James, D.R. (1987). Development at cold hardening temperatures. The structure and composition of purified rye light harvesting complex II. *Plant Physiology*, **84**, 19–24.

Lemieux, B. (1996). Molecular genetics of epicuticular wax synthesis. *Trends in Plant Science*, **1**, 312–18.

Luhs, W. & Friedt, W. (1994). The major oil crops. In *Designer Oil Crops*, ed. D.J. Murphy, pp. 5–71. Weinheim: VCH.

Lynch, D.V. & Phinney, A.J. (1995). The transbilayer distribution of glucosylceramide in plant plasma membrane. In *Plant Lipid Metabolism*, ed. J-C. Kader & P. Mazliak, pp. 239–41. Dordrecht: Kluwer.

Mazliak, P. (1977). Glyco- and phospholipids of biomembranes in higher plants. In *Lipids and Lipid Polymers in Higher Plants*, ed. M. Tevini & H.K. Lichtenthaler, pp. 48–74. Berlin: Springer-Verlag.

Mazliak, P. (1997). Lipids in cell signalling: a review. In *Physiology,*

Biochemistry and Molecular Biology of Plant Lipids, ed. J.P. Williams, M.U. Khan & N.W. Lem, pp. 151–3. Dordrecht: Kluwer.

Mudd, J.B. (1980). Phospholipid biosynthesis. In *The Biochemistry of Plants*, vol. 4, ed. P.K.Stumpf, pp. 250–82. New York: Academic Press.

Mudd, J.B. & Kleppinger-Sparace, K.F. (1987). Sulfolipids. In *The Biochemistry of Plants*, vol. 9, ed. P.K. Stumpf, pp. 275–89. New York: Academic Press.

Munnik, T., Arisz, S.A., De Vrije, T. & Musgrave, A. (1995). G protein activation stimulates phospholipase D signalling in plants. *The Plant Cell*, **7**, 2197–210.

Murata, N. (1983). Molecular species composition of phosphatidylglycerol from chilling-sensitive and chilling-resistant plants. *Plant and Cell Physiology*, **24**, 81–6.

Murphy, D.J. (1986). The molecular organisation of the photosynthetic membranes of higher plants. *Biochimica et Biophysica Acta*, **864**, 33–95.

Murphy, D.J. (ed.) (1994*a*). *Designer Oil Crops*. Weinheim: VCH.

Murphy, D.J. (1994*b*). Biotechnology of oil crops. In *Designer Oil Crops*, ed. D.J. Murphy, pp. 219–51. Weinheim: VCH.

Nawrath, C., Poirier, Y. & Somerville, C. (1994). Targeting of the polyhydroxybutyrate biosynthetic pathway to the plastids of *Arabidopsis thaliana* results in high levels of polymer accumulation. *Proceedings of the National Academy of Sciences, USA*, **91**, 12 760–4.

Norberg, P. & Liljenberg, C. (1991). Lipids of plant membranes prepared from oat root cells. *Plant Physiology*, **96**, 1136–41.

Packer, L. & Douce, R. (eds.) (1987). Plant cell membranes. *Methods Enzymology*, series ed. S.P. Colowick & N.O. Kaplan, vol. 148. New York: Academic Press.

Poirier, Y., Dennis, D.E., Klomparens, K. & Somerville, C. (1992). Polyhydroxybutyrate, a biodegradable thermoplastic produced in transgenic plants. *Science*, **256**, 520–3.

Pugh, C.E., Roy, A.B., Hawkes, T. & Harwood, J.L. (1995). A new pathway for the synthesis of the plant sulpholipid, sulphoquinovosyldiacylglycerol. *The Biochemical Journal*, **309**, 513–19.

Quinn, P. & Williams, W.P. (1983). The structural role of lipids in photosynthetic membranes. *Biochimica et Biophysica Acta*, **737**, 223–66.

Remy, R., Trémolières, A. & Ambard-Bretteville, G. (1984). Formation of oligomeric light-harvesting chlorophyll a/b protein by interaction between its monomeric form and liposomes. *Photobiochemistry and Photobiophysics*, **7**, 267–76.

Russell, N.J. & Harwood, J.L. (1979). Changes in the acyl lipid composition of photosynthetic bacteria grown under photosynthetic and non-photosynthetic conditions. *The Biochemical Journal*, **181**, 339–45.

Schmidt, H., Sperling, P. & Heinz, E. (1995). PCR-based cloning of membrane-bound desaturases. In *Plant Lipid Metabolism*, ed. J-C. Kader & P. Mazliak, pp. 21–3. Dordrecht: Kluwer.

Sen, A., Mannock, B.A., Collins, B.J., Quinn, P.J. & Williams, W.P. (1983). Thermotropic phase properties and structure of 1,2-distearoyl-galactosylglycerols in aqueous systems. *Proceedings of the Royal Society, Series B*, **218**, 349–64.

Shanklin, J., Whittle, E.J. & Fox, B.G. (1995). Membrane bound desaturases and hydroxylases: structure function studies. In *Plant Lipid Metabolism*, ed. J-C. Kader & P. Mazliak, pp. 18–20. Dordrecht: Kluwer.

Siedow, J.N. (1991). Plant lipoxygenase: structure and function. *Annual Review of Plant Physiology and Plant Molecular Biology*, **42**, 145–88.

Stymne, S. & Stobart, A.K. (1987). Triacylglycerol biosynthesis. In *The Biochemistry of Plants*, vol. 9, ed. P.K. Stumpf, pp. 175–214. New York: Academic Press.

Telfer, A., Hodges, M. & Barber, J. (1983). Analysis of chlorophyll fluorescence induction curves in the presence of dichlorophenyldimethylurea as a function of magnesium concentration and NADPH-activated light-harvesting chlorophyll a/b protein phosphorylation. *Biochimica et Biophysica Acta*, **724**, 167–75.

Testi, R. (1996). Sphingomyelin breakdown and cell fate. *Trends in Biochemical Sciences*, **21**, 468–71.

Trémolières, A., Garnier, J. & Dubertrot, G. (1992). Lipid protein interactions in relation to light energy distribution in photosynthetic membranes of eukaryotic organisms. In *Metabolism, Structure and Utilisation of Plant Lipids*, ed. A. Cherif, D.B. Miled-Daoud, B. Marzouk, A. Smaoui & M. Zarrouk, pp. 289–92. Tunis: Centre National Pedagogique.

Tuquet, C., Guillot-Salomon, T.D., de Lubac, M.F. & Signol, M. (1977). Granum formation and the presence of phosphatidylglycerol containing trans-3-hexadecenoic acid. *Plant Science Letters*, **8**, 59–64.

Uemura, M. & Steponkus, P.L. (1994). A contrast of the plasma membrane composition of oat and rye leaves in relation to freezing tolerance. *Plant Physiology*, **104**, 479–96.

Vick, B.A. & Zimmerman, D.C. (1987). Oxidative systems for modification of fatty acids: the lipoxygenase pathway. In *The Biochemistry of Plants*, vol. 9, ed. P.K. Stumpf, pp. 53–90. New York: Academic Press.

von Wettstein-Knowles, P.M. (1993). Waxes, cutin and suberin. In *Lipid Metabolism in Plants*, ed. T.S. Moore, pp. 127–66. Boca Raton: CRC Press.

Walton, T.E. (1990). Waxes, cutin and suberin. In *Methods in Plant*

Biochemistry, vol. 4, ed. J.L. Harwood & J.R. Bowyer, pp. 105–58. London: Academic Press.

Wang, X. (1993). Phospholipases. In *Lipid Metabolism in Plants*, ed. T.S. Moore, pp. 499–520. Boca Raton: CRC Press.

Wang, X., Xu, L. & Zheng, L. (1994). Cloning and expression of phosphatidylcholine-hydrolyzing phospholipase D from *Ricinus communis* L. *Journal of Biological Chemistry*, **269**, 20 312–17.

Wang, X., Zheng, S., Pappan, K. & Zheng, L. (1997). Characterisation of phospholipase D-overexpressed and expressed transgenic tobacco and *Arabidopsis*. In *Physiology, Biochemistry and Molecular Biology of Plant Lipids*, ed. J.P. Williams, M.U. Khan & N.W. Lem, pp. 345–7. Dordrecht: Kluwer.

Williams, J.P., Khan, M.U. & Lem, N.W. (eds.) (1997). *Physiology, Biochemistry and Molecular Biology of Plant Lipids*. Dordrecht: Kluwer.

Williams, W.P. (1994). The role of lipids in the structure and function of photosynthetic membranes. *Progress in Lipid Research*, **33**, 119–28.

PART I: *De novo* fatty acid biosynthesis

C. ALBAN and R. DOUCE

2 Biotin-dependent carboxylases and their biotinylation in higher plants

Introduction

Biotin is a water-soluble vitamin (vitamin H or B8), essential for life. Biosynthesis is limited to most bacteria, some fungi and plants. In all organisms, biotin is an essential cofactor for a small number of enzymes involved in the transfer of CO_2 during carboxylation, decarboxylation and transcarboxylation reactions (Knowles, 1989). Biotinylated proteins are rare in nature. For example, the only biotin-dependent carboxylase present in *Escherichia coli* is acetyl-CoA carboxylase (ACCase; EC 6.4.1.2), a multisubunit enzyme, in which one of the subunits is biotinylated and corresponds to the biotin carboxyl carrier protein (BCCP). Other bacteria contain one to three biotinylated proteins. Eukaryotic cells appear to contain a slightly greater number of biotinylated protein species. Thus, *Saccharomyces cerevisiae* contains four or five biotinylated proteins depending on growth conditions, whereas mammals are reported to contain four biotinylated proteins (Samols *et al.*, 1988; Knowles, 1989). The four biotin-dependent carboxylases found in mammals play crucial cellular housekeeping functions. Thus, ACCase that catalyses the ATP-dependent carboxylation of acetyl-CoA is recognised as the regulatory enzyme of lipogenesis; methylcrotonoyl-CoA carboxylase (MCCase; EC 6.4.1.4) catalyses the conversion of methylcrotonoyl-CoA to methylglutaconyl-CoA, a key reaction in the degradation pathway of Leu; propionyl-CoA carboxylase (PCCase; EC 6.4.1.3) is a key enzyme in the catabolic pathway of odd-numbered fatty acids and the amino acids, Ile, Thr, Met and Val; pyruvate carboxylase (PyrCase; EC 6.4.1.1) has an anaplerotic role in the formation of oxaloacetate (Fig. 1). The common feature of these reactions is the transfer of a carboxyl group from bicarbonate to an acceptor, utilising biotin as an intermediate carboxyl carrier. The reactions catalysed by these enzymes actually take place in two steps (1) and (2), resulting in the overall reaction (3):

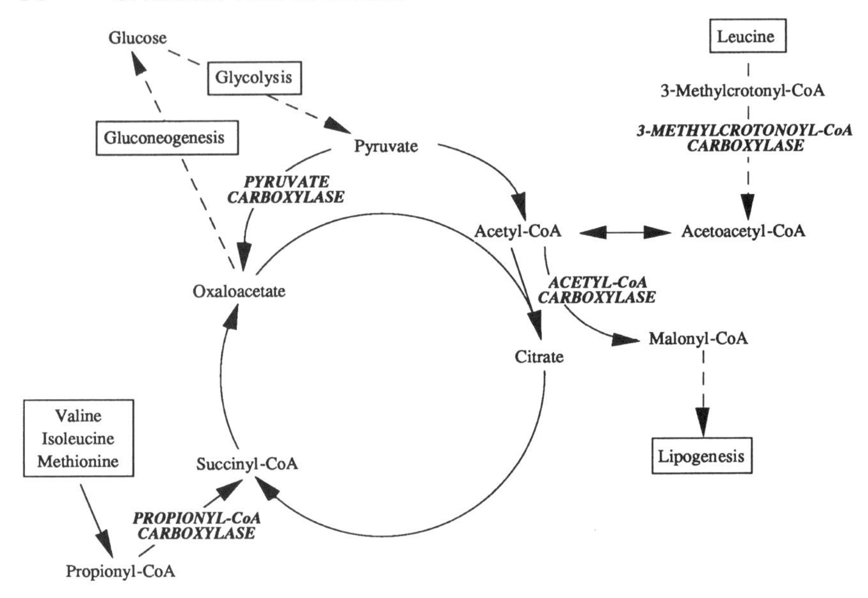

Fig. 1. Metabolic functions of biotin-dependent carboxylases in mammalian cells.

$$HCO_3^- + \text{Enzyme-Biotin} + \text{ATP-Mg} \rightarrow \text{Enzyme-Biotin-}CO_2^- + \text{ADP-Mg} + \text{Pi} \quad (1)$$

$$\text{Enzyme-Biotin-}CO_2^- + \text{Acceptor} \rightarrow \text{Acceptor-}CO_2^- + \text{Enzyme-Biotin} \quad (2)$$

$$HCO_3^- + \text{Acceptor} + \text{ATP-Mg} \rightarrow \text{Acceptor-}CO_2^- + \text{ADP-Mg} + \text{Pi} \quad (3)$$

The features that distinguish the reactions of each of these enzymes are the acceptor substrates. The family of biotin enzymes also includes oxaloacetate, methylmalonyl-CoA and glutaconyl-CoA decarboxylases that are involved in sodium transport in anaerobic prokaryotes as well as transcarboxylase (EC 2.1.3.1) that participates in propionic acid fermentation in *Propionibacterium shermanii*. These two last classes of enzymes do not require ATP as a substrate (Knowles, 1989). In all biotin enzymes described to date, the biotin is covalently linked to the ε amino group of a specific Lys residue located within a highly conserved (Ala/Val)-Met-Lys-(Met/Leu) tetrapeptide motif (found at the extremity of an hairpin loop) by the action of a biotin ligase (called biotin holocarboxylase synthetase in eukaryotes; EC 6.3.4.10) (for a review, see Samols *et al.*, 1988). Mutagenesis studies with the biotinyl domain from human propionyl-CoA carboxylase and the 1.3-S subunit of *Propionibacterium shermanii* transcarboxylase indicate that the methionine residues flanking the biotinylated lysine are not required for biotinylation despite their high conservation in biotin enzymes (Shenoy *et al.*, 1988; León-Del-Rio & Gravel, 1994). Rather, these amino acids, particularly -1 Met facilitate the carboxylation reaction and have a critical role in the carboxyl transfer reaction. A similar functional role was previously demonstrated for the corresponding -1 Met of the biotinylated subunit of *E. coli* ACCase (Kondo *et al.*, 1984). Consistent with this finding, Schatz (1993) reported the isolation from a peptide library of synthetic peptides that did not contain the consensus sequence found in natural biotinylated proteins but were efficiently biotinylated in *E. coli*, suggesting that these peptides somehow mimic the folded structure formed by the natural substrates. Finally, Duval *et al.* (1994*a*, *b*) recently discovered a novel seed-specific biotinyl protein (called SBP 65) devoid of biotin-dependent carboxylase activity, which binds covalently biotin to a Lys residue in a atypical biotinylation domain. The absence of the two Met residues flanking the biotinylated Lys found in biotin enzymes, thus account for the lack of known biotin-dependent carboxylase activity associated with SBP 65.

ACCase has been documented in plants since 1961 (Hatch & Stumpf, 1961). Investigations of plant ACCases increased in the late 1980s because of its regulatory role in fatty acid biosynthesis and because this enzyme represents the molecular target of powerful herbicides (the cyclohexanediones and the aryloxyphenoxypropionates) in use since the early 1980s and effective against grasses (the *Graminaceae*) including grass weeds (Harwood, 1988). The first direct evidence of the occurrence of biotin enzymes other than ACCase in plants was provided in 1990 by Wurtele and Nikolau. Using cell-free extracts of various monocot and dicot plant species, these authors detected in addition to ACCase activity, MCCase, PCCase and PyrCase activities, i.e. the four biotin-dependent activities found in mammals. Since 1990, numerous contributions to the knowledge of the structure, regulation and function of plant biotin-dependent carboxylases, as well as biotin biosynthesis and protein biotinylation processes in higher plants, were obtained from several laboratories over the world, including our group.

In this chapter, an attempt has been made to summarise these advances and their possible implications for industry, as for example the rational design of new herbicides.

Methylcrotonoyl-CoA carboxylase

MCCase catalyses the ATP-divalent cation-dependent carboxylation of 3-methylcrotonyl-CoA to form 3-methylglutaconyl-CoA, a reaction essential for Leu catabolism in animals and microbes. In addition, MCCase has been implicated in the recycling of carbon from mevalonic acid, a process called 'mevalonate shunt' (for a review, see Wood & Barden, 1977). In humans, the fatal, inherited disorder 3-methylcrotonylglycinuria has been shown to be the result of a deficiency of MCCase (Gompertz, Goodey & Barlett, 1973), thus explaining the further extensive investigations made about the functions, structures and chemical mechanism of catalysis of MCCase in bacteria and animals. In plants, the first significant characterisation of MCCase was obtained from Baldet *et al.* (1992) who demonstrated that all the biotin present in the matrix space of the plant mitochondrion (to a concentration of about 13 μM) was protein bound and associated with MCCase. Using subcellular fractionation procedures of isolated pea leaf protoplasts, these authors also demonstrated that plant MCCase was mostly, if not exclusively, located into mitochondria (Baldet *et al.*, 1992). Shortl after, MCCase activity was purified to homogeneity from pea leaf and potato tuber Percoll-purified mitochondria using avidin affinity chromatography (Alban *et al.*, 1993). Indeed, avidin binds biotin specifically and extremely tightly, with a dissociation constant (K_d) of 10^{-15}. Because the avidin–biotin interaction is so tight, it was necessary prior to chromatography to denature avidin to its monomeric form to reduce its affinity for biotin (K_d of 10^{-7}; Kohansky & Lane, 1985). The use of purified mitochondria and affinity chromatography appeared to be the most convenient way to purify plant MCCase to homogeneity (Alban *et al.*, 1993). Indeed, plant MCCase is very labile. Attempts to purify MCCase from crude extracts of carrot somatic embryos and from maize leaves led to the obtention of less pure and less active fractions, probably because of the presence of other biotin proteins in these extracts and the lack of protection of the enzyme by the mitochondrial membranes from proteases released from the vacuolar space during the course of tissue grinding (Chen *et al.*, 1993; Diez, Wurtele & Nikolau, 1994). Structure and biochemical properties of pure pea and potato enzymes were determined (Table 1) and were found to be similar to those reported for bacterial and mammal MCCases. Pea leaf and potato tuber MCCases are composed of two non-identical subunits. The larger subunit is biotinylated and has a molecular mass of 74–76 kDa,while the smaller subunit is biotin free and has an apparent molecular mass

Table 1. *Summary of the physicochemical and kinetic properties of mitochondrial MCCase from pea leaves and potato tubers*

	Pea MCCase	Potato MCCase
Native *M*r (gel filtration)	530 000	500 000
Subunit *M*r is:	β: 54 000	β: 53 000
	α: 76 000	α: 74 000
Biotin content (μg/mg protein)	1.83	1.9
Structure	α4β4	α4β4
K_m for 3-methylcrotonyl-CoA	0.1 mM	0.1 mM
K_m for ATP	0.1 mM	0.07 mM
K_m for HCO_3^-	0.9 mM	0.34 mM
Optimal pH	8.3	8.0–8.2
Optimal temperature	35 °C	38 °C

of 53–54 kDa (Table 1). The native enzyme was determined to have an octameric structure (α4β4). Indeed, gel filtration experiments disclosed a M_r in the range 500 000 to 530 000 for both purified pea and potato MCCases. Secondly, the biotin content of plant MCCase was consistent with a stoichiometry of four molecules per α4β4 tetramer, presumably with each α subunit containing one biotin prosthetic group (Alban *et al.*, 1993; Table 1). In this respect, pea and potato MCCases resemble bacterial MCCase (Apitz-Castro, Rehn & Lynen, 1970) but differs from mammalian, carrot and maize MCCases, which seems to exhibit a dodecameric structure (α6β6) (Lau, Cochran & Fall, 1980; Chen *et al.*, 1993; Diez *et al.*, 1994). Steady-state kinetic experiments demonstrated that the plant MCCase-catalysed reaction proceeds by a double-displacement mechanism ('Bi Bi Uni Uni ping–pong' mechanism) where ATP and bicarbonate bind to the enzyme in a first half reaction and react to form carboxybiotinyl enzyme. Then, ADP and Pi produced by hydrolysis of ATP are released. In a second half-reaction, 3-methyl-crotonyl-CoA binds to the carboxybiotinyl enzyme and is carboxylated to form 3-methylglutaconyl-CoA (Alban *et al.*, 1993; Diez *et al.*, 1994). Nothing is known concerning factors regulating MCCase activity in plants. However, *in vitro*, MCCase activity is modulated by several metabolites. Purified enzyme is significantly inhibited by the reaction end products, ADP and Pi (Alban *et al.*, 1993) and by acetoacetyl-CoA, an end product of Leu degradation (Chen *et al.*, 1993). Plant MCCase is also sensitive to N-ethylmaleimide, a sulphydryl reagent, and to phenylglyoxal, a reagent known to modify Arg residues. The inhibition

by these products is prevented by preincubation of MCCase with ATP and 3-methylcrotonyl-CoA suggesting that Cys and Arg residues may be involved in catalysis or regulation of the enzyme activity (Alban *et al.*, 1993; Diez *et al.*, 1994). Recently, it was shown that, in tomato, MCCase activity is lower in leaves than in roots (Wang, Wurtele & Nikolau, 1995). However, the steady-state levels of the biotinyl subunit of MCCase and its mRNA are approximately equal in both roots and leaves. Instead, labelling experiments with ^{125}I-streptavidin suggested that the lower activity of MCCase in leaves could be attributed to the reduced biotinylation of the biotin-containing subunit of the enzyme, and consistent with this hypothesis, a pool of non-biotinylated enzyme was found in leaves (Wang *et al.*, 1995). These data strongly suggest that the relative biotinylation of a biotin-containing enzyme is a significant mechanism for regulating its activity. Nevertheless, by directly measuring the biotin holocarboxylase synthetase activity, using a very sensitive radioactive test, Tissot *et al.* (1996) were not able to detect any biotin incorporation in pea leaf extracts at any stage of the plant development, until they add exogenous apo-carboxylase substrate to the reaction medium, indicating that the level of apo-carboxylases in pea leaves was negligible. The isolation of cDNAs and genes coding for the biotinylated subunit of MCCase revealed that this subunit contains the functional domains for the first half-reaction catalysed by all biotin-dependent carboxylases namely, the carboxylation of biotin (Song, Wurtele & Nikolau, 1994). These domains are arranged serially on the polypeptide, with the biotin carboxylase domain at the N terminus and the biotin-carboxyl carrier domain at the C terminus. Cloning of the mRNA and gene coding for the biotin-free smaller subunit of this enzyme will be very useful to study the regulation of the coordinate expression of both MCCase subunits, to understand how these polypeptides associate together to form an active MCCase and if this association plays any role in regulation of MCCase activity in plants.

Despite these biochemical and molecular characterisations, the metabolic function of MCCase in plants is still unclear. As in mammals, MCCase may be involved in the mitochondrial catabolism of Leu and/or in the 'mevalonate shunt' by which mevalonate can be metabolised to non-isoprenoid compounds. Radiotracer studies of plants provided with either ^{14}C-Leu (Stewart & Beevers, 1967) or ^{14}C-mevalonate (Nes & Bach, 1985) indicate that Leu catabolism and the mevalonate shunt operate *in vivo*. Whether these two pathways implicate MCCase remains to be demonstrated. Indeed, little is known concerning the catabolism of amino acids in plants. In particular, the degradation pathway(s) of branched-chain amino acids (Leu, Ile and Val) in plants are poorly understood. The first step leads by transamination to branched-chain 2-oxo acids (2-oxoisocaproate in the case of Leu), whose oxidative decarboxylation (to isovaleryl-CoA in the case of Leu) and further oxidation (to 3-methylcrotonyl-CoA in the case of Leu) have been shown to take place in peroxisomes (Gerbling & Gerhardt, 1989). These authors suggested that further steps could also involve peroxisomal enzymes. However, the fate of 3-methylcrotonyl-CoA was not clearly identified. In mammalian mitochondria and in bacteria this metabolite is carboxylated to 3-methylglutaconyl-CoA by MCCase. Gerbling and Gerhardt (1989) suggested that an extraperoxisomal pathway is involved in Leu catabolism, since peroxisomes were unable to carboxylate 3-methylcrotonyl-CoA. However, despite the lack of direct evidence for the implication of plant MCCase in Leu catabolism, a number of circumstantial evidences indicate that the pathway may be similar to that in animals and bacteria. Plant MCCase is constitutively expressed, but its activity increases strongly during leaf senescence where intense degradation of proteins occurs (Alban *et al*., 1993). In addition, MCCase activity is also induced in cotyledons during the mobilisation of storage proteins for seedling growth (Duval *et al*., 1994*a*). Finally, Leu after Asn, is the second most abundant amino acid which transiently accumulates in higher plant cells during prolonged carbon starvation, in relation with the induction of proteolytic activities leading to a massive breakdown of proteins (Genix *et al*., 1990). These amino acids, as well as the fatty acids deriving from the lipid breakdown, are used in place of sugar to fuel the respiration of the mitochondria spared by autophagy. Interestingly, MCCase activity is induced in sycamore starved cells (up to sixfold increase in specific activity after four days of sucrose starvation) in correlation with an accumulation of the enzyme in mitochondria (Aubert *et al*., 1996). Altogether, these observations are consistent with the idea that the plant MCCase, as for its animal counterpart, is involved in the degradation of Leu derived from protein mobilisation. Indeed, in these developmental situations (senescence and germination), as well as during carbohydrate starvation, proteins are massively degraded and amino acids are used as carbon and energy sources.

Acetyl-CoA carboxylase

ACCase (acetyl-CoA: carbon dioxide ligase (ADP-forming)) catalyses the ATP-Mg-dependent carboxylation of acetyl-CoA to form malonyl-CoA, with bicarbonate as a source of carbon. The reaction is the first committed step in the synthesis of fatty acids. Indeed, in plant cells, large amounts of malonyl-CoA are needed in the plastids to sustain *de novo* fatty acid synthesis ($C_{16:0}$, $C_{18:0}$, $C_{18:1}$), but malonyl-CoA is also needed in the cytosol in a variety of reactions including elongation of very long chain fatty acids, the synthesis of secondary metabolites, such as flavonoids, anthocyanins, stilbenoids and malonylation of some amino acids and secondary metabolites (for a review, see Harwood, 1988). Because of the occurrence of cytosolic reactions needing malonyl-CoA, and because plastid is not permeable to malonyl-CoA, this led to the hypothesis that at least two isoenzymes of ACCase, a plastidic isoenzyme and a cytosolic isoenzyme are present in plants (for a review, see Harwood, 1991).

Interest in the plant ACCase has increased with the discovery that it is the primary target site of several major classes of grass herbicides, affecting fatty acid synthesis, principally the cyclohexanediones (CHD) and the aryloxyphenoxypropionates (APP) (reviewed by Harwood, 1988). CHD and APP herbicides are reversible and linear inhibitors of grass ACCase, and different kinetic analyses indicate that these compounds inhibit the carboxyltransferase partial reaction by acting as a transition-state analogue of the complex formed at this site (for a review, see Harwood, 1996). This last observation seems logical since this is the partial reaction unique to individual biotin-dependent carboxylases, and inhibition here is in keeping with the specificity of these compounds for plant ACCase. The other partial reaction, the carboxylation of biotin, is common to all biotin enzymes, and inhibition here would render sensitive these other carboxylases, which is not the case (Alban *et al.*, 1993). Double-inhibition kinetic analysis revealed that the CHD and the APP are mutually exclusive inhibitors of ACCase, i.e. the binding of a herbicide from one class prevents binding of a herbicide from the other class and vice versa (Burton *et al.*, 1991). This suggests that the two classes bind in overlapping domains, or alternatively that they bind in separate locations, but in that case the binding of one class, through allosteric effects, would prevent the binding of the other. Most grass species are sensitive to CHD and APP herbicides whereas broad-leaved plants and monocotyledonous species other than grasses are resistant (Harwood, 1991). With few exceptions, the selectivity of these compounds is expressed at the target site. This situation is uncommon among herbicide classes since selectivity is usually determined by the ability of plants to metabolise herbicides, or by differences in uptake and/or translocation of the active compound. In the few cases where grass species exhibit tolerance to these herbicides, it is principally due to the ability of the plant, wheat for example, to metabolise the herbicide (Devine & Shimabukuro, 1994). There are, however, several examples where herbicide resistance in grasses is due to the presence of a tolerant form of ACCase, as in natural populations of red fescue (*Festuca rubra*) and annual meadow grass (*Poa annua*) (see, for example, Herbert *et al.*, 1996*a*). Also, widespread herbicide use has led to the selection of resistant biotypes in weeds that were previously susceptible (for a review, see Devine & Shimabukuro, 1994). The main resistance mechanism in these cases involves modification of the target site ACCase. Recently, resistance to ACCase-inhibiting herbicides of a biotype of *Lolium rigidum* following selection with diclofop-methyl for 10 consecutive years, was found to correlate with possession of a modified resistant form of ACCase, controlled by a single major gene (Tardif *et al.*, 1996). Nevertheless, the structural or molecular basis for the difference in herbicide sensitivity of the enzymes from grasses and all the other plants, particularly dicotyledons remained unknown for a long time. The recent elucidation of the two different types of ACCase structures in plants, together with the discovery of ACCase isoforms in different cell and tissue compartments has allowed, at least in a large part, to answer this intriguing question.

The molecular organisation of ACCase differs according to the source of the enzyme. In *Escherichia coli*, heteromeric ACCase is composed of four distinct subunits that readily dissociate into three components: a dimer of 49 kDa subunits (biotin carboxylase module; BC), a dimer of 17 kDa subunits (biotin carboxyl carrier module; BCCP) to which biotin is covalently bound, and a carboxyltransferase tetramer (CT) containing two 33 kDa and two 35 kDa ($\alpha2\beta2$) subunits. These are encoded by four separate genes, which in *E. coli* are named *acc*C, *acc*B, *acc*A and *acc*D, respectively (Alix, 1989; Kondo *et al.*, 1991; Li & Cronan, 1992*a*, *b*). In contrast, in animals, fungi and yeasts, these entities are located on a single, multifunctional polypeptide having a molecular mass exceeding 200 kDa (Lopez-Casillas *et al.*, 1988; Al-Feel, Chirala & Wakil, 1992; Bailey *et al.*, 1995). The functional ACCase enzyme in these organisms is composed of multimers of this large polypeptide. The structure of the plant ACCase has been a subject of considerable confusion in the past. Early experiments indicated that ACCase in chloroplasts have a structure similar to that found in prokaryotes or represented an intermediate form (Kannangara & Stumpf, 1972). It was also postulated that plants may contain distinct isozymes of ACCase in separate spatial and temporal compartments for the generation of malonyl-CoA needed for a variety of phytochemicals (Stumpf, 1980). However, the subsequent inclusion in purification media of effective proteinase inhibitors and the development of avidin-affinity columns allowing a rapid purification of the enzyme led to the isolation of a more stable homomeric ACCase comparable to that found in other eukaryotes (for reviews, see Harwood, 1988, 1991). Then, it was claimed and generally agreed that the occurrence of low-molecular-mass biotinyl polypeptides in purified ACCase fractions was essentially, if not exclusively, due to a severe degradation of the high-molecular-mass polypeptide form by endogenous proteinases during the course of ACCase isolation (for reviews, see Harwood, 1988, 1991). Consequently, the concept of a prokaryotic form of ACCase in plants fell into disfavour. On the other hand, as in the same time, evidence for compartmentation of ACCase was not obtained, this last hypothesis was rejected as well. And yet, it has now become evident that both of these hypotheses are perfectly correct. The confusion arose in large part because first, plants contain structurally different forms of the enzyme, one of which easily loses activity during attempts to characterise it, not due to proteases inactivation, but rather because it readily dissociates into its constituents, and secondly because the enzymes from Gramineae were mainly studied. Indeed, it now appears that most plants other than Gramineae have a multisubunit plastidial ACCase (frequently referred to as prokaryotic form) and a multifunctional extra-plastidial ACCase (also called eukaryotic form), while members of the Gramineae have only the multifunctional-type ACCase in both plastids and cytosol (Fig. 2) (Egli *et al.*, 1993; Alban *et al.*, 1994; Sasaki, Konishi & Nagano,

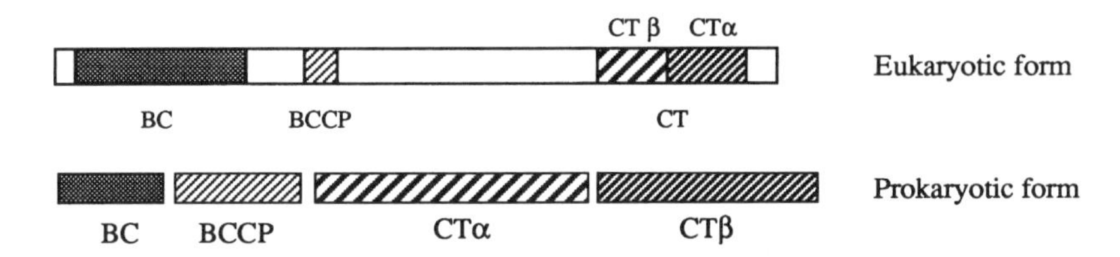

Plant \ Location	Plastids	Cytosol of epidermal cells
Gramineae	Eukaryotic form	Eukaryotic form
other plants	Prokaryotic form	Eukaryotic form

Fig. 2. Structure and compartmentation of the two forms of ACCase in plants. BC: biotin carboxylase; BCCP: biotin carboxyl carrier protein; CT, carboxyltransferase.

1995; Konishi *et al.*, 1996*a*). The first significant feature that revived the idea that plant ACCase consisted of several different subunits was the discovery of an *acc*D homologue (coding for a putative polypeptide exhibiting sequence similarities with the β subunit of the carboxyltransferase component of *E. coli* ACCase) in pea chloroplast genome (Li & Cronan, 1992*c*). Indeed, it was further shown that the product of this gene in pea bound at least two other polypeptides, one of which was biotinylated (Sasaki *et al.*, 1993). Shortly after, direct biochemical evidences definitively assessed the presence of a prokaryotic form of ACCase in pea chloroplasts (Alban, Baldet & Douce, 1994; Sasaki *et al.*, 1995). Indeed, chloroplasts of pea mesophyll cells contain an ACCase that is a heteromeric enzyme consisting of subunits of different sizes, one of which of 38 kDa is biotinylated, for a native molecular mass >600 kDa. This form of the enzyme represents about 80% of total ACCase activity in the whole leaf (Alban *et al.*, 1994) and is totally insensitive to herbicides of the CHD and APP classes (Alban *et al.*, 1994; Sasaki *et al.*, 1995; Dehaye *et al.*, 1994). The enzyme subunits are organised in two functional domains interacting through ionic interactions, that are readily dissociable and reassociable (Alban *et al.*, 1994; Shorrosh *et al.*, 1996). The chloroplastic ACCase is able to carboxylate free D-biotin as an alternate substrate in lieu of biotin carboxyl carrier protein and in the absence of acetyl-CoA. This specific property of ACCase was used to purify the biotin carboxylase component of the enzyme (Alban *et al.*, 1995). The minimal structure of the active biotin carboxylase domain corresponds to a complex of two polypeptides in tight interaction including the 38 kDa biotinylated subunit and a 32 kDa biotin-free polypeptide (Alban *et al.*, 1995). This contrasts with bacterial ACCase organisation where all the components are known to be freely separable (Guchhait *et al.*, 1974). Steady-state kinetic results with pea biotin carboxylase were compatible with an ordered mechanism in which MgATP binds first, followed by free biotin and then bicarbonate. Consistent with this mechanism, bicarbonate-dependent ATP hydrolysis by the enzyme could only be observed in the presence of added D-biotin (Alban *et al.*, 1995). These data suggested that functional differences may also exist between the multisubunit isoform of ACCase found in plants and that in bacteria (Knowles, 1989). cDNAs encoding BCCP and BC subunits from various plants have been cloned (Shorrosh *et al.*, 1995; Choi *et al.*, 1995; Elborough *et al.*, 1996). These subunits are nuclear encoded and share substantial sequence similarity with the bacterial ACCase subunits. Furthermore, their expression is coordinated during seed and leaf development (Roesler *et al.*, 1996). The second functional domain of ACCase found in pea chloroplasts is composed of the α and

β subunits of CT (Shorrosh *et al.*, 1996). The α subunit (91 kDa) which was recently characterised is nuclear encoded and presents a structural motif similar to a prokaryotic membrane lipoprotein lipid attachment site. As this polypeptide was previously identified as an inner envelope membrane constituent of pea chloroplasts, this observation suggests that the prokaryotic form of ACCase in pea may be at least partly associated with chloroplast membranes (Shorrosh *et al.*, 1996). The gene encoding the β subunit (65 to 80 kDa) has been identified in the plastid genome by its similarity to one of the CT subunits of *E. coli* ACCase (Li & Cronan, 1992*c*; Sasaki *et al.*, 1993; Elborough *et al.*, 1996). This is the only component of plant lipid metabolism known to be encoded by the plastid genome.

In addition to the major prokaryotic form, pea leaves also contain a minor eukaryotic form of ACCase mainly if not exclusively concentrated in the epidermis (Alban *et al.*, 1994). In contrast, in pea seeds this ACCase contributes the major portion of the enzyme activity (Bettey, Ireland & Smith, 1992; Dehaye *et al.*, 1994). This enzyme is a homodimer of a single biotinylated polypeptide of about 220 kDa and is sensitive to the herbicides of the APP class, such as diclofop and quizalofop (Alban *et al.*, 1994; Dehaye *et al.*, 1994). Even though this was not demonstrated directly, it is reasonable to think that the eukaryotic form of ACCase in pea has an extraplastidial location, probably the cytosol. Indeed, several genes and cDNA clones have been isolated for this type of ACCase from other dicotyledonous plants, that do not seem to present typical organelle-targeting presequences (for reviews see Harwood, 1996; Ohlrogge & Browse, 1995). All of these clones encode a single polypeptide with the biotin carboxylase domain at the N-terminus, the biotin carboxyl carrier domain in the middle, and the carboxyltransferase domain at the C-terminus. A non-negligible portion of the enzyme situated between the two last domains is not assigned by any component of *E. coli* ACCase, and may play a role in bringing the N-terminal active region close to the C-terminal active region by its suitable folding. Grasses like maize and wheat also contain two ACCase isoforms, the major one in the chloroplasts of mesophyll cells, the other in an extraplastidial compartment (probably cytosol) of other cell type (Egli *et al.*, 1993; Gornicki & Haselkorn, 1993; Herbert *et al.*, 1996*b*). But, in contrast to dicotyledonous plants, both isoforms are composed of a single type of high-molecular-mass biotin-containing polypeptide of 227 kDa (ACCase 1, mesophyll chloroplast enzyme) or 219 kDa (ACCase 2, non-mesophyll-chloroplast enzyme) (Egli *et al.*, 1993; Ashton *et al.*, 1994). Antibodies raised to the 227 kDa form were poorly recognised by the minor 219 kDa isoform. Indeed, both ACCase isoforms are encoded by distinct nuclear genes (Ashton, Jenkins & Whitfeld, 1994). Finally, both enzymes are inhibited by CHD and APP herbicides, although ACCase 1 is much more sensitive than ACCase 2 (Egli *et al.*, 1993; Herbert *et al.*, 1996*b*). Thus, it now appears that all plants contain two ACCase isoforms in two subcellular and tissue different locations. However, the Gramineae family of plants is different in that both the plastid and cytosolic ACCase isozymes are of the eukaryotic type. Coincident with this evolutionary difference, it is interesting to note the recent finding that probably all the plants except for those in Gramineae contain the *acc*D gene, encoding for the β subunit of carboxyltransferase, in their plastid genome (Konishi *et al.*, 1996a). Consistent with this finding, both Western and Southern blot analyses were unable to detect subunits of the prokaryotic form of ACCase in all grass species studied, while prokaryotic and eukaryotic forms of ACCase were detected in all the other plants analysed including monocotyledons and dicotyledons (Konishi *et al.*, 1996*a*; Roesler *et al.*, 1996). Thus, the difference in ACCase organisation in the Gramineae has now provided an explanation for the action of grass-specific herbicides, which inhibit only the eukaryotic form of the enzyme and not the prokaryotic one (Alban *et al.*, 1994; Sasaki *et al.*, 1995). Indeed, the chloroplastic ACCase which is believed to play a key role in *de novo* fatty acid biosynthesis (the target of graminicides), is strongly inhibited by these compounds in the Gramineae (eukaryotic form) and not in other plants (prokaryotic form). On the other hand, the cytosolic ACCase (eukaryotic form) is only a minor form in leaves, probably involved in secondary pathways, and is much less affected by herbicides.

In addition to these differences in sensitivity to herbicides, the two structurally distinct ACCase isozymes have important differences in their biochemical properties. The multifunctional enzyme has a much lower K_m for acetyl-CoA than the multisubunit complex and has the ability to carboxylate propionyl-CoA at substantial rates, which the multisubunit complex does not (Alban *et al.*, 1994; Dehaye *et al.*, 1994; Egli *et al.*, 1993). Furthermore, no other propionyl-CoA carboxylase activity, different from that catalysed by the eukaryotic forms of ACCase, can be detected from either reproductive and vegetative organs of pea plants and from maize leaves at any stage of development (Dehaye *et al.*, 1994; Herbert *et al.*, 1996*b*). Dehaye *et al.* (1994) investigated therefore, steady-state kinetics of the multifunctional form of ACCase as purified from mature pea seeds with respect to the substrate specificity and inhibition by quizalofop. These authors demonstrated that both reactions catalysed by the enzyme (acetyl-CoA and propionyl-CoA carboxylations) occur at separate sites on the enzyme and are inhibited by quizalofop. One site binds either acetyl-CoA or propionyl-CoA and is inhibited by quizalofop. The other is specific for acetyl-CoA and is much less affected by quizalofop. Owing to the existence of these two catalytically different sites, the enzyme obeyed Michaelis–Menten kinetics with propionyl-CoA, but exhibited kinetic co-operativity in the presence of acetyl-CoA. In addition, kinetics of propionyl-CoA carboxylase activity exhibited hyperbolic inhibition in the presence of quizalofop, but co-operative inhibition when following the ACCase activity of the enzyme. These results suggested that the higher the substrate specificity the lower the quizalofop sensitivity of the active site, indicating that the nature of the interactions between the two identical enzyme subunits may play some role in governing the strength of herbicide sensitivity. Presumably, such interactions introduce some distortions in the structure of the individual active sites, for example, generating kinetic co-operativity with substrates and inhibitors (Dehaye *et al.*, 1994). Similar conclusions were drawn from kinetic studies with maize ACCases (Herbert *et al.*, 1996*b*). Also, the apparent correlation between the substrate specificity and the sensitivity of ACCase toward quizalofop was confirmed by kinetic analyses of the prokaryotic form of ACCase from pea leaf chloroplasts. This enzyme, which is insensitive to quizalofop inhibition, was not able to carboxylate propionyl-CoA (Dehaye *et al.*, 1994).

The discovery of the occurrence of ACCase isoforms in separate plant cell compartments raises the question of their physiological significance. The main, if not the exclusive, role of ACCase in plastids is to provide the precursor, malonyl-CoA, for *de novo* fatty acid biosynthesis. Indeed, plastids represent the major site for fatty acid biosynthesis in plants (Ohlrogge *et al.*, 1979). Furthermore, the level of the chloroplastic form of ACCase in plants shows considerable variations during development, being maximum in young leaves, presumably reflecting a strong demand in fatty acids required for the biosynthesis of chloroplastic glycerolipids (Dehaye *et al.*, 1994). Indeed, during the biogenesis of thylakoid membranes, chloroplasts are the most important sites of glycerolipids biosynthesis in green leaves. Considerable *in vivo* evidence suggests that chloroplastic ACCase is involved in regulation of plant fatty acid synthesis. Thus, analysis of substrate and product pool size implicated ACCase in the light/dark regulation of fatty acid synthesis in spinach leaves and chloroplasts (Post-Beittenmiller, Jaworski & Ohlrogge, 1991; Post-Beittenmiller, Roughan & Ohlrogge, 1992). Herbicide inhibition of ACCase was used to determine flux control coefficients, and led to the conclusion that chloroplastic ACCase exerts major control over fatty acid synthesis rates in barley and maize leaves (Page, Okada & Harwood, 1994). ACCase was also the apparent site of feedback inhibition of fatty acid synthesis in tobacco suspension cells supplemented with exogenous fatty acids (Shintani & Ohlrogge, 1995). Finally, recent *in vivo* evidence for a regulatory role of ACCase in oilseeds was obtained by means of genetic engineering approaches (Roesler *et al.*, 1997). Thus, now that clones are available for chloroplastic ACCases, transgenic plant experiments will provide crucial *in vivo* tests of the role of this enzyme in controlling flux through the fatty acid biosynthetic pathway. At present, there is no evidence that *de novo* fatty acid synthesis occurs in the cytosol of plant cell. It is interesting to note, at this level, the recent finding that plant mitochondria contain not only ACP, but also all the enzymes required for *de novo* fatty acid synthesis, from malonate but not from acetate. Thus, in the present case, ACCase is probably not required (Wada, Shintani & Ohlrogge, 1997). Indeed, plant mitochondria do not present detectable levels of ACCase activity (Baldet *et al.*, 1992; Alban *et al.*, 1993).

Although it is recognised that malonyl-CoA is required, in addition to fatty acid synthesis, in a number of biosynthetic pathways, the physiological role of the multifunctional cytosolic ACCase is not yet clearly established, and it can be involved in the biosynthesis of very-long-chain fatty acids required for cuticular waxes, or the biosynthesis of flavonoids via chalcone synthase. In this context, it is interesting that, in pea leaves, both processes occur in the cytosol of epidermal cells, thus matching the tissue localisation of the eukaryotic extraplastidial form of pea leaf ACCase (Alban *et al.*, 1994). Cuticular wax and flavonoids are important in the interaction of plants with their environment, for example, for protection against UV light and pathogens. Interestingly, two recent findings indicate that this isozyme of ACCase probably helps to control the synthesis of these protective compounds, when necessary: first the transcript for the eukaryotic form of alfalfa ACCase is induced by yeast or fungal elicitors of isoflavonoid phytoalexin synthesis (Shorrosh *et al.*, 1994), secondly, the cytosolic eukaryotic form but not the chloroplastic prokaryotic form of ACCase is induced by UV-B irradiation of fully expanded pea leaves, in parallel to the induction of some of the enzymes involved in flavonoid synthesis, including chalcone synthase (Konishi *et al.*, 1996*b*). The mammalian and yeast ACCases are regulated by phosphorylation/dephosphorylation through the action of an AMP-activated protein kinase. There is as yet no evidence that such a mechanism may play a significant role in regulation of both forms of plant ACCases.

Biotin biosynthesis pathway

Plants, like microorganisms, have the ability to synthesise biotin. In contrast, other multicellular eukaryotic organisms are biotin auxotrophs. Biotin biosynthesis has been well characterised in *E. coli*, and more recently in other bacteria such as *Bacillus sphaericus*, by combined biochemical and genetic studies. In these bacteria, the *bio(ABFCD)* locus contains genes that are required for biotin biosynthesis (Eisenberg, 1987; Ifuku *et al.*, 1994). These genes encode for the enzymes catalysing the synthesis of biotin from pimeloyl-CoA and alanine: 7-keto-8-aminopelargonic acid synthase (KAPA synthase), coded by *bioF*; 7,8-diaminopelargonic acid synthase (DAPA) aminotransferase, coded by *bioA*; dethiobiotin synthase, coded by *bioD*, and biotin synthase, coded by *bioB*. Until recently, little was known about biotin metabolism in plants. Initial information on biotin synthesis and transport in higher plants came from analysis of the *bio1* biotin auxotroph of *A. thaliana* discovered by Schneider *et al.* (1989). Seeds homozygous for the mutation failed to develop unless exogenous biotin, dethiobiotin or DAPA but not KAPA was supplied to the plant (Shellhamer & Meinke, 1990). Recently, Patton, Volrath and Ward (1996*a*) have shown that the *E. coli bioA* gene, which codes for DAPA aminotransferase, can genetically complement the *bio1* mutation, demonstrating that *bio1/bio1* mutant plants are defective in this enzyme. By treatment of a biotin overexpressing strain of lavender cells with ^{3}H-pimelic acid, Baldet *et al.* (1993*a*), showed that all the intermediates of biotin synthesis established in bacteria plus the novel metabolite 9-mercaptodethiobiotin (9-mDTB) accumulate in plants, demonstrating that the pathway of biotin synthesis in bacteria is conserved in higher plants and that the reaction catalysed by biotin synthase may proceed in two distinct steps involving 9-mDTB as an intermediate. The catalytic mechanism of the last step of biotin biosynthesis is still unclear. Thus, it has been demonstrated in *E. coli* that the conversion of dethiobiotin to biotin is catalysed by a complex involving two (or three) proteins in addition to biotin synthase which alone is not able to catalyse this reaction (Ifuku *et al.*, 1994). By functional complementation of a *bioB* biotin auxotroph mutant of *E. coli*, Baldet and Ruffet (1996) have isolated a cDNA encoding *A. thaliana* biotin synthase. This cDNA shows specific regions of similarity with biotin synthase gene from bacteria and yeast. The predicted amino acid sequence of the plant protein contains the consensus region GX*C*XED*C*XY*C*XQ involved in a (2Fe-2S) cluster binding. Interestingly, the plant sequence contains an N-terminal extension of about 40 amino acids that is not found in the bacterial counterparts, suggesting an organellar location for this enzyme (Baldet & Ruffet, 1996). Indeed, computer analysis of the primary sequence of the *A. thaliana* biotin synthase predicted that this protein was targeted to mitochondria (Weaver *et al.*, 1996; Patton *et al.*, 1996*b*). Also, Western blot analyses using antibodies raised against the plant recombinant enzyme confirmed its mitochondrial location (P. Baldet, C. Alban and R. Douce, unpublished data). This subcellular location for plant biotin synthase is intriguing since most of the free biotin pool in plant mesophyll cells have been shown to be concentrated in the cytosol to a level of about 11 μM (Baldet *et al.*, 1993*b*). Thus, if biotin is well synthesised into mitochondria, it would need to be exported to accumulate in the cytosol. The precise role of the free-biotin pool in cytosol is not known. One possibility would be that, as in bacteria, the level of free biotin might control the expression of genes encoding the biotin-containing enzymes and/or the enzymes involved in biotin synthesis. In support of this suggestion, it was demonstrated that the expression of the gene encoding biotin synthase in *A. thaliana* is strongly induced during biotin-limited conditions (Patton *et al.*, 1996*b*). Finally, the fact that the *E. coli bioB* mutant used to clone the plant biotin synthase is unable to grow on minimal medium supplemented with dethiobiotin, but addition of 9-mDTB restores the growth of the mutant could indicate that the isolated cDNA rather corresponds to the enzyme which is involved in the insertion of sulphur atom to dethiobiotin, and not to the enzyme which converts 9-mDTB into biotin (Baldet & Ruffet, 1996).

Biotin holocarboxylase synthetase

Biotinylation of biotin-dependent carboxylases is an important process that allows the transformation of inactive apo-carboxylases in active holo-forms. Thus, as suggested by Wang *et al.* (1995), the relative biotinylation of a biotin-containing enzyme may be a significant mechanism for regulating its activity. In *E. coli*, the biotinylation of apo ACCase (the unique biotin-enzyme found in this bacteria) is realised by biotin ligase (EC 6.3.4.10). The enzyme catalyses the post-translational incorporation of D-biotin to a specific Lys residue of newly synthesised apoenzyme, via an amide linkage between the biotin carboxyl group and a unique ε-amino group of Lys residue (Samols *et al.*, 1988). This covalent attachment, essential for enzymatic activation of apo-ACCase into holo-ACCase, occurs in two steps (1) and (2) as follows:

$$\text{D-biotin} + \text{ATP} \rightarrow \text{D-biotinyl 5'-AMP} + \text{PPi} \quad (1)$$

D-biotinyl 5′-AMP + apo-ACCase → holo-ACCase + AMP (2)

The first step is the activation of D-biotin by ATP, which yields D-biotinyl 5′-AMP. This is then followed by the covalent attachment of the biotinyl group to the ε-amino group of a specific Lys residue of the apo-ACCase, with release of AMP. Biotin ligase has been purified from *E. coli* and its gene cloned (Howard, Shaw & Otsuka, 1985). This enzyme, also called Bir A, is a 33.5 kDa protein that also acts as a repressor of the biotin operon. The corresponding enzymes from various mammalian species have been purified and were referred to as biotin holocarboxylase synthetase (HCS) (Xia, Zhang & Ahmad, 1994; Chiba *et al.*, 1994). Recently, clones encoding *Saccharomyces cerevisiae* HCS gene (Cronan & Wallace, 1995) and human HCS cDNAs have been obtained (Suzuki *et al.*, 1994; León-Del-Rio *et al.*, 1995). These latter enzymes show some sequence similarity to the biotin ligases from bacteria, but are more than twice their size. The first direct evidence for the existence of HCS activity in plants was provided by Tissot *et al.* (1996). These authors partially purified and characterised biochemically HCS activity from pea leaves. The enzyme activity was assayed using ^{3}H-biotin and bacterial apo-biotin carboxyl carrier protein from a *birA E. coli* mutant lacking *in vitro* biotin ligase activity, as substrates. These data indicate that plant HCS acts across species barriers, being able to biotinylate bacterial apo-carboxylase. This cross-species activity reveals a molecular mechanism common to these enzymes. In contrast, plant HCS showed a very high specificity for its substrate biotin, exhibiting an apparent K_m value of 28 nM (Tissot *et al.*, 1996). Fractionation of pea leaf protoplasts and purification of chloroplasts and mitochondria from this tissue clearly indicated that HCS activity is associated with several subcellular compartments (Tissot, Douce & Alban, 1997). Enzyme activity was mainly located in cytosol (approximately 90% of total cellular activity). Significant activity was also identified in the soluble phase of both mitochondria and chloroplasts. Two enzyme forms were separated by anion-exchange chromatography. The major form was found to be specific for the cytosol compartment, while the minor form was present in mitochondria as well as in chloroplasts. The great purity and the higher latency values of HCS activity measured in Percoll-purified chloroplasts and mitochondria, together with the protection of the enzyme activity in these organelles during thermolysin treatment, assessed that HCS is a genuine constituent of chloroplasts and mitochondria (Tissot *et al.*, 1997). Thus, the existence of HCS isoforms in different cell compartments suggests that the different biotin-dependent carboxylases in plants (localised in chloroplasts, mitochondria and cytosol, respectively) are biotinylated in the cell compartment within which they are localised. One possible explanation for this feature is that the active site of HCS recognises a three-dimensional folded structure of apo-carboxylases rather than a primary amino acid sequence. By functional complementation of the *birA E. coli* mutant with an *A. thaliana* cDNA expression library, a full-length HCS cDNA encoding a 41 kDa polypeptide was isolated (Tissot *et al.*, 1997). Plant HCS shows specific enclosed regions of similarity with other known biotin ligases of bacterial, yeast and human origins. These similarities are restricted to the ATP binding domain and the biotin binding domain, respectively. These motifs, located in the central part of the protein are interconnected, thus reflecting the requirement for ATP and biotin to be spatially close to permit formation of biotinyl-5′-AMP (Fig. 3). Interestingly, the eight amino acid residues involved in direct contact with biotin in BirA from *E. coli*, as determined by X-ray crystallography (Wilson *et al.*, 1992) are strictly conserved in *A. thaliana* HCS. The N-terminal region of this *A. thaliana* HCS exhibits the characteristic features of an organelle transit peptide. Also, the occurrence of two methionine residues close together in this region may suggest the existence of cytosolic and 'organelle targeted' HCS, synthesised from a single species of mRNA by alternative translational initiation. Such a possibility has been recently suggested to account for the synthesis of human mitochondrial and cytosolic HCS isoforms, together with an alternative splicing mechanism (León-Del-Rio *et al.*, 1995). The obtention of two translation products of the expected sizes by *in vitro* transcription–translation experiments using the plant cDNA is consistent with this hypothesis (Tissot *et al.*, 1997). However, further studies

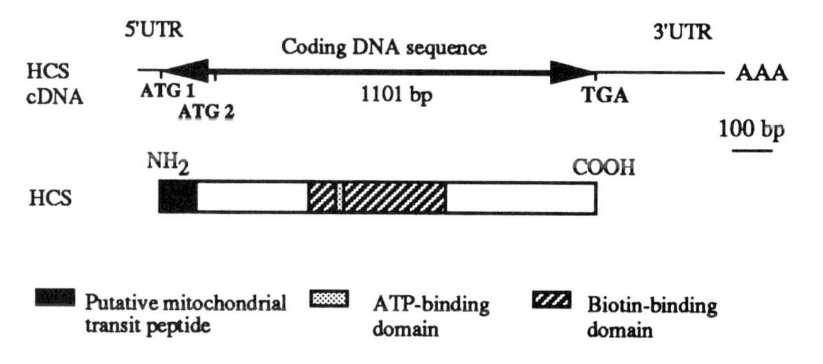

Fig. 3. Representation of cDNA and predicted amino acid sequences encoding *A. thaliana* HCS. UTR: untranslated region; HCS: biotin holocarboxylase synthetase; AAA: Poly A tail.

concerning the uptake of the cloned plant HCS by mitochondria and/or chloroplasts, as well as *in situ* localisation experiments of plant cells overexpressing this clone are needed to definitively assign the cellular localisation of this clone to a specific compartment. Finally, an exhaustive characterisation of HCS activities in the different cell compartments, and particularly the identification of a possible structurally distinct HCS isoform, will be determinant for understanding the mechanism of biotinylation of apocarboxylases in plants and to elucidate the question of why HCS activity has to be compartmentalised in plant cells.

Conclusion

The past few years have seen important advances in our understanding of the enzymes that manipulate biotin in plants, including biotin-containing carboxylases, biotin synthesising enzymes and biotin ligases. Most of these enzymes have been, or soon will be, purified and/or their genes cloned. With these achievements, a better understanding of the regulation and the interconnection between these different pathways will soon be possible. This knowledge will allow for more rational and directed efforts at their manipulation by genetic engineering. Indeed, some of the processes that involve biotin generate biochemicals that serve a broad range of nutritional and industrial purposes. For example, plant storage oils, for which biosynthesis requires ACCase, are a major resource both for human and animal nutrition, but also for a number of non-food uses including pharmaceutical, cosmetics, detergents and even fuels. Finally, the future elucidation of the structure of these enzymes, will certainly allow one to envisage, particularly by the means of molecular modelisation techniques, creation of new inhibitor families with herbicidal activities, affecting specifically plants and therefore having a lower impact on the environment.

Note added in proof: We have now demonstrated that the cloned *A. thaliana* HCS cDNA encodes the chloroplastic isoform.

References

Alban, C., Baldet, P., Axiotis, S. & Douce, R. (1993). Purification and characterization of 3-methylcrotonyl-coenzyme A carboxylase from higher plant mitochondria. *Plant Physiology*, **102**, 957–65.

Alban, C., Baldet, P. & Douce, R. (1994). Localization and characterization of two structurally different forms of acetyl-CoA carboxylase in young pea leaves, of which one is sensitive to aryloxyphenoxypropionate herbicides. *Biochemical Journal*, **300**, 557–65.

Alban, C., Jullien, J., Job D. & Douce R. (1995). Isolation and characterization of biotin carboxylase from pea chloroplasts. *Plant Physiology*, **109**, 927–35.

Al-Feel, W., Chirala, S.S. & Wakil, S.J. (1992). Cloning of the yeast *FAS3* gene and primary structure of yeast acetyl-CoA carboxylase. *Proceedings of the National Academy of Sciences, USA*, **89**, 4534–8.

Alix, J.H. (1989). A rapid procedure for cloning genes from lambda libraries by complementation of *E. coli* defective mutants. Application to the FabE region of the *E. coli* chromosome. *DNA*, **8**, 779–89.

Apitz-Castro, R., Rehn, K. & Lynen, F. (1970). β-Methylcrotonyl-CoA carboxylase. Kristallisation und einige physikalische Eigenschaften. *European Journal of Biochemistry*, **16**, 71–9.

Ashton, A.R., Jenkins, C.L.D. & Whitfeld, P.R. (1994). Molecular cloning of two different cDNAs for maize acetyl-CoA carboxylase. *Plant Molecular Biology*, **24**, 35–49.

Aubert, S., Alban, C., Bligny, R. & Douce, R. (1996). Induction of β-methylcrotonyl coenzyme A carboxylase in higher plant cells during carbohydrate starvation: evidence for the role of MCCase in leucine catabolism. *FEBS Letters*, **383**, 175–80.

Bailey, A., Keon, J., Owen, J. & Hargreaves, J. (1995). The *ACC1* gene, encoding acetyl CoA carboxylase, is essential for growth in *Ustilago maydis*. *Molecular General Genetics*, **249**, 191–201.

Baldet, P. & Ruffet, M.L. (1996). Biotin synthesis in higher plants: isolation of a cDNA encoding *Arabidopsis thaliana bioB*-gene product equivalent by functional complementation of a biotin auxotroph mutant *bioB105* of *Escherichia coli* K12. *Compte Rendus de l'Académie des Sciences Paris*, **309**, 99–106.

Baldet, P., Alban, C., Axiotis, S. & Douce, R. (1992). Characterization of biotin and 3-methylcrotonyl-coenzyme A carboxylase in higher plant mitochondria. *Plant Physiology*, **99**, 450–5.

Baldet, P., Gerbling, H., Axiotis, S. & Douce, R. (1993*a*). Biotin biosynthesis in higher plant cells. Identification of intermediates. *European Journal of Biochemistry*, **217**, 479–85.

Baldet, P., Alban, C., Axiotis, S. & Douce, R. (1993*b*). Localization of free and bound biotin in cells from green pea leaves. *Archives of Biochemistry and Biophysics*, **303**, 67–73.

Bettey, M., Ireland, R.J. & Smith, A.M. (1992). Purification and characterization of acetyl-CoA carboxylase from developing pea embryos. *Journal of Plant Physiology*, **140**, 513–20.

Burton, J.D., Gronwald, J.W., Keith, R.A., Somers, D.A., Gengenbach, B.G. & Wyse, D.L. (1991). Kinetics of inhibition of acetyl-coenzyme A carboxylase by Sethoxydim and Haloxyfop. *Pesticide Biochemistry and Physiology*, **39**, 100–9.

Chen, Y., Wurtele, E.S., Wang, X. & Nikolau, B.J. (1993). Purification and characterization of 3-methylcrotonyl-CoA carboxylase from somatic embryos of *Daucus carota*. *Archives of Biochemistry and Biophysics*, **305**, 103–9.

Chiba, Y., Suzuki, Y., Aoki, Y., Ishida, Y. & Narisawa, K. (1994). Purification and properties of bovine liver holocarboxylase synthetase. *Archives of Biochemistry and Biophysics*, **313**, 8–14.

Choi, J.K., Yu, F., Wurtele, E.S. & Nikolau, B.J. (1995). Molecular cloning and characterization of the cDNA coding for the biotin-containing subunit of the chloroplastic acetyl-CoA carboxylase. *Plant Physiology*, **109**, 619–25.

Cronan, J.E., Jr. & Wallace, J.C. (1995). The gene encoding the biotin-apoprotein ligase of *Saccharomyces cerevisiae*. *FEMS Microbiology Letters*, **130**, 221–30.

Dehaye, L., Alban, C., Job, C., Douce, R. & Job, D. (1994). Kinetics of the two forms of acetyl-CoA carboxylase from *Pisum sativum*. Correlation of the substrate specificity of the enzymes and sensitivity towards aryloxyphenoxypropionate herbicides. *European Journal of Biochemistry*, **225**, 1113–23.

Devine, M.D. & Shimabukuro, R.H. (1994). Resistance to acetyl coenzyme A carboxylase inhibiting herbicides. In *Herbicide Resistance in Plants: Biology and Biochemistry*, ed. S.B. Powles & J.A.M. Holtum, pp. 141–69. Boca Raton: Lewis Publishers.

Diez, T.A., Wurtele, E.S. & Nikolau, B.J. (1994). Purification and characterization of 3-methylcrotonyl-Coenzyme A carboxylase from leaves of *Zea mays*. *Archives of Biochemistry and Biophysics*, **310**, 64–75.

Duval, M., Job, C., Alban, C., Douce, R. & Job, D. (1994*a*). Developmental patterns of free and protein-bound biotin during maturation and germination of seeds of *Pisum sativum*. Characterization of a novel seed-specific biotinylated protein. *Biochemical Journal*, **299**, 141–50.

Duval, M., DeRose, R.T., Job, C., Faucher, D., Douce, R. & Job, D. (1994*b*). The major biotinyl protein from *Pisum sativum* seeds covalently binds biotin at a novel site. *Plant Molecular Biology*, **26**, 265–73.

Egli, M.A., Gengenbach, B.G., Gronwald, J.W., Somers, D.A. & Wyse, D.L. (1993). Characterization of maize acetyl-CoA carboxylase. *Plant Physiology*, **101**, 499–506.

Eisenberg, M.A. (1987). Biosynthesis of biotin and lipoic acid. In Escherichia coli *and* Salmonella typhimurium. *Cellular and Molecular Biology*, ed. F.C. Neidhardt, J.L. Ingraham, K.B. Low, B. Magasanik, M. Schaechter & M.E. Umbarger, pp. 544–50. New York: American Society of Microbiology.

Elborough, K.M., Winz, R., Deka, R.K., Markham, J.E., White, A.J., Rawsthorne, S. & Slabas, A.R. (1996). Biotin carboxyl carrier protein and carboxyltransferase subunit form of acetyl-CoA carboxylase from *Brassica napus*: cloning and analysis of expression during oilseed rape embryogenesis. *Biochemical Journal*, **315**, 103–12.

Genix, P., Bligny, R., Martin, J.B. & Douce, R. (1990). Transient accumulation of asparagine in sycamore cells after a long period of sucrose starvation. *Plant Physiology*, **94**, 717–22.

Gerbling, H. & Gerhardt, B. (1989). Peroxisomal degradation of branched-chain 2-oxo acids. *Plant Physiology*, **91**, 1387–92.

Gompertz, D., Goodey, P.A. & Barlett, K. (1973). Evidence for the enzymatic defect in beta-methylcrotonylglycinuria. *FEBS Letters*, **32**, 13–14.

Gornicki, P. & Haselkorn, R. (1993). Wheat acetyl-CoA carboxylase. *Plant Molecular Biology*, **22**, 547–52.

Guchhait, R.B., Polakis, S.E., Dimroth, P., Stall, E., Moss, J. & Lane, M.D. (1974). Acetyl coenzyme A carboxylase system of *Escherichia coli*. Purification and properties of the biotin carboxylase, carboxyltransferase, and carboxyl carrier protein components. *Journal of Biological Chemistry*, **249**, 6633–45.

Harwood, J.L. (1988). Fatty acid metabolism. *Annual Review of Plant Physiology and Plant Molecular Biology*, **39**, 101–38.

Harwood, J.L. (1991). Lipid synthesis. In *Target Sites for Herbicide Action*, ed. R.C. Kirkwood, pp. 57–94. New York: Plenum Press.

Harwood, J.L. (1996). Recent advances in the biosynthesis of plant fatty acids. *Biochimica et Biophysica Acta*, **1301**, 7–56.

Hatch, M.D. & Stumpf, P.K. (1961). Fat metabolism in higher plants. XVI. Acetyl coenzyme A carboxylase and acyl coenzyme A-malonyl coenzyme A transcarboxylase from wheat germ. *Journal of Biological Chemistry*, **236**, 2879–85.

Herbert, D., Cole, D.J., Pallett, K.E. & Harwood, J.L. (1996*a*). Susceptibilities of different test systems from maize (*Zea mays*), *Poa annua*, and *Festuca rubra* to herbicides that inhibit the enzyme acetyl-coenzyme A carboxylase. *Pesticide Biochemistry and Physiology*, **55**, 129–39.

Herbert, D., Price, L.J., Alban, C., Dehaye, L., Job, D., Cole, D.J., Pallett, K.E. & Harwood, J.L.(1996*b*). Kinetic studies on two isoforms of acetyl-CoA carboxylase from maize leaves. *Biochemical Journal*, **318**, 997–1006.

Howard, P.K., Shaw, J. & Otsuka, A.J. (1985). Nucleotide sequence of the *birA* gene encoding the biotin operon repressor and biotin holoenzyme synthetase of *Escherichia coli*. *Gene*, **35**, 321–31.

Ifuku, O., Koga, N., Haze, S., Kishimoto, J. & Wachi, Y. (1994). Flavodoxin is required for conversion of dethiobiotin to biotin in *Escherichia coli*. *European Journal of Biochemistry*, **224**, 173–8.

Kannangara, C.G. & Stumpf, P.K. (1972). Fat metabolism in higher plants. A prokaryotic type acetyl-CoA carboxylase in spinach chloroplasts. *Archives of Biochemistry and Biophysics*, **152**, 83–91.

Knowles, J.R. (1989). The mechanism of biotin-dependent enzymes. *Annual Review of Biochemistry*, **58**, 195–221.

Kohansky, R.A. & Lane, D.L. (1985). Homogeneous functional insulin receptor from 3T3-L1 adipocytes. Purification using $N^{\alpha B1}$-(biotinyl-ε-aminocaproyl)insulin and avidin-Sepharose. *Journal of Biological Chemistry*, **260**, 5014–25.

Kondo, H., Uno, S., Komizo, Y. & Sunamoto, J. (1984). Importance of methionine residues in the enzymatic carboxylation of biotin-containing peptides representing the local biotinyl site of *E. coli* acetyl-CoA carboxylase. *International Peptide and Protein Research*, **23**, 559–64.

Kondo, H., Shiratsuchi, K., Yoshimoto, T., Masuda, T., Kitazono, A., Tsuru, D., Anai, M., Sekiguchi, M. & Tanabe, T. (1991). Acetyl-CoA carboxylase from *Escherichia coli:* gene organization and nucleotide sequence of the biotin carboxylase subunit. *Proceedings of the National Academy of Sciences, USA*, **88,** 9730–3.

Konishi, T., Shinohara, K., Yamada, K. & Sasaki, Y. (1996*a*). Acetyl-CoA carboxylase in higher plants: most plants other than gramineae have both the prokaryotic and the eukaryotic forms of this enzyme. *Plant Cell Physiology*, **37**, 117–22.

Konishi, T., Kamoi, T., Matsuno, R. & Sasaki, Y. (1996*b*). Induction of cytosolic acetyl-coenzyme A carboxylase in pea leaves by ultraviolet-B irradiation. *Plant Cell Physiology*, **37**, 1197–200.

Lau, E.P., Cochran, B.C. & Fall, R.R. (1980). Isolation of 3-methylcrotonyl-Coenzyme A carboxylase from bovine kidney. *Archives of Biochemistry and Biophysics*, **205**, 352–9.

León-Del-Rio, A. & Gravel, R.A. (1994). Sequence requirements for the biotinylation of carboxyl-terminal fragments of human propionyl-CoA carboxylase α subunit expressed in *Escherichia coli. Journal of Biological Chemistry*, **269**, 22 964–8.

León-Del-Rio, A., Leclerc, D., Akerman, B., Wakamatsu, N. & Gravel, R.A. (1995). Isolation of a cDNA encoding human holocarboxylase synthetase by functional complementation of a biotin auxotroph of *Escherichia coli. Proceedings of the National Academy of Sciences, USA*, **92**, 4626–30.

Li, S. & Cronan, J.E. (1992*a*). The gene encoding the biotin carboxylase subunit of *Escherichia coli* acetyl-CoA carboxylase. *Journal of Biological Chemistry*, **267**, 855–63.

Li, S. & Cronan, J.E. (1992*b*). The genes encoding the two carboxyltransferase subunits of *E. coli* acetyl-CoA carboxylase. *Journal of Biological Chemistry*, **267**, 16 841–7.

Li, S. & Cronan, J.E. (1992*c*). Putative zinc finger protein encoded by a conserved chloroplast gene is very likely a subunit of a biotin-dependent carboxylase. *Plant Molecular Biology*, **20**, 759–61.

Lopez-Casillas, F., Bai, D.H., Luo, X., Kong, I.S., Hermodson, M.A. & Kim, K.H. (1988). Structure of the coding sequence and primary amino acid sequence of acetyl-CoA carboxylase. *Proceedings of the National Academy of Sciences, USA*, **85**, 5784–8.

Nes, W.D. & Bach, T.J. (1985). Evidence for a mevalonate shunt in a tracheophyte. *Proceedings of the Royal Society of London. Series B. Biological Sciences*, **225**, 425–44.

Ohlrogge, J.B. & Browse, J. (1995). Lipid biosynthesis. *The Plant Cell*, **7**, 957–70.

Ohlrogge, J.B., Kuhn, D.N. & Stumpf, P.K. (1979). Subcellular localisation of acyl carrier protein in leaf protoplasts of *Spinacia oleracea*. *Proceedings of the National Academy of Sciences, USA*, **76**, 1194–8.

Page, R.A., Okada, S. & Harwood, J.L. (1994). Acetyl-CoA carboxylase exerts strong flux control over lipid synthesis in plants. *Biochimica et Biophysica Acta*, **1210**, 369–72.

Patton, D.A., Volrath, S. & Ward, E.R. (1996*a*). Complementation of an *Arabidopsis thaliana* biotin auxotroph with an *Escherichia coli* biotin biosynthetic gene. *Molecular General Genetics*, **251**, 261–6.

Patton, D.A., Johnson, M. & Ward, E.R. (1996*b*). Biotin synthase from *Arabidopsis thaliana*. cDNA isolation and characterization of gene expression. *Plant Physiology*, **112**, 371–8.

Post-Beittenmiller, D., Jaworski, J.G. & Ohlrogge, J.B. (1991). *In vivo* pools of free and acylated acyl carrier proteins in spinach: evidence for sites of regulation of fatty acid biosynthesis. *Journal of Biological Chemistry*, **266**, 1858–65.

Post-Beittenmiller, D., Roughan, G. & Ohlrogge, J.B. (1992). Regulation of plant fatty acid biosynthesis: analysis of acyl-CoA and acyl-ACP substrate pools in spinach and pea chloroplasts. *Plant Physiology*, **100**, 923–30.

Roesler, K.R., Savage, L.J., Shintani, D.K., Shorrosh, B.S & Ohlrogge, J.B. (1996). Co-purification, co-immunoprecipitation, and coordinate expression of acetyl-coenzyme A carboxylase activity, biotin carboxylase, and biotin carboxyl carrier protein of higher plants. *Planta*, **198**, 517–25.

Roesler, K., Shintani, D., Savage, L., Boddupalli, S. & Ohlrogge, J.B. (1997). Targeting of the *Arabidopsis* homomeric acetyl-CoA carboxylase to plastid rapeseeds. *Plant Physiology*, **113**, 75–81.

Samols, D., Thornton, C.G., Murtif, V.L., Kumar, G.K., Hasse, F.C. & Wood, H.G. (1988). Evolutionary conservation among biotin enzymes. *Journal of Biological Chemistry*, **263**, 6461–4.

Sasaki, Y., Hakamada, K., Suama, Y., Nagano, Y., Furusawa, I. & Matsuno, R. (1993). Chloroplast-encoded protein as a subunit of acetyl-CoA carboxylase in pea plant. *Journal of Biological Chemistry*, **268**, 25 118–23.

Sasaki, Y., Konishi, T. & Nagano, Y. (1995). The compartmentation of acetyl-coenzyme A carboxylase in plants. *Plant Physiology*, **108**, 445–9.

Schatz, P.J. (1993). Use of peptide libraries to map the substrate specificity of a peptide-modifying enzyme: a 13 residue consensus peptide specifies biotinylation in *Escherichia coli*. *Bio/Technology*, **11**, 1138–43.

Schneider, T., Dinkins, R., Robinson, K., Shellhammer, J. & Meinke, D.W. (1989). An embryo-lethal mutant of *Arabidopsis thaliana* is a biotin auxotroph. *Developmental Biology*, **131**, 161–7.

Shellhammer, J. & Meinke, D. (1990). Arrested embryos from the *bio1* auxotroph of *Arabidopsis thaliana* contain reduced levels of biotin. *Plant Physiology*, **93**, 1162–7.

Shenoy, B.C., Paranjape, S., Murtif, V.L., Kumar, G.K., Samols, D. & Wood, H.G. (1988). Effect of mutation at Met-88 and Met-90 on the biotinylation of Lys-89 of the apo 1.3S subunit of transcarboxylase. *FASEB Journal*, **2**, 2505–11.

Shintani, D.K. & Ohlrogge, J.B. (1995). Feedback inhibition of fatty acid synthesis in tobacco suspension cells. *The Plant Journal*, **7**, 577–87.

Shorrosh, B.S., Dixon, R.A. & Ohlrogge, J.B. (1994). Molecular cloning, characterization, and elicitation of acetyl-CoA carboxylase from alfalfa. *Proceedings of the National Academy of Sciences, USA*, **91**, 4323–7.

Shorrosh, B.S., Roesler, K.R., Shintani, D., van de Loo, F.J. & Ohlrogge, J.B. (1995). Structural analysis, plastid localization, and expression of the biotin carboxylase subunit of acetyl-Coenzyme A carboxylase from tobacco. *Plant Physiology*, **108**, 805–12.

Shorrosh, B.S., Savage, L.J., Soll, J. & Ohlrogge, J.B. (1996). The pea chloroplast membrane-associated protein, IEP96, is a subunit of acetyl-CoA carboxylase. *The Plant Journal*, **10**, 261–8.

Song, J., Wurtele, E.S. & Nikolau, B.J. (1994). Molecular cloning and characterization of the cDNA coding for the biotin-containing subunit of 3-methylcrotonyl-CoA carboxylase: Identification of the biotin carboxylase and biotin carrier domains. *Proceedings of the National Academy of Sciences, USA*, **91**, 5779–83.

Stewart, C.R. & Beevers, H. (1967). Gluconeogenesis from amino acids in germinating castor bean endosperm and its role in transport to the embryo. *Plant Physiology*, **42**, 1587–95.

Stumpf, P.K. (1980). Biosynthesis of saturated and unsaturated fatty acids. In *The Biochemistry of Plants*, ed. P.K. Stumpf & E.E. Conn, pp. 177–204, New York: Academic Press.

Suzuki, Y., Aoki, Y., Ishida, Y., Chiba, Y., Iwamatsu, A., Kishino, Y., Niikawa, N., Matsubara, Y. & Narisawa, K. (1994). Isolation and characterization of mutations in the human holocarboxylase synthetase cDNA. *Nature Genetics*, **8**, 122–8.

Tardif, F.J., Preston, C., Holtum, J.A.M. & Powles, S.B. (1996). Resistance to acetyl-coenzyme A carboxylase-inhibiting herbicides endowed by a single major gene encoding a resistant target site in a biotype of *Lolium rigidum. Australian Journal of Plant Physiology*, **23**, 15–23.

Tissot, G., Job, D., Douce, R. & Alban, C. (1996). Protein biotinylation in higher plants: characterization of biotin holocarboxylase synthetase activity from pea (*Pisum sativam*) leaves. *Biochemical Journal*, **314**, 391–5.

Tissot, G., Douce, R. & Alban, C. (1997). Evidence for multiple forms of biotin holocarboxylase synthetase in pea (*Pisum sativum*) and in *Arabidopsis thaliana*: subcellular fractionation studies and isolation of a cDNA clone. *Biochemical Journal*, **323**, 179–88.

Wada, H., Shintani, D. & Ohlrogge, J.B. (1997). Why do mitochondria synthesize fatty acids? Evidence for involvement in lipoic acid production. *Proceedings of the National Academy of Sciences, USA*, **94**, 1591–6.

Wang, X., Wurtele, E.S. & Nikolau, B.J. (1995). Regulation of β-methylcrotonyl-coenzyme A carboxylase activity by biotinylation of the apoenzyme. *Plant Physiology*, **108**, 1133–9.

Weaver, L.M., Yu, F., Wurtele, E.S. & Nikolau, B.J. (1996). Characterization of the cDNA and gene coding for the biotin synthase of *Arabidopsis thaliana. Plant Physiology*, **110**, 1021–8.

Wilson, K.P., Shewchuk, L.M., Brennan, R.G., Otsuka, A.J. & Matthews, B.W. (1992). *Escherichia coli* biotin holoenzyme synthetase-*bio* repressor crystal structure delineates the biotin and DNA-binding domains. *Proceedings of the National Academy of Sciences, USA*, **89**, 9257–61.

Wood, H.G. & Barden, R.E. (1977). Biotin enzymes. *Annual Review of Biochemistry*, **46**, 385–413.

Wurtele, E.S. & Nikolau, B.J. (1990). Plants contains multiple biotin enzymes: discovery of 3-methylcrotonyl-CoA carboxylase, propionyl-CoA carboxylase and pyruvate carboxylase in the plant kingdom. *Archives of Biochemistry and Biophysics*, **278**, 179–86.

Xia, W-L., Zhang, J. & Ahmad, F. (1994). Biotin holocarboxylase synthetase: purification from rat liver cytosol and some properties. *Biochemistry and Molecular Biology International*, **34**, 225–32.

A.J. WHITE, K.M. ELBOROUGH,
S.Z. HANLEY and A.R. SLABAS

3 Down-regulation of lipid synthesis in plants implies higher orders of regulation in fatty acid biosynthesis

Introduction

The biosynthesis of fatty acids in plants has undergone considerable advancement over the last 15 years. This has been due to commercial considerations, coupled with the increasing realisation of their biological importance in a variety of areas encompassing environmental stress, signalling, carbon storage, and the structure of membranes. Several excellent reviews have previously been published in this area (Topfer, Norbert & Schell, 1995; Somerville & Browse, 1991; Ohlrogge & Browse, 1995) and advances since then have been recently updated (Slabas, Carey & White, 1997). In this chapter we consider *de novo* fatty acid biosynthesis in seeds of oilseed rape, concentrating on recent results from our laboratory aimed at down-regulating lipid metabolism. Specifically, we compare the resulting phenotypes of plants down-regulated in fatty acid synthase (FAS) activity with those plants that have been down-regulated for the high molecular weight, single polypeptide form of acetyl CoA carboxylase (ACCase). We hope that, rather than merely providing a description of the data, we can also convey how this work has evolved, and how it might give further insights into the roles of ACCases within plant tissue.

Central pathway of fatty acid biosynthesis in plants

Two main enzymatic components are central to *de novo* fatty acid biosynthesis in plants: (a) ACCase, and (b) FAS. The former enzyme generates malonyl CoA from acetyl CoA while the latter catalyses the progressive elongation of fatty acyl groups starting with an acetyl primer. In its simplest form this can be summarised by the equation below and further details can be found in these volume by Baldock *et al.* (Chapter 4).

$$\text{Acetyl CoA} + CO_2 + \text{ATP} \xleftrightarrow{\text{ACCase}} \text{Malonyl CoA} + \text{ADP} + \text{Pi}$$

$$\text{Acetyl CoA} + 16\text{NADH} + 16\text{NADPH} \xleftrightarrow{\text{FAS}} \text{Stearoyl ACP} + 16\text{NAD}^+$$
$$+32H^+ + 16 \text{ Malonyl CoA} + 16\text{NADPH} \qquad + 16\text{NADP}^+ + 16\text{CoASH}$$

In higher plants FAS exists as a freely dissociable multipolypeptide enzyme system, which performs a cyclic series of condensation, reduction, dehydration and further reduction reactions within the plastid. This results in the elongation of an acyl chain by two carbons for each series of reactions. Although a brief description of the individual components is given below, each has been well documented in numerous reviews and articles and the reader is referred to these for a detailed description (see Slabas & Fawcett, 1992 and references therein). Briefly, acetyl CoA ACP transacylase (ACAT) and malonyl CoA ACP transacylase (MCAT) convert acetyl CoA and malonyl CoA, respectively, to their ACP derivatives. However, the involvement of ACAT in fatty acid synthesis has been questioned, following the identification of a plant condensing enzyme that catalyses the condensation of malonyl ACP directly with acetyl CoA to form a ketoacyl ACP (Jaworski, Clough & Barnum, 1989). It is worthwhile noting that an *E.coli* strain severely deficient in KAS III activity *in vitro* does not show any major change in fatty acid composition (Tsay *et al.*, 1992). The condensation reactions are performed by three different ketoacyl synthases (KAS I, II, and III). KAS III is involved in the initial elongation from C2 to C6, KAS I in the elongation of C6 to C16 and KAS II from C16 to C18. The first reduction step is catalysed by β-ketoacyl ACP reductase, an NADPH-dependent reaction, and produces hydroxyacyl-ACP. Hydroxyacyl-ACP is dehydrated by a dehydratase and then further reduced by enoylacyl ACP reductase to form the acyl ACP, this last reaction being NADH dependant. Elongation of the acyl chain then either continues by further rounds of the above reactions or is terminated by a thioesterase. Many of these components have now been characterised and cloned from various plant species including oilseed rape, where they are present as small multigene families.

We have previously documented the molecular weights of proteins and clones which have been obtained for plant FAS components (Slabas *et al.*, 1993). Three components still have to be cloned to complete the full set of cDNAs encompassing plant FAS. These are ACAT, MCAT, and dehydratase. In the proceedings of this conference the isolation of the cDNA encoding the dehydratase from oilseed rape was described by Doig *et al.* (1997), and progress is being made in our laboratory on the cloning of MCAT. The cloning of ACAT is likely to be more complicated, as no clone coding for this protein has been isolated from any plant or bacterial species. A complementation approach as used to isolate the membrane-bound plant 2-acyltransferase (Brown *et al.*, 1994) is likely to hold the most promise.

ACCase in plants

In both yeast and animal systems acetyl CoA carboxylase has three functional domains within one high molecular weight polypeptide of over 200 kDa. These are (a) biotin carboxylase (BC) (b) biotin carboxyl carrier protein (BCCP) and (c) carboxyl transferase (CT). In *E.coli* the organisation of this enzyme is different: it is composed of four separate dissociable polypeptide chains BC, BCCP, CTα and CTβ. These range in size from 17 kDa for the BCCP polypeptide up to 37 kDa for the CT domains. In this chapter the different structural forms of ACCase are referred to as high molecular weight (HMW), in which all the polypeptides are on one chain, or low molecular weight (LMW), in which the components are on multiple polypeptides. Nomenclatures based on prokaryotic and eukaryotic considerations are likely to prove erroneous as is the case with the FAS from *Brevibacterium ammoniagenes* (Kawaguchi & Okuda, 1977). A schematic of both these forms is shown below.

HMW ACCase

BC BCCP CT

LMW ACCase

BC	BCCP
CTα	CTβ

All components except for CTα are nuclear encoded in plants.

Initial studies on plant ACCase from a number of different laboratories indicated that plants possessed a multiple polypeptide form of ACCase. Kannangara and Stumpf (1973) demonstrated that, in barley, the BCCP component was associated with the thylakoid membrane; however, in the light of more recent studies from Sasaki and Konishi (1997), this result warrants reinvestigation. Subsequent studies also demonstrated the presence of a single polypeptide form of ACCase in a number of plant species, including wheat, parsley and oilseed rape (Hellyer, Bambridge & Slabas, 1986). At that time, the existence of a multiple polypeptide form of ACCase was being dismissed as a probable artefact brought about by proteolysis; comments were made from my own laboratory to this effect. However, in 1991 an open reading frame with considerable similarity to the CT domain of the *E. coli* LMW ACCase was identified in the pea chloroplast DNA by Nagano, Matsuno & Sasaki (1991). This had previously been designated a zinc finger protein. Subsequent experimentation confirmed that both the high and low molecular weight forms of ACCase were present in pea. However, the LMW form has not been found in pea seeds (Alban *et al.*, this volume). Furthermore, a detailed study of the distribution of the LMW form of ACCase in plants has now indicated that this form is present in all plants examined with the exception of the Graminaceae (Sasaki & Konishi, 1997). This provides an explanation for the differential action of the monocot-specific acyloxyphenoxy-proprionate and cyclohexone-diose group of herbicides, whose main target is the HMW form of ACCase present in the plastid (Rendina *et al.*, 1988). During the purification of the major herbicide sensitive form of ACCase away from the minor insensitive form of ACCase in maize, Ashton, Jenkins and Whitfeld (1994) commented 'as yet, it is unclear why ACCase activity in leaves of monocotyledonous species is generally higher than in leaves of dicotyledonous species'. This can probably now be explained as an effect of the differential compartmentalisation of the two ACCase forms in different plant species, and the effect of dilution on measurcments of their *in vitro* activity. Experiments have been performed in dicots by Alban *et al.* (1994), which clearly demonstrate a plastidic LMW form of ACCase and a cytoplasmic HMW form in pea leaves. It is worthwhile noting here the difference in identified forms of ACCase in leaves and seeds of pea. This could point to a fundamental difference in the structural organisation of the lipid synthesis machinery of these tissues. In maize it appears that there is a HMW ACCase in the plastid responsible for *de novo* fatty acid biosynthesis which is herbicide sensitive, and a further HMW ACCase in the cytoplasm responsible for elongation which is insensitive (Fig. 1). The lower biological activity of ACCase in dicots vs. monocots could therefore be explained by the dissociable LMW form of ACCase being dilution sensitive in the assay and hence being underestimated whilst the HMW form is not. Indeed, there is a new assay from the Norwich group of Rawsthorne *et al.*, which can be explained on this basis. This differential localisation has been used to suggest different functions for the two forms of ACCase in dicots. The LMW form found in the plastid would be mainly involved in the production of malonyl CoA for *de novo* lipid biosynthesis, whereas the HMW cytosolic form would produce malonyl CoA which could then be utilised for flavanoid production and fatty acid elongation from C18 upwards (Sasaki & Konishi, 1997).

It is perhaps now apparent that the early findings suggesting the presence of a LMW ACCase in plants should not have been so readily

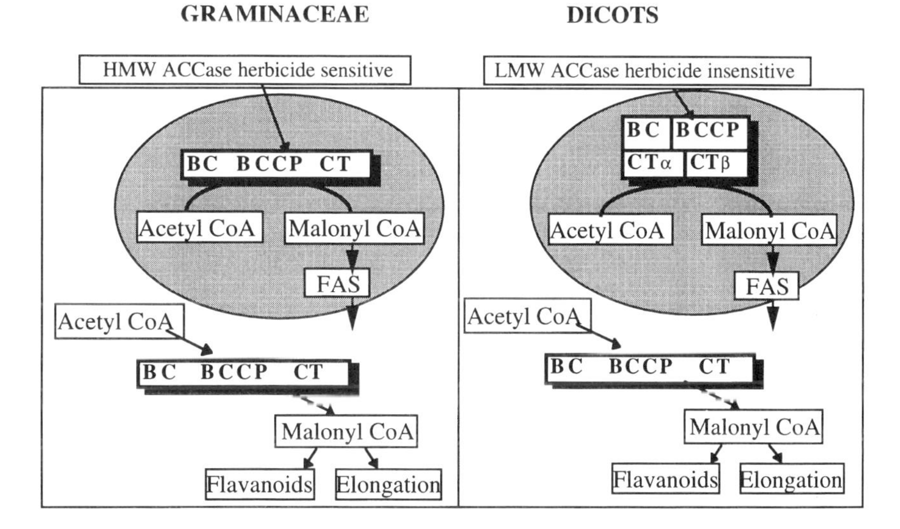

Fig. 1. Current proposed model of cellular localisation and possible functions for the different forms of ACCase. The shaded area represents the plastid, and the clear area the cytoplasm. ACCase in the cytoplasm is responsible for the elongation of C18 fatty acids to longer chain lengths, and is also believed to be involved in flavanoid production, while the plastidic ACCase participates in the synthesis of fatty acids up to C18 only.

dismissed as purification artefacts. They act as a reminder that Nature has more than one way of doing things and the biochemical structures of enzymes in all plants are not the same. Perhaps the best example of this is the difference between C3 and C4 plants in the details of CO_2 fixation (Leegood, 1993). To avoid misinterpretation of the scientific literature surrounding ACCase in plants, we are deliberately restricting ourselves in the following comments to oilseed rape, a major industrial crop, and to *Arabidopsis*, a close relative. This is in order to clarify what is really known, as opposed to what could have been extrapolated by comparisons with other plant species. The following four points are important here.

Correlation of ACCase activity and lipid synthesis in seeds

Early studies by Turnham and Northcote (1983) on ACCase activity in rape embryos during seed maturation indicated that the induction of ACCase activity closely resembled that of lipid deposition. By analogy with lipid synthesis in animal systems, the authors indicated that ACCase could therefore be a rate-limiting step in *de novo* lipid biosynthesis. However, more recent studies by Kang *et al.* (1994) reported no justification for considering ACCase to be a rate-limiting enzyme. These authors, using an intact plastid extraction procedure, showed a five-fold increase in ACCase activity levels over those previously reported. This difference can, of course, be explained by differences in the extraction procedures affecting the dilution-dependent activity of the LMW ACCase.

Existence of mRNA for HMW and LMW forms of ACCase in rape embryos

Both the HMW and LMW forms of ACCase exist in rape embryos. Partial cDNAs encoding the CT and BCCP domains of HMW ACCase have been cloned from a rape embryo library, together with a genomic clone containing the ACCase gene (Elborough *et al.*, 1994; Schulte, Schell & Topfer, 1994). Sequence comparisons between the cDNA and genomic clone suggests that these may encode two different isoforms of this gene. A small multigene family for the HMW ACCase in oilseed rape had previously been predicted (Elborough *et al.*, 1994). More recently the plastid-encoded βCT, and the nuclear-encoded BCCP cDNAs were also isolated from rape embryo libraries (Elborough *et al.*, 1996). Developmental studies on both forms of ACCase show that expression increases during the early stages of embryogenesis prior to maximal lipid deposition and then rapidly decreases. However, according to the model described in Fig. 1, the LMW ACCase should be the major plastidic form, while the HMW ACCase would be located within the cytosol. Similarly, in *Arabidopsis* both full-length genes and cDNAs have been cloned in addition to components of the LMW form (Roesler, Shorrosh & Ohlrogge, 1994; Choi *et al.*, 1995).

Nucleotide sequence comparison of LMW and HMW forms of ACCase

Sequence comparisons at both the nucleotide and amino acid levels for the LMW ACCase components CTα, CTβ and BCCP show no similarity to the corresponding domains on the HMW ACCase (see Fig. 2) in rape. However, strong similarity exists between the HMW ACCases from various plant sources and even with the animal ACCase (Elborough *et al.*, 1994). The LMW ACCase components also show some degree of similarity with the equivalent component from other plant species although these appear to have diverged to a greater extent.

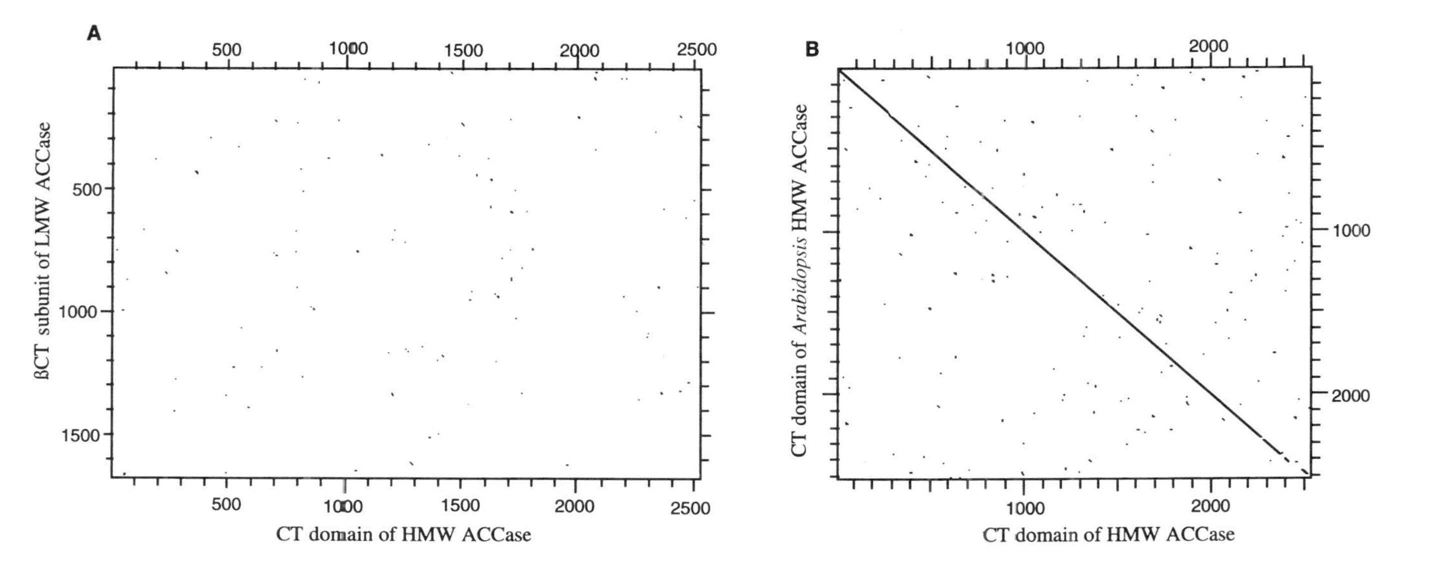

Fig. 2. Dot matrix of homology between (A) CT domain of HMW ACCase from *Brassica napus* against the bCT subunit of the LMW ACCase also from rape. No similarity is evident. Stringency 15 window 23. (B) CT domain of HMW ACCase from *Brassica napus* against the same domain from the *Arabidopsis* HMW ACCase. Stringency 15, window 23. In comparison to the first example, strong homology is present.

It is crucial to our arguments concerning the use of antisense to down-regulate the CT domain of HMW ACCase since there is no *a priori* reason based on nucleotide sequence similarity to expect a reduction in either of the CT components of LMW ACCase.

Are LMW and HMW proteins of ACCase present in both rape seed and leaf?

Roesler *et al.* (1997), as part of a study into overexpression of the HMW ACCase, has indicated, by the use of both immunoblots and propionyl-CoA assays, that the HMW ACCase could be present in plastids of rape embryos, albeit at low levels. In addition, results from our laboratory have shown the presence of the BCCP component within chloroplasts of rape leaves (Markham *et al.*, unpublished data). However, to date, no experiments have been performed to quantify the subcellular compartmentalisation and the role of LMW and HMW ACCase in seed storage synthesis in the plastids of rape embryos. Whilst it might be convenient to assume that what occurs in the leaf is occurring within the embryo, there is no experimental evidence to suggest that this is the case.

Considerations and design of antisense lipid experiments

Initially, we had two main reasons for wanting to down-regulate lipid biosynthesis.
(a) To study the multitude of effects that could result from down-regulation of such an important pathway.

Several basic questions could be asked by such an experiment. Amongst these are:

(i) Are there any changes in the phenotype of the plant at both the gross and cellular level?
(ii) Is there a hierarchy of lipids which must be preserved for normal cellular function?
(iii) How important an *in vivo* role does any one enzyme play in metabolic control and what level of plasticity is there within this metabolic pathway?
(iv) Could insights be gained into higher orders of regulation in lipid metabolism enabling us to understand fundamental mechanisms responsible for the perception of

metabolic signals and conservation of stoichiometry of metabolic components?

(b) To allow for the diversion of acetyl CoA to other engineered pathways by reducing the accumulation of fatty acids in storage compounds.

Whilst descriptions were present in the literature of mutations in desaturase genes no mutations had been reported for central FAS metabolism. Therefore, we had no idea what they would look like and it was possible they could be lethal. In addition, the enzymatic steps we were interested in were encoded by multigene families in rape and therefore the technique chosen would have to be able to reduce the levels of all isoforms of the protein. The use of antisense technology was an obvious way forward which had previously been used extensively by Stitt and colleagues to study the importance of Rubisco (Quick *et al.*, 1991; Stitt *et al.*, 1991).

Down-regulation could be affected at two levels, either FAS or ACCase. β-Ketoacyl reductase, a component of FAS, was chosen as a target; similarly, the ACCase also seemed appropriate. At the time, the existence of two forms of ACCase was unknown, and we had just cloned the HMW ACCase from rape. If the model presented in Fig. 1 on the compartmentalisation and predicted functions of the ACCase forms had been available at the time, the constructs we eventually designed would never have been logically arrived at. Both antisense constructs were placed individually under the control of two promoters:

(i) The 35S CaMV promoter to allow rapid analysis of leaf material prior to segregation.
(ii) The seed-specific ACP promoter kindly supplied by Unilever. This of course would be subject to segregation problems.

The constructs are shown in Fig. 3. Transformation of the rape was carried out by Zeneca who were collaborating with us on this project.

A total of 96 lines were generated. Primary transformants were analysed for the presence of both the kanamycin gene and the antisense constructs. Plants which were PCR positive were then further analysed for copy number and expression of the antisense construct. Two main problems can be encountered when carrying out such studies. The first problem is the inherent variability found within the parental cultivar, a feature that was recently strongly emphasised by Still, Dahal and Bradford (1997), who were working on tomatoes. The second problem concerns the positive identification of changes being brought about by an

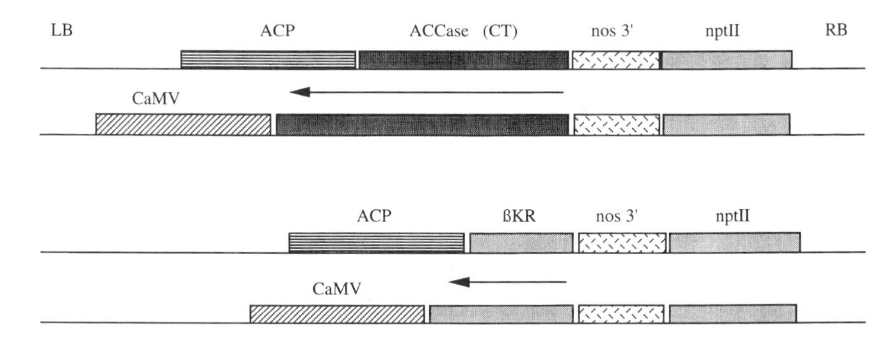

Fig. 3. Schematic diagram of antisense constructs used for transformation of rape plants. ACP and CaMV refer to the respective promoters chosen to provide general expression and seed-specific expression. LB and RB refer to the left- and right-hand borders of the T-DNA respectively. The *npt*II gene provides kanamycin resistance following transformation for selection purposes.

antisense effect, as opposed to insertional inactivation of a structural or regulatory gene other than the target. Appropriate controls using vectors lacking the coding area of interest allow for identification of somaclonal variation.

The parental cultivar used in our study is Westar, which is known to be heterogeneous. In order to reduce the level of variability, doubled haploid lines can be used, and have been in this study. In addition, we carried out preliminary studies on the developmental changes of the various enzymes of interest in both the seed and leaf to quantify the variation in Westar. This revealed that small (5%) changes in enzyme levels would be statistically unidentifiable.

Antisense FAS plants

Various lines of the antisense βKR plants showed several common phenotypic differences in comparison to the Westar parental cultivar. These were observed at differing degrees of severity in different transgenic lines. Following elimination of phenotypes which could be attributed to insertional events a number of characteristics emerged. These included reduced seed set, decreased germination rates (a feature that might be predicted when reducing the carbon energy source used in germination), and deformation of the leaves as manifested by the leaf curling in on itself. Additionally, the seeds rather than forming a spherical shape were crinkled, and on removal of the testa showed spots of discoloration on the embryo. These phenotypes generally followed a pattern of being more severe in those plants containing multiple inserts and in addition were identified in successive generations of these lines. The characterisation of the phenotypes, particularly the crinkling of seed, enabled us to rapidly identify plants that were down-regulated in fatty acid synthesis and was crucial during the analysis of antisense HMW ACCase plants.

Antisense HMW ACCase plants

Following the identification of the multisubunit ACCase in plants, initially reported by Nagano *et al.* (1991) and confirmed by results from Alban *et al.* (1994), our immediate thoughts were to discard the antisense ACCase plants as a waste of time and to concentrate on making new antisense constructs to the LMW ACCase. According to the current model we had down-regulated the wrong ACCase, if we wished to modulate *de novo* fatty acid biosynthesis. However, after producing these plants, we were reluctant to discard them in case they might reveal something unexpected. After characterising a few lines, several general similarities were observed between the antisense HMW ACCase and the antisense FAS plants. Again, lines were identified in which the leaves would curl up. In fact, in one line, which contained eight inserts, the most severe form of this characteristic phenotype was observed with the leaves almost forming tubes. In addition, reduced germination rates, reduced seed set, and a crinkled appearance to the seed were also present. This posed a difficult question: why should reducing the level of the HMW ACCase produce similar phenotypes to FAS antisense? Considerations of the proposed role of HMW ACCase would predict a reduction in flavanoids and very long chain fatty acids (VLCFAs). We had not predicted, however, that this would affect the growth of the plant. More surprising are the marked similarities to the phenotype seen in the antisensed FAS plants. Analysis of lipid levels in seeds has shown that reduced levels of HMW ACCase cause a reduction in total seed lipid levels.

Further analyses of the biochemical differences in the down-regulated HMW ACCase plants

It was of interest to us to study whether the observed similarity in antisensed HMW ACCase and FAS phenotypes is due to pleiotrophic effects on the FAS enzymes. Accordingly, we have measured the level of βKR protein in these plants immunologically. There is clearly observable down-regulation of βKR which accompanies HMW ACCase down-regulation. However, it cannot be directly occurring as a consequence of nucleotide sequence homology to FAS components, or more importantly by any similarity to the LMW ACCase. If the regulation is not occurring directly at the nucleotide level, then other methods of regulation need to be invoked. Two mechanisms can be put forward: either a physical (mechanical) or a signalling (chemical) one. Physical events such as the alteration of membrane fluidity are known to affect gene transcription (Vigh *et al.*, 1993) and a whole cascade of metabolic signals has been described in the literature including jasmonic acid (Farmer & Ryan, 1990; Franceschi & Grimes, 1991). Clearly, the generation of these antisense plants could raise more questions than answers.

Acknowledgements

We would like to thank Unilever for the gift of the ACP promoter, Zeneca for their help throughout the project, and BBSRC for financial support.

References

Alban, C., Baldet, P. & Douce, R. (1994). Localization and characterization of two structurally different forms of acetyl CoA carboxylase in young pea leaves, of which one is sensitive to aryloxyphenoxypropionate herbicides. *Biochemical Journal*, **300**, 557–65.

Ashton, A.R., Jenkins, C.L.D. & Whitfeld, P.R. (1994). Molecular cloning of two different cDNAs for maize acetyl-CoA carboxylase. *Plant Molecular Biology*, **24**, 35–49.

Brown, A.P., Coleman, J., Tommey, A.M., Watson, M.D. & Slabas, A.R. (1994). Isolation and characterisation of a maize cDNA that complements a 1-acyl sn-glycerol-3-phosphate acyltransferase mutant of *Escherichia coli* and encodes a protein which has similarities to other acyltransferases. *Plant Molecular Biology*, **26**, 211–23.

Choi, J.K., Yu, F., Wurtele, E.S. & Nikolau, B.J. (1995). Molecular cloning and characterization of the cDNA coding for the biotin-containing subunit of the chloroplastic acetyl-coenzyme A carboxylase. *Plant Physiology*, **109**, 619–25.

Doig, S., Slabas, A.R. & Hawkes, T.R. (1997). β-hydroxyl ACP dehydratase from type II fatty acid synthases. *Journal of Experimental Botany*, **48**, suppl. P9.7.

Elborough, K.M., Swinhoe, R., Winz, R., Kroon, J.T.M., Farnsworth, L., Fawcett, T., Martinez-Rivas, J.M. & Slabas, A.R. (1994). Isolation of cDNAs from *Brassica napus* encoding the biotin-binding and transcarboxylase domains of acetyl CoA carboxylase: assignment of the domain structure in a full length *Arabidopsis thaliana* genomic clone. *Biochemical Journal*, **301**, 599–605.

Elborough, K.M., Winz, R., Deka, R.K., Markham, J.E., White, A.J., Rawsthorne, S. & Slabas, A.R. (1996). Biotin carboxylase carrier protein and carboxyltransferase subunits of the multi-subunit form of acetyl CoA carboxylase from *Brassica napus*: cloning and analysis of expression during oilseed rape embryogenesis. *Biochemical Journal*, **315**, 103–12.

Farmer, E.E. & Ryan, C.L. (1990). Interplant communication: airborne methyl jasmonate induces synthesis of proteinase inhibitors in plant cells. *Proceedings of the National Academy of Sciences, USA*, **87**, 7713–16.

Franceschi, V.R. & Grimes, H.D. (1991). Induction of soybean vegetative storage proteins and anthocyanins by low-level atmospheric methyl jasmonate. *Proceedings of the National Academy of Sciences, USA*, **88**, 6745–9.

Hellyer, A., Bambridge, H.E. & Slabas, A.R. (1986). Plant acetyl Co-A carboxylase. *Biochemical Society Transactions*, **14**, 565–8.

Jaworski, J.G., Clough, R.C. & Barnum, S.R. (1989). A cerulenin insensitive short chain 3-ketoacyl-acyl carrier protein synthase in *Spinacia oleracea* leaves. *Plant Physiology*, **90**, 41–4.

Kang, F., Ridout, C.J., Morgan, C.L. & Rawsthorne, S. (1994). The activity of acetyl-CoA carboxylase is not correlated with the rate of lipid synthesis during development of oilseed rape (*Brassica napus L.*) embryos. *Planta*, **193**, 320–5.

Kannangara, C.G. & Stumpf, P.K. (1973). Fat metabolism in higher plants. Distribution and nature of biotin in chloroplasts of different plant species. *Archives of Biochemistry and Biophysics*, **155**, 391–9.

Kawaguchi, A. & Okuda, S. (1977). Fatty acid synthetase from *Brevibacterium ammoniagenes*: formation of monounsaturated fatty acids by a multienzyme complex. *Proceedings of the National Academy of Sciences, USA*, **74**, 3180–3.

Leegood, R.C. (1993). Carbon dioxide concentrating mechanisms. In *Plant Biochemistry and Molecular Biology*, ed. P.J. Lea & R.C. Leegood. Chichester, UK: John Wiley.

Nagano, Y., Matsuno, R. & Sasaki, Y. (1991). Sequence and transcriptional analysis of the gene cluster trnQ-zfpA-psaI-ORF231-petA in pea chloroplasts. *Current Genetics*, **20**, 431–6.

Ohlrogge, J. & Browse, J. (1995). Lipid biosynthesis. *Plant Cell*, **7**, 957–70.

Quick, W.P., Schurr, U., Scheibe, R., Schulze, E.D., Rodermel, S.R., Bogorad, L. & Stitt, M. (1991). Decreased ribulose-1,5-bisphosphate carboxylase-oxygenase in transgenic tobacco transformed with 'antisense' *rbcS* .I. Impact on photosynthesis in ambient growth conditions. *Planta*, **183**, 542–54.

Rendina, A.R., Felts, J.M., Beaudoin, J.D., Craig-Kennard, A.C.,

Look, L.L., Paraskos, S.L. & Hagenah, J.A. (1988). Kinetic characterization, stereoselectivity, and species selectivity of the inhibition of plant acetyl-CoA carboxylase by the aryloxyphenoxypropionic acid grass herbicides. *Archives of Biochemistry and Biophysics*, **265**, 219–25.

Roesler, K.R., Shorrosh, B.S. & Ohlrogge, J.B. (1994). Structure and expression of an *Arabidopsis* acetyl-coenzyme A carboxylase gene. *Plant Physiology*, **105**, 611–17.

Roesler, K.R., Shintani, D., Savage, L., Boddupalli, S. & Ohlrogge, J.B. (1997). Targeting of the *Arabidopsis* homomeric acetyl-coenzyme A carboxylase to plastids of rapeseeds. *Plant Physiology*, **113**, 75–81.

Sasaki, Y. & Konishi, T. (1997). Two forms of acetyl -CoA carboxylase in higher plants and effects of UV-B on the enzyme levels. In *Physiology, Biochemistry and Molecular Biology of Plant Lipids*, ed. J.P.Williams, M.U.Khan & N.W.Lem. Dordrecht, Netherlands: Kluwer Academic Publishers.

Schulte, W., Schell, J. & Topfer, R. (1994). A gene encoding acetyl-coenzyme A carboxylase from *Brassica napus*. *Plant Physiology*, **106**(2), 793–4.

Slabas, A.R. & Fawcett, T. (1992). The biochemistry and molecular biology of plant lipid biosynthesis. *Plant Molecular Biology*, **19**, 169–91.

Slabas, A.R., Fawcett, T. & Griffiths, G. & Stobard, K. (1993). Biochemistry and molecular biology of lipid biosynthesis in plants: potential for genetic manipulation. In *Plant Biotechnology Series Vol. 3. Biosynthesis and Manipulation of Plant Products*, ed. D. Grierson. Glasgow, UK: Blackie Academic and Professional.

Slabas, A.R., Carey, A.T. & White, A.J. (1997). Manipulation of seed oils for industrial use. *The Biochemist*, **19**(1), 11–15.

Somerville, C. & Browse, J. (1991). Plant lipids: metabolism, mutants, and membranes. *Science*, **252**, 80–7.

Stitt, M., Quick, W.P., Schurr, U., Schulze, E-D., Rodermel, S.R. & Bogorad, L. (1991). Decreased ribulose-1,5-bisphosphate carboxylase-oxygenase in transgenic tobacco transformed with 'antisense' *rbcS* . *Planta*, **183**, 555–66.

Still, D.W., Dahal, P. & Bradford, K.J. (1997). A single-seed assay for endo-β-mannanase activity from tomato endosperm and radical tissues. *Plant Physiology*, **113**, 13–20.

Topfer, R., Norbert, M. & Schell, J. (1995). Modification of plant lipid synthesis. *Science*, **268**, 681–6.

Tsay, J-T., Oh, W., Larson, T.J., Jackowski S. & Rock, C.O. (1992). Isolation and characterization of the β-ketoacyl-acyl carrier protein synthase III gene (*fabH*) from *Escherichia coli* K-12. *Journal of Biological Chemistry*, **267**, 6807–14.

Turnham, E. & Northcote, D.H. (1983). Changes in the activity of acetyl CoA carboxylase during rapeseed formation. *Biochemical Journal*, **212**, 223–9.

Vigh, L., Los, D.A., Horvath, I. & Murata, N. (1993). The primary signal in the biological perception of temperature: Pd-catalyzed hydrogenation of membrane lipids stimulated the expression of the *desA* gene in *Synechocystis* PCC6803. *Proceedings of the National Academy of Sciences, USA*, **90**, 9090–4.

C. BALDOCK, J.B. RAFFERTY, A.R. STUITJE
and D.W. RICE

4 Molecular structure of a reductase component of fatty acid synthase

Introduction

With the increasing availability of structural information on proteins, recent attention has focused on the use of such data in strategies for the design of new enzyme inhibitors. One of the most important features of any inhibitor is its ability to discriminate between the target enzyme and its counterpart in the host. To this end, significant differences in enzymes which carry out the same reaction afford particularly attractive opportunities for drug development and one area of metabolism where this has proved to be important is in fatty acid biosynthesis. Lipids are synthesised by the multifunctional enzyme complex fatty acid synthetase (FAS) through the extension of the acetyl group attached to an acyl carrier protein (ACP) by two carbon units derived from malonyl-CoA in a stepwise reaction (McCarthy & Hardie, 1984). The components of the pathway have been studied intensively and include an acetyl CoA: ACP transacylase, a malonyl CoA:ACP transacylase, a β-ketoacyl ACP synthetase, a β-ketoacyl ACP reductase, a β-hydroxyacyl ACP dehydratase and an enoyl ACP reductase (Fig. 1). The FAS complex has been classified into types: type I FAS complex, found in eukaryotes and yeast in which all the catalytic domains reside on one or two polypeptides; type II FAS, found in plants and most prokaryotes, in which the enzymes that catalyse the individual steps are found on separate polypeptides (McCarthy & Hardie, 1984).

Enoyl ACP reductase (ENR) catalyses the final reaction of FAS, the reversible reduction of a carbon–carbon double bond in the enoyl moiety of its substrate, which is covalently linked to an acyl carrier protein.

$$R - CH = CH - CO - ACP + NADH + H^+ \leftrightarrow R - CH_2 - CH_2 - CO - ACP + NAD^+$$

Investigation in *B. napus*, which accumulates up to 40% lipid by weight

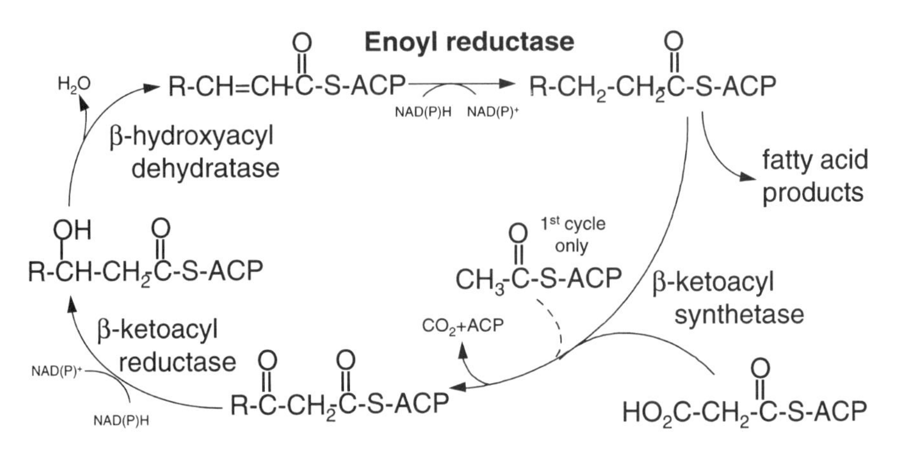

Fig. 1. A schematic representation of the fatty acid synthase cycle.

in its seeds (Slabas *et al.*, 1987), has demonstrated that both the activity and level of enoyl reductase increases throughout the entire lipid deposition phase in maturing seeds (Slabas *et al.*, 1990). Thus there is, in seeds, a continued strong demand for the enzyme during lipid biosynthesis, which is not satisfied by a single elevated catalytic rate.

The *B. napus* ENR is tetrameric (Slabas *et al.*, 1986) and has been cloned and overexpressed in *E. coli* using the bacteriophage T7 polymerase and promoter system (Kater *et al.*, 1991) and crystallised in the presence of cofactor (Fig. 2) (Rafferty *et al.*, 1994). ENR from *E. coli* is also tetrameric (Wagner *et al.*, 1994) and accepts both NADH and NADPH as cofactors (Bergler *et al.*, 1996) and is inhibited by a range of diazaborines (Bergler *et al.*, 1994), which are heterocyclic boron-containing compounds, whose action is thought to lead to the inhibition of cell growth by preventing lipopolysaccharide synthesis (Högenauer & Woisetschläger, 1981). Biochemical studies on the *E. coli* enzyme have shown that NAD^+ is required for diazaborine binding, and this has led to the suggestion that the drug either binds to ENR in association with NAD^+ or that NAD^+ converts the drug to an active form (Kater *et al.*, 1994). There are no reports of inhibition of *B. napus* ENR by diazaborine. The ENR from *Mycobacterium tuberculosis* is the target for a metabolite of isoniazid (Banerjee *et al.*, 1994), a potent drug that is used in the front-line chemotherapeutic treatment of tuberculosis. However, strains of *M. tuberculosis* are emerging that are isoniazid resistant (Cole, 1994) with consequent problems in treatment.

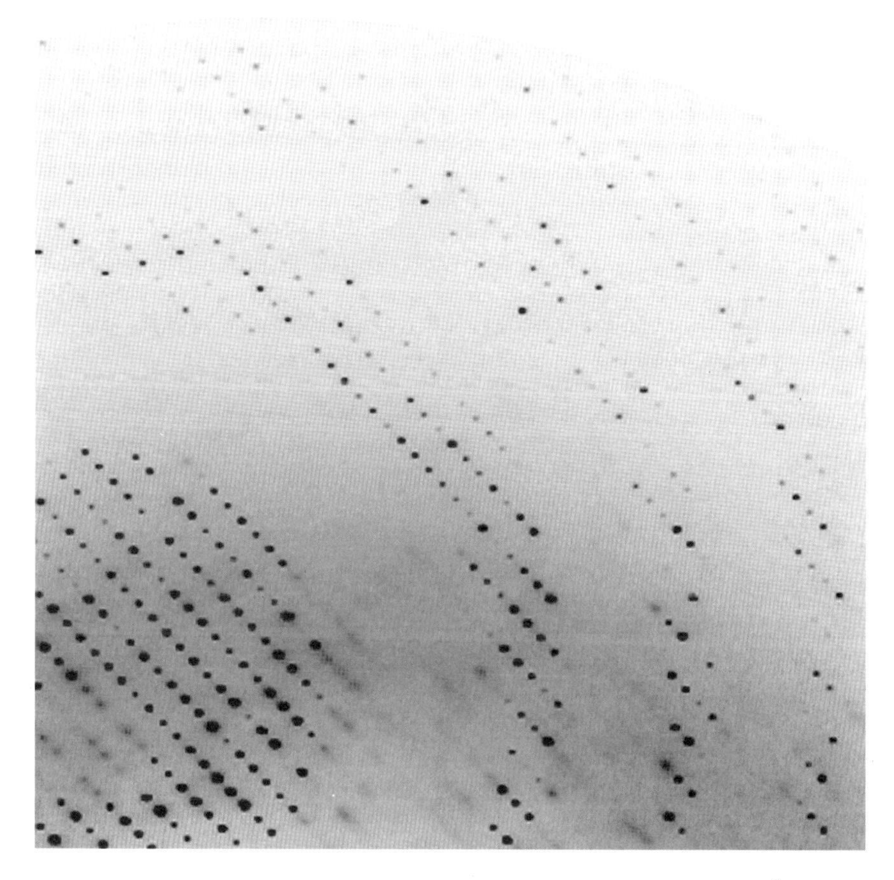

Fig. 2. A diffraction pattern of *Brassica napus* ENR to 1.9 Å.

Structure of *B. napus* ENR

The structure of a binary complex of *B. napus* ENR with NAD^+ was determined by X-ray crystallography to 1.9 Å resolution (Rafferty *et al.*, 1995). A model composed of 297 out of the total of 312 residues and 123 solvent molecules has been constructed. The additional 15 residues not seen in the electron density map correspond to the first 11 and final 4 residues. The electron density for the NAD^+ is interpretable only for the adenine ring and its associated ribose sugar moiety, but difference Fourier studies using data from crystals soaked in NADH have enabled the unambiguous location of the entire cofactor. The ENR subunit comprises a single domain of dimensions 55 × 55 × 50 Å and contains seven β-strands (β1–β7), creating a parallel β-sheet, and seven α-helices (α1–α7). The parallel β-sheet is flanked on one side by helices α1, α2 and α7 and, on the other, by helices α3, α4 and α5 with α6 sitting along the 'top edge' of the β-sheet above the COOH-terminal ends of strands β6 and β7 (Fig. 3). This fold is highly reminiscent of the Rossmann fold commonly found in dinucleotide binding enzymes (Birktoft & Banaszak, 1984).

ENR has been shown to be a tetramer by gel filtration (Slabas *et al.*, 1986), and the tetramer can be clearly seen in the crystal packing with approximate dimensions 90 × 80 × 80 Å (Fig. 4). Each monomer in the tetramer makes contact with all three symmetry-related partners on for

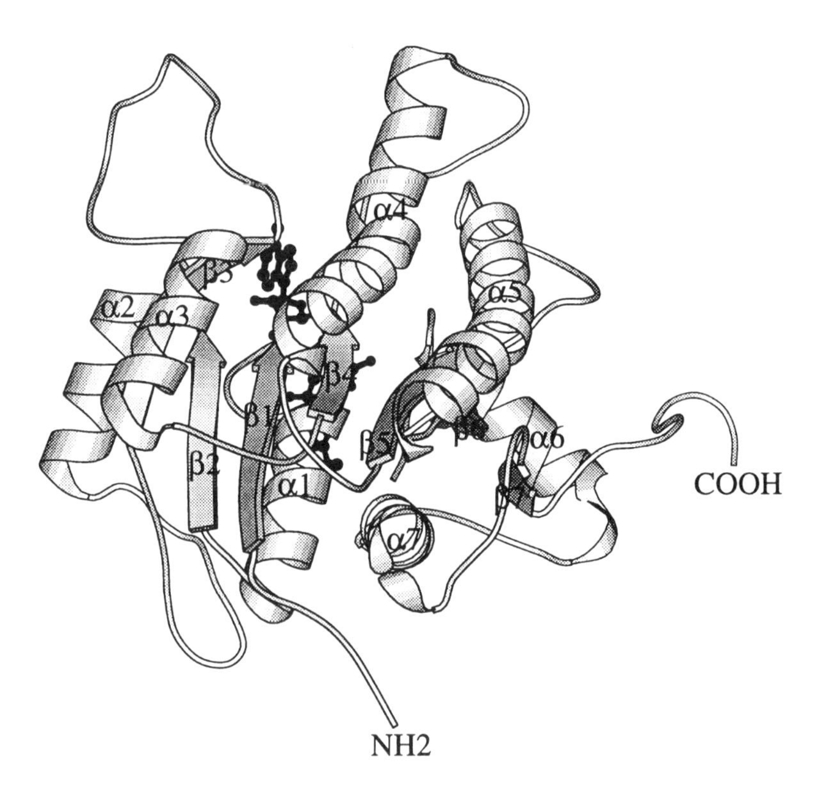

Fig. 3. View of one subunit of the *B. napus* ENR tetramer. The cofactor is shown in all atom representation, α-helices and β-strands are shown as coiled ribbons and flattened arrows, respectively, and numbered as in the text. (Produced using the program MOLSCRIPT, Kraulis, 1991.)

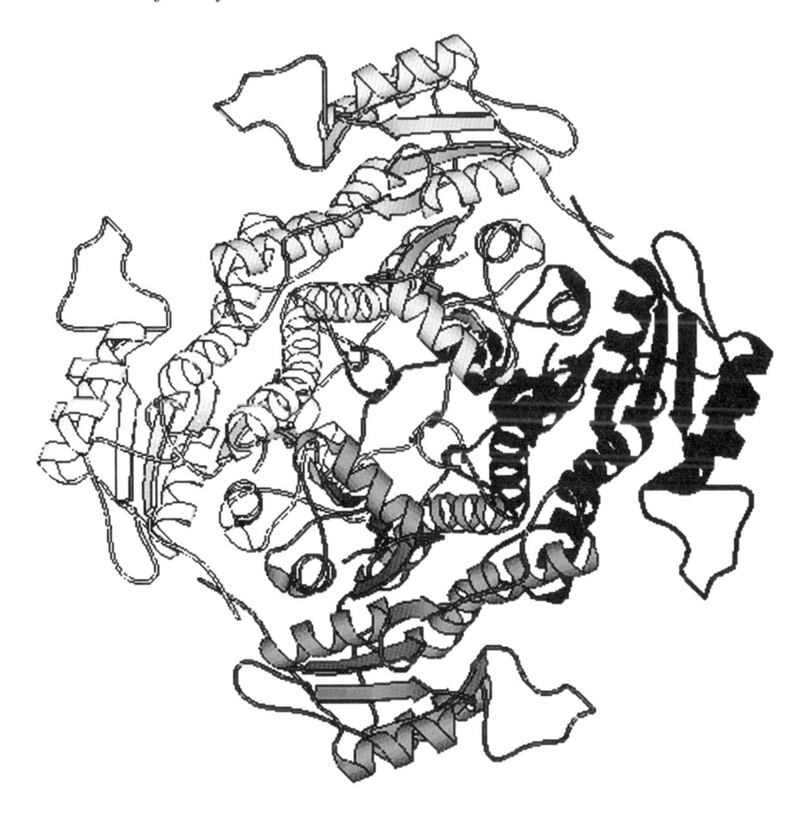

Fig. 4. View of the *B. napus* ENR tetramer along a molecular two-fold axis. α-Helices and β-strands shown as coiled ribbons and flattened arrows, respectively. (Produced using the program MOLSCRIPT, Kraulis, 1991.)

mation of the tetramer and approximately 4000 Å^2 of the solvent accessible surface is buried per monomer.

Nucleotide binding site

Difference Fourier experiments on crystals soaked in stabilising solutions containing NADH have resulted in maps with good, interpretable density for the entire cofactor. Thus, unlike NAD^+, the nicotinamide ring in NADH is ordered on the protein surface. Analysis of the ENR/NADH binary complex has shown that the cofactor is bound in an extended conformation and both ribose sugar rings are found as the

C2′-endo conformers. The nicotinamide ring is in the *syn* conformation with its B-face accessible from the proposed substrate binding site.

The 2′ hydroxyl group of the adenine ribose sits in a shallow depression on the enzyme surface in a region flanked by Ala27, Trp52 and Ala55. The lack of space and absence of positively charged side chains provide an explanation for the strict NADH-dependence of the enzyme (Slabas *et al.*, 1986). The pyrophosphate moiety of the NADH lies close to the COOH-terminal end of the β-sheet, where it interacts with residues which form the loop between strand β1 and helix α1 and residues from the first turn of helix α1. There are no positively charged side chains close to the pyrophosphate moiety, and thus its stabilisation on the enzyme is via the series of hydrogen bonds and the helix dipole of α1 alone. Interactions made by the nicotinamide ribose hydroxyl groups involve hydrogen bonds to the side chain amino group of Lys206 and to an ordered solvent molecule which is in turn contacted by the side chain imidazole ring ND1 nitrogen of His135 and the peptide nitrogen of Leu186. Hydrogen bonds are formed between the oxygen and nitrogen of the carboxyamide moiety of the nicotinamide ring, and the peptide nitrogen of Leu236 and the cofactor pyrophosphate moiety, respectively, and stabilise the glycosidic bond in the *syn* conformation burying the A-face of the ring against the protein and exposing the B-face to the active site. Thus, for this enzyme, stereospecific hydride transfer would appear to be via the *pro*-4S hydrogen of the NADH, in agreement with the previous studies on *E. coli* ENR (Saito *et al.*, 1980).

A comparison of the structures of ENR with NAD^+ or NADH bound shows that a significant shift occurs in the position of the side chain of Tyr32 between the two structures. In the complex with NADH, the tyrosine side chain can be unambiguously located and the edge of the phenolic ring packs against the A-face of the nicotinamide ring. In contrast, in the NAD^+ bound structure there is evidence in the electron density for one predominant side chain conformation for Tyr32 plus other lower occupancy conformations amongst which is that observed in the NADH bound structure. In the major conformation in the NAD^+ complex the phenolic ring occupies part of the binding pocket for the nicotinamide moiety and hence partially fills it. The movement of this side chain is therefore essential for the localisation on the enzyme of the nicotinamide ring.

Location of the active site

A multiple sequence alignment of ENR from *B. napus*, *E. coli*, *M. tuberculosis* and *Anabena sp.* reveals that there are 36 completely conserved residues. When these residues are mapped onto the structure of the *B. napus* enzyme, Tyr198, Met202, Lys206, Ala233, Gly234 and Pro235 lie immediately adjacent to the nicotinamide ring of the NADH and are therefore likely to be critical for either the substrate recognition or the enzyme chemistry. Of the remaining completely conserved residues, Gly25 lies close to the adenine ribose and the others are dispersed throughout the structure occurring mainly at tetramer subunit interfaces or in positions which help to stabilise the orientation of secondary structure elements. Studies on *E. coli* enoyl reductase have shown the reduction to proceed by a *syn* addition of hydrogen via a 2-*Re*, 3-*Si* attack on the double bond (Saito *et al.*, 1981), but little is currently known about the molecular details of the mechanism of ENR. However, a plausible mechanism for the catalytic activity of ENR proceeds through attack of a hydride ion from NADH upon position C-3 at the double bond in the substrate followed by the formation of an enolate anion intermediate (Fig. 5). In subsequent steps a proton could then be added to the oxygen of the enolate anion to form an enol which would then tautomerise to give the reduced acyl product. Examination of the region close to the C4 position of the NADH focuses attention on the conservation of Tyr198 and Lys206. A possible role for Tyr198 might be to act as the base which donates the proton to the enolate anion in the catalytic mechanism, and Lys206 might act to stabilise a transition state.

Similarity of ENR with HSD

Structural comparisons have shown that ENR is similar to numerous dehydrogenases, but a very strong similarity was shown between ENR and 3α,20β-hydroxysteroid dehydrogenase (HSD) from *Streptomyces hydrogenans* (Protein Data Bank (PDB) entry 2HSD) (Edwards & Orr, 1978). HSD catalyses the NADH-dependent reversible oxidation of the 3α-hydroxyl and 20β-hydroxyl groups of derivatives of the steroids androstane and pregnane. The similarity between HSD and ENR extends over virtually the entire fold of the two enzymes and includes nearly all the secondary structure elements. As shown in Fig. 6, the region of common structure between HSD and ENR comprises 7 β-strands and 5 α-helices (for purposes of the comparison search helix α4 in ENR and helix α4 in HSD were split by the program into two smaller helices) with identical topology and sequence order. Both enzymes are tetrameric, and the similarity is also observed for the association of their subunits in formation of the quaternary structures. Alignment of the HSD and ENR sequences shows a 22% identity over the 254 residues of HSD (Fig. 7).

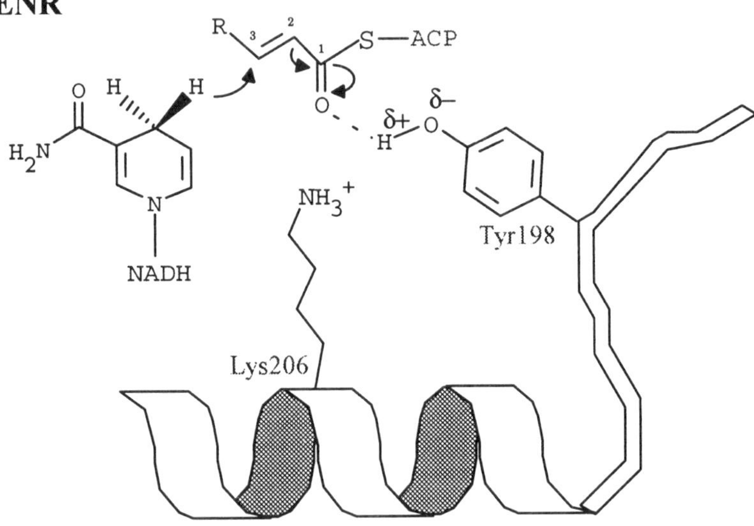

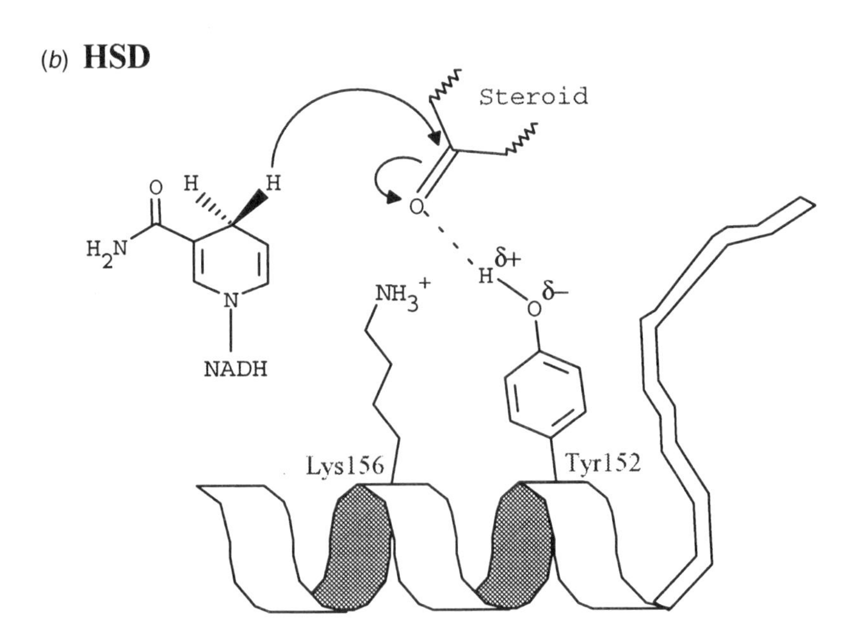

Fig. 5. Proposed catalytic mechanisms of substrate reduction by ENR and HSD. (*a*) Reduction of the double bond in an enoyl substrate by ENR. (*b*) Reduction of the keto group in a steroid substrate by HSD (Edwards & Orr, 1978).

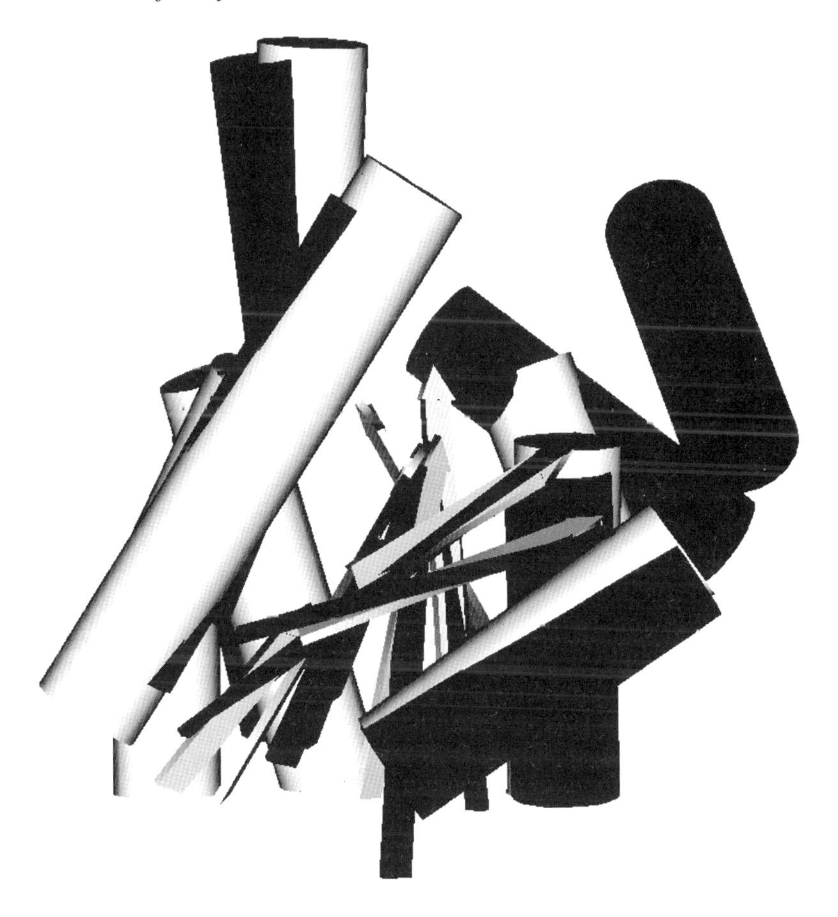

Fig. 6. View of the structural overlap between ENR and HSD, the α-helices shown as solid cylinders and the β-strands shown as arrows are black for ENR and grey for HSD (produced using the program PETO, P.J. Artymiuk, unpublished program).

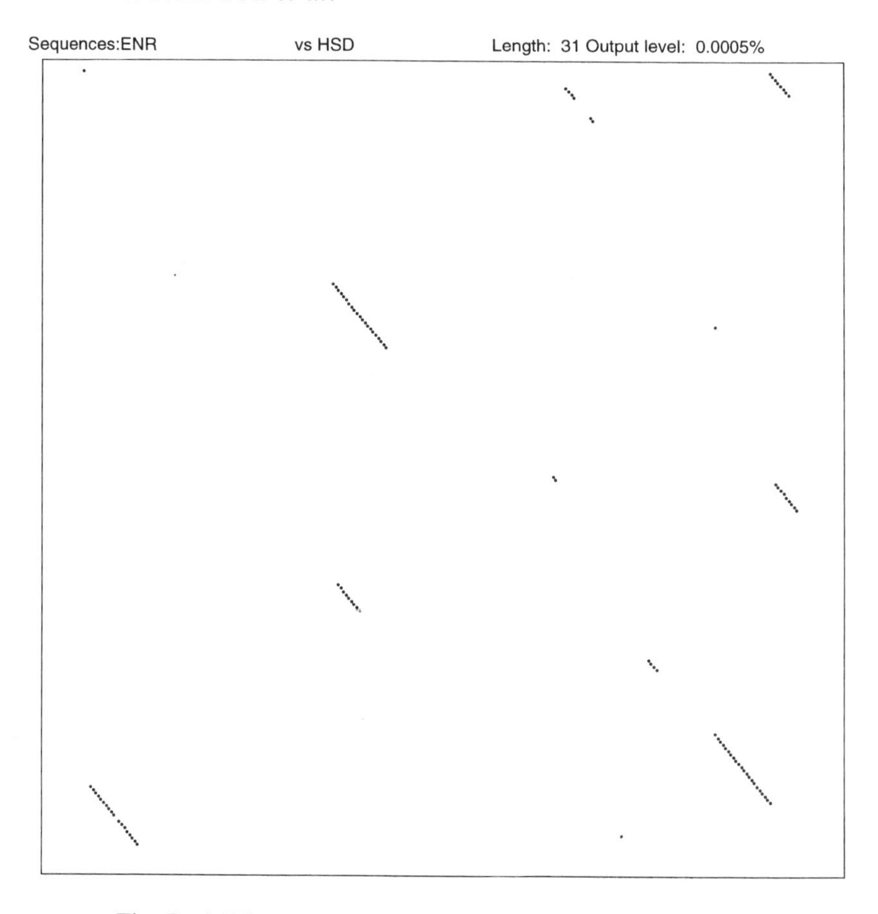

Fig. 7. A Diagon plot with the sequence of ENR on the vertical axis and HSD on the horizontal axis.

A catalytic mechanism for steroid keto-group reduction by HSD involves stereospecific hydride transfer from the NADH cofactor to the carbon atom of the keto group followed by donation of a proton to the oxygen of the resultant alkoxide ion intermediate from the phenolic oxygen of Tyr152 (Fig. 5). The positively charged amino group of Lys156 has been proposed in some way to stabilise the transition state. HSD is a member of the family of homologous short chain dehydrogenases, oxidoreductases and epimerases (Persson, Krook & Jörvall, 1991; Labesse *et al.*, 1994), including rat liver dihydropteridine reductase (Varughese *et al.*, 1992), *E. coli* UDP-galactose 4-epimerase (Bauer *et al.*, 1992) and *B. napus* β-keto acyl-carrier protein reductase (BKR). The family has been identified as carrying a signature motif YxxxK, where x is any residue, and where the tyrosine and lysine residues are those highlighted in the proposed catalytic mechanism of HSD.

Closer inspection of the superimposition shows that the similarities extend to the location and conformation of the bound NADH cofactors, where the C4 carbons of the nicotinamide rings are less than 1 Å apart, and to the equivalence of the putative catalytic Lys156 in HSD and Lys206 from helix α5 in ENR. Furthermore, although the equivalent residue to the catalytic Tyr152 in HSD is Met202 in ENR, the position occupied by the phenolic oxygen of Tyr152 in HSD is approximately taken up by the phenolic oxygen of Tyr198 in ENR. The similarity in side chain position of Tyr152 in HSD and Tyr198 in ENR lends further support for the role of the latter as a base in the ENR reaction mechanism. However, it is clear that there is a definite difference in the position of the phenolic oxygen with respect to the C4 position of the nicotinamide ring. In HSD the separation of the phenolic oxygen of Tyr152 and the C4 position of the nicotinamide ring is approximately 4 Å, which is consistent with the direct attack of the hydride on the carbon atom of the keto group of the steroid substrate. In contrast, in the reaction catalysed by ENR, where the carbon atom attacked by the NADH is two bonds removed from the keto group thought to be protonated by the base, the phenolic oxygen of Tyr198 is located approximately 6 Å from the C4 position. Thus the shift in the relative positions of the base and NADH may be a direct consequence of the difference in the chemistry. Nevertheless, the striking similarity in the active sites shows how the same structural framework has been adapted to carry out quite different enzyme chemistry.

Previous studies have shown that the sequence of BKR, an enzyme which carries out the other reductive step in fatty acid biosynthesis, must also be related to HSD (sequence identity of 35%) (Fig. 8). Therefore, the recognition that ENR and HSD adopt similar structures further implies that ENR is also related to BKR. This further implies that the two enzymes may have diverged from a common ancestor during the evolution of the biosynthetic pathway.

The structure of *E. coli* ENR

In tandem with our structure determination of *B. napus* ENR, we have also been studying this enzyme from *E. coli*. This has led to the

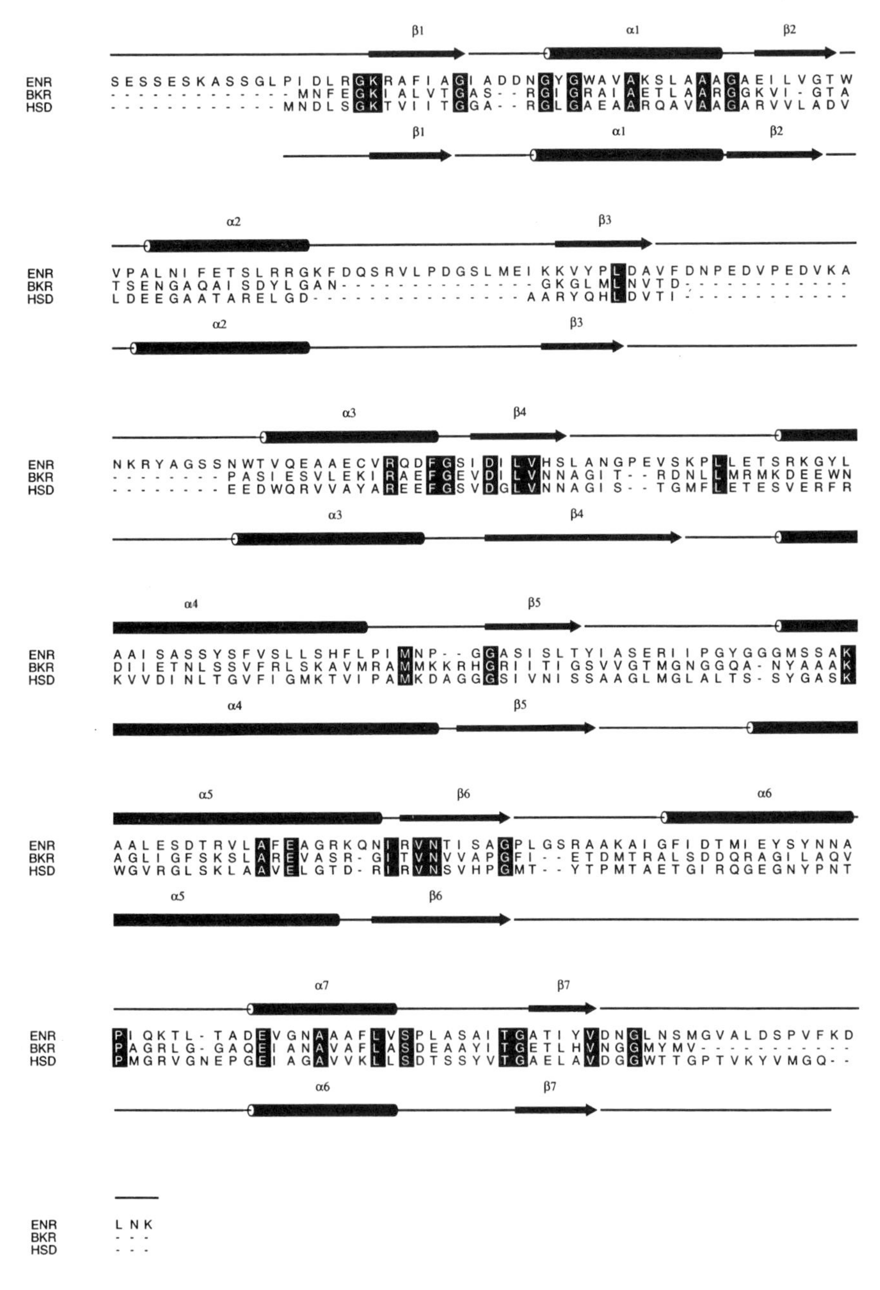
β1 α1 β2
ENR S E S S E S K A S S G L P I D L R G K R A F I A G I A D D N G Y G W A V A K S L A A A G A E I L V G T W
BKR - - - - - - - - - - - - - M N F E G K I A L V T G A S - - R G I G R A I A E T L A A R G G K V I - G T A
HSD - - - - - - - - - - - - - M N D L S G K T V I I T G G A - - R G L G A E A A R Q A V A A G A R V V L A D V
β1 α1 β2
α2 β3
ENR V P A L N I F E T S L R R G K F D Q S R V L P D G S L M E I K K V Y P L D A V F D N P E D V P E D V K A
BKR T S E N G A Q A I S D Y L G A N - - - - - - - - - - - - - - G K G L M L N V T D - - - - - - - - - - -
HSD L D E E G A A T A R E L G D - - - - - - - - - - - - - - - - A A R Y Q H L D V T I - - - - - - - - - -
α2 β3
α3 β4
ENR N K R Y A G S S N W T V Q E A A E C V R Q D F G S I D I L V H S L A N G P E V S K P L L E T S R K G Y L
BKR - - - - - - - - P A S I E S V L E K I R A E F G E V D I L V N N A G I T - - R D N L L M R M K D E E W N
HSD - - - - - - - - E E D W Q R V V A Y A R E E F G S V D G L V N N A G I S - - T G M F L E T E S V E R F R
α3 β4
α4 β5
ENR A A I S A S S Y S F V S L L S H F L P I M N P - - G G A S I S L T Y I A S E R I I P G Y G G G M S S A K
BKR D I I E T N L S S V F R L S K A V M R A M M K K R H G R I I T I G S V V G T M G N G G Q A - N Y A A A K
HSD K V V D I N L T G V F I G M K T V I P A M K D A G G G S I V N I S S A A G L M G L A L T S - S Y G A S K
α4 β5
α5 β6 α6
ENR A A L E S D T R V L A F E A G R K Q N I R V N T I S A G P L G S R A A K A I G F I D T M I E Y S Y N N A
BKR A G L I G F S K S L A R E V A S R - G I T V N V V A P G F I - - E T D M T R A L S D D Q R A G I L A Q V
HSD W G V R G L S K L A A V E L G T D - R I R V N S V H P G M T - - Y T P M T A E T G I R Q G E G N Y P N T
α5 β6
α7 β7
ENR P I Q K T L - T A D E V G N A A A F L V S P L A S A I T G A T I Y V D N G L N S M G V A L D S P V F K D
BKR P A G R L G - G A Q E I A N A V A F L A S D E A A Y I T G E T L H V N G G M Y M V - - - - - - - - - -
HSD P M G R V G N E P G E I A G A V V K L L S D T S S Y V T G A E L A V D G G W T T G P T V K Y V M G Q - -
α6 β7
ENR L N K
BKR - - -
HSD - - -

(Fig. 8.)

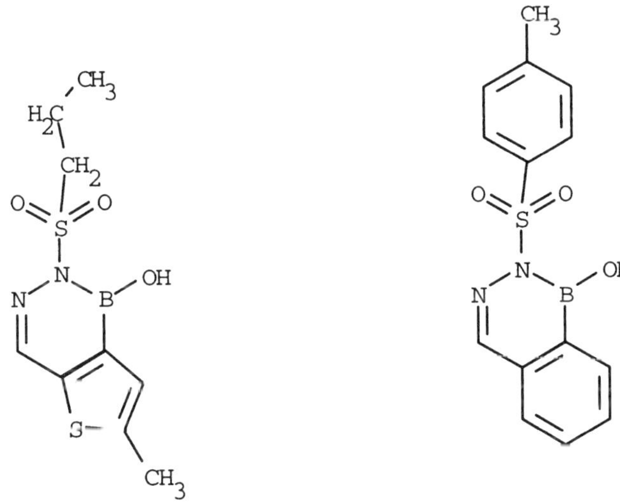

Thieno-diazaborine Benzo-diazaborine

Fig. 9. Structural formulae of thieno-diazaborine and benzo-diazaborine.

three-dimensional structure determination of *E. coli* ENR in complex with NAD^+ to 2.1 Å resolution and in complex with NAD^+ and either thieno-diazaborine or benzo-diazaborine (Fig. 9) (Baldock *et al.*, 1996). The structure of *E. coli* ENR, was found to be very similar to that of *B. napus* ENR, with 208 Cα atoms superimposing with a root mean square deviation (rmsd) of 1.4 Å as might be expected from the 33% sequence identity. In particular, in the region of the active site, the putative catalytic tyrosine and lysine occupy very similar positions

---

Fig. 8. A multiple sequence alignment of the enoyl reductase from *B. napus* (ENR), β keto reductase from *E. coli* (BKR) and 3α–20β hydroxy steroid dehydrogenase from *Streptomyces hydrogenans* (HSD). The secondary structure in ENR is shown above and HSD below the sequences (produced using the program ALSCRIPT, Barton, 1993).

(Fig. 10). Both diazaborine compounds bind in a similar manner in a pocket on the protein surface, adjacent to the nicotinamide ring of the cofactor. The bicyclic rings of the diazaborines form extensive π–π stacking interactions with the nicotinamide ring. The difference in binding of the two diazaborines is that their respective tosyl and propyl groups occupy subtly modified positions. The tosyl moiety lies perpendicular to the bicyclic ring and interacts with the main chain peptide between Gly93 and Ala95 and the side chain of Leu100, whereas the propyl moiety folds back onto the planar bicyclic ring system in a manner reminiscent of a scorpion's tail and forms interactions with the side chain of Met159 and Ile200 and the main chain peptide of both Gly93 and Phe94.

Surprisingly, analysis of the drug complex showed that the distance between the boron atom of the diazaborine and the 2′OH of the nicotinamide ribose was approximately 1.7 Å, comparable with a B–O covalent bond length of 1.6 Å and implying that the interaction between these two atoms is covalent. This is further supported by the unambiguous identification of the position of the hydroxyl oxygen to which the

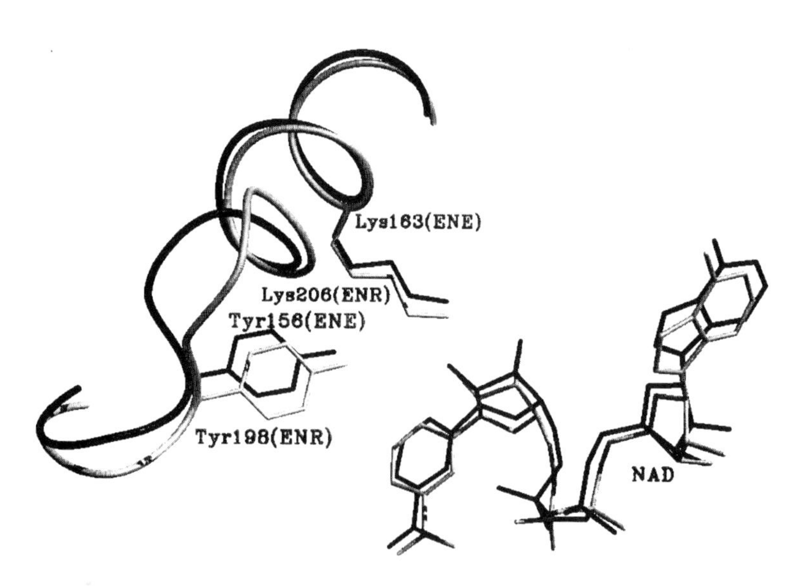

Fig. 10. View of the superimposition of ENR and ENE near the nicotinamide ring of the NADH cofactor. The alpha carbon backbone traces, NADH cofactors and key tyrosine and lysine residues (see text) are shown in grey for ENR and black for ENE (produced using the program MIDAS, Ferrin *et al.*, 1988).

boron is linked, which can be seen to form part of a tetrahedral, rather than a trigonal arrangement as required if the boron forms four covalent bonds (Fig. 11). This finding provides a clear explanation for the strong inhibitory properties of the diazaborines and for the requirement of NAD^+ for diazaborine binding. We note that the formation of the covalent bond in this case resembles the manner in which the boronic acid inhibitors of serine proteases act via the chemical modification of the active site serine to give a covalently bound tetrahedral adduct (Zhong *et al.*, 1991).

Examination of the structure of *E. coli* ENR shows that the G93S mutation which leads to resistance to diazaborine (Turnowsky *et al.*, 1989) maps to a region close to the nucleotide binding site. Modelling

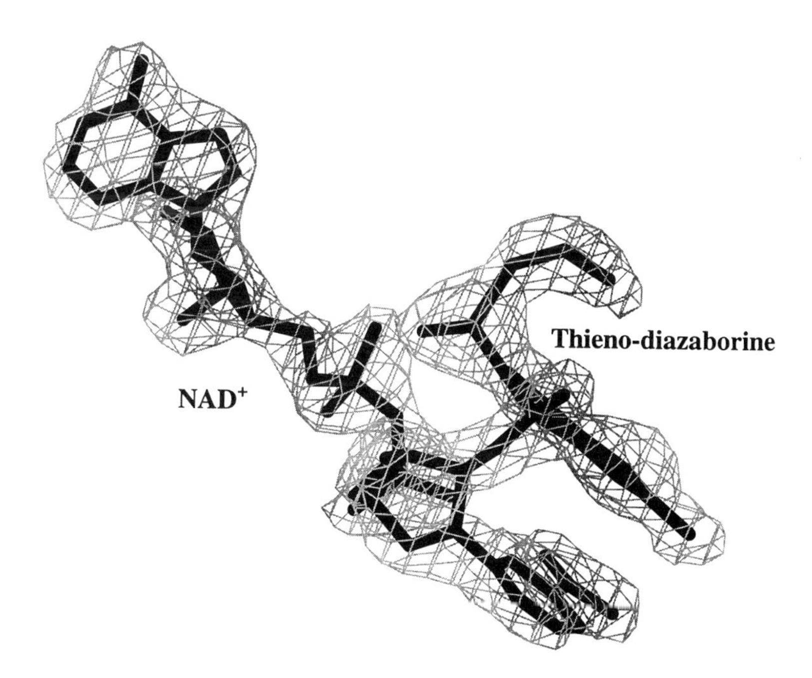

Fig. 11. Initial Fourier map of the NAD^+-thieno-diazaborine complex at 2.2 Å resolution with the final refined structures superimposed. The density (contoured at 1.2σ) was calculated with coefficients $2|F_o| - |F_c|$ and calculated phases using the refined structure from the molecular replacement solution using the model for the *E. coli* ENR–NAD^+ complex which contained no information about the inhibitor (produced using BOBSCRIPT, R. Esnouf, Oxford, personal communication, a modified version of MOLSCRIPT, Kraulis, 1991).

studies show that the Cβ atom of the serine side chain would be unacceptably close to the two oxygens of the sulphonyl group of the diazaborine. Therefore, resistance to diazaborine is probably explained by the serine side chain of the G93S mutant encroaching into the drug binding site and causing severe steric hindrance.

The position of the aromatic bicyclic ring of the diazaborine above the nicotinamide ring, strongly resembles the proposed model for the binding of the enoyl substrate suggested from studies on *B. napus* ENR (Rafferty *et al.*, 1995) with the proposed position for the negatively charged oxygen of the enolate anion of the substrate close to that of the boron atom in the drug. Thus, the formation of a covalent bond between

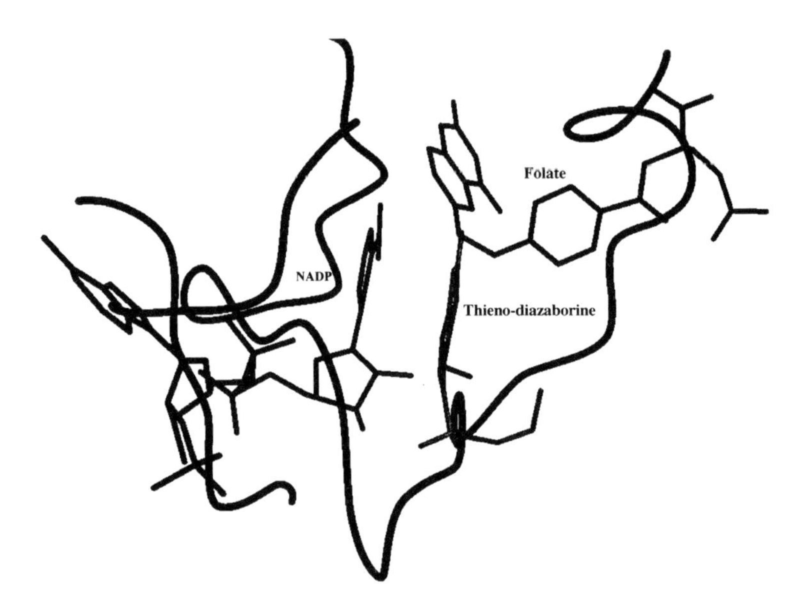

Fig. 12. The superposition (based on the nicotinamide and its associated ribose) of the nucleotide/inhibitor complex of ENR into the active site of the substrate complex of DHFR (PDB entry 7DFR, Bystroff, Oatley & Kraut, 1990). The NAD^+, thieno-diazaborine and folate are shown in all atom representation and the Cα backbone of DHFR is shown as a worm trace. When the thieno-diazaborine/NAD^+ complex is fitted into the active site of DHFR, there are some steric clashes between the sulphonyl group and the propyl tail of the diazaborine with parts of the enzyme surface. Nevertheless, there is sufficient space around the 2OH of the nicotinamide ribose to envisage the formation of a linker between the ribose and a folate analogue.

E. coli ENR and diazaborine generates a tight, non-covalently bound bisubstrate analogue. In the light of this, an important feature of the *E. coli* ENR–diazaborine complex structure is that it provides clear evidence for the type of linkage that may need to be created in order to synthesise a bisubstrate analogue with the necessary geometry to occupy the active site cleft, potentially guiding the design of new ENR inhibitors. A number of NAD(P)-dependent oxidoreductases are known to be drug targets, including dihydrofolate reductase (DHFR), the target for the anti-cancer agent, methotrexate (Bolin *et al.*, 1982). Structural analysis of members of this family of dehydrogenases has shown that the relative position of the nicotinamide ring, its associated ribose and the enzyme active site are closely related. Furthermore, in each of these enzymes the catalytic cycle involves the presentation of a π electron system of a substrate to the face of the nicotinamide ring. Therefore, there is an excellent opportunity to mimic the chemistry seen in the diazaborine/NAD^+ complex in the synthesis of new enzyme inhibitors. For example, the superposition of the thieno-diazaborine inhibitor of *E. coli* ENR into the active site of the nucleotide/substrate complex of DHFR (PDB entry 7DFR (Bystroff, Oatley & Kraut, 1990)) indicates that linking a folate analogue to the nicotinamide ribose is a distinct possibility and might be utilised for the design of new anti-cancer agents (Fig. 12). Therefore, this structural comparison suggests that the utilisation of the ribose hydroxyl to create a bisubstrate analogue might find important applications in other areas of medicinal chemistry.

References

Baldock, C., Rafferty, J.B., Sedelnikova, S.E., Baker, P.J., Stuitje, A.R., Slabas, A.R., Hawkes, T.R. & Rice, D.W. (1996). A mechanism of drug action revealed by structural studies of enoyl reductase. *Science*, **274**, 2107–10.

Banerjee, A., Dubnau, E., Quemard, A., Balasubramanian, V., Sun Um, K., Wilson, T., Collins, D., de Lisle, G. & Jacobs, Jr, W.R. (1994). *inhA*, a gene encoding a target for isoniazid and ethionamide in *Mycobacterium tuberculosis*. *Science*, **263**, 227–30.

Barton, G.J. (1993). ALSCRIPT a tool to format multiple alignments. *Protein Engineering*, **6**, 37–40.

Bauer, A.J., Rayment, I., Frey, P.A. & Holden, H.M. (1992). The molecular structure of UDP-galactose 4-epimerase from *Escherichia coli* determined at 2.5 Å resolution. *Proteins: Structure, Function and Genetics*, **12**, 372–81.

Bergler, H., Wallner, P., Ebeling, A., Leitinger, B., Fuchsbichler, S., Aschauer, H., Kollenz, G., Hogenauer, G. & Turnowsky, F. (1994). Protein EnvM is the NADH-dependent enoyl-ACP reductase

(FabI) of *Escherichia coli. Journal of Biological Chemistry*, **269**, 5493–6.

Bergler, H., Fuchsbichler, S., Hogenauer, G. & Turnowsky, F. (1996). The enoyl-[acyl-carrier-protein] reductase (FabI) of *Escherichia coli*, which catalyses a key regulatory step in fatty acid biosynthesis, accepts NADH and NADPH as cofactors and is inhibited by palmitoyl-CoA. *European Journal of Biochemistry*, **242**, 689–94.

Birktoft, J.J. & Banaszak, L.J. (1984). Structure-function relationships among nicotinamide-adenine dinucleotide dependent oxidoreductases. In *Peptide and Protein Reviews*, ed. M.T.W. Hearn, vol. 4, pp. 1–46. New York: Dekker.

Bolin, J.T., Filman, D.J., Matthews, D.A., Hamlin, R.C. & Kraut, J. (1982). Crystal structures of *Escherichia coli* and *Lactobacillus casei* dihydrofolate reductase refined at 1.7 Å resolution. *Journal of Biological Chemistry*, **257**, 13 650–62.

Bystroff, C., Oatley S.J. & Kraut, J. (1990). Crystal structures of *Escherichia coli* dihydrofolate reductase: the NADP holoenzyme and the folate-NADP ternary complex. Substrate binding and a model for the transition state. *Biochemistry*, **29**, 3263–77.

Cole, S.T. (1994). *Mycobacterium tuberculosis:* drug-resistance mechanisms. *Trends in Microbiology*, **2**, 411–15.

Edwards, C.A.F. & Orr, J.C. (1978). Comparison of the 3α- and 3β-hydroxysteroid dehydrogenase activities of the cortisone reductase of *Streptomyces hydrogenans. Biochemistry*, **17**, 4370–6.

Ferrin, T.E., Huang, C.C., Jarvis, L.E. & Langridge, R. (1988). The MIDAS display system. *Journal of Molecular Graphics*, **6**, 13–27.

Högenauer, G. & Woisetschläger, M., (1981). A diazaborine derivative inhibits lipopolysaccharide biosynthesis. *Nature*, **293**, 662–4.

Kater, M.M., Koningstein, G.M., Nijkamp, H.J.J. & Stuitje, A.R. (1991). cDNA cloning and expression of *Brassica napus* enoyl-acyl carrier protein reductase in *Escherichia coli. Plant Molecular Biology*, **17**, 895–909.

Kater, M.M., Koningstein, G.M., Nijkamp, H.J.J. & Stuitje, A.R. (1994). The use of a hybrid genetic system to study the functional relationship between prokaryotic and plant multi-enzyme fatty acid synthetase complexes. *Plant Molecular Biology*, **25**, 771–90.

Kraulis, P.J. (1991). MOLSCRIPT: a program to produce both detailed and schematic plots of protein structures. *Journal of Applied Crystallography*, **24**, 946–50.

Labesse, G., Vidal-Cros, A., Chomilier, J., Gaudry, M. & Mornon, J-P. (1994). Structural comparisons lead to the definition of a new superfamily of NAD(P)(H)-accepting oxidoreductases: the single-domain reductases/epimerases/ dehydrogenases (the 'RED' family). *Biochemical Journal*, **304**, 95–9.

McCarthy, A.D. & Hardie, D.G. (1984). Fatty acid synthase – an example of protein evolution by gene fusion. *Trends in Biochemical Sciences*, **9**, 60–3.

Persson, B., Krook, M. & Jörvall, H. (1991). Characteristics of short-chain dehydrogenases and related enzymes. *European Journal of Biochemistry*, **200**, 537–43.

Rafferty, J.B., Simon, W.J., Stuitje, A.R., Slabas, A.R., Fawcett, T. & Rice, D.W. (1994). Crystallization of the NADH-specific enoyl acyl carrier protein reductase from *Brassica napus*. *Journal of Molecular Biology*, **237**, 240–2.

Rafferty, J.B., Simon, J.W., Baldock, C., Artymiuk, P.J., Baker, P.J., Stuitje, A.R., Slabas, A.R. & Rice, D.W. (1995). Common themes in redox chemistry emerge from the X-ray structure of oilseed rape (*Brassica napus*) enoyl acyl carrier protein reductase. *Structure*, **3**, 927–38.

Saito, K., Kawaguchi, A., Okuda, S., Seyama, Y. & Yamakawa, T. (1980). Incorporation of hydrogen atoms from deuterated water and stereospecifically deuterium-labelled nicotinamide nucleotides into fatty acids with the *Escherichia coli* fatty acid synthetase system. *Biochimica et Biophysica Acta*, **618**, 202–13.

Saito, K., Kawaguchi, A., Seyama, Y., Yamakawa, T. & Okuda, S. (1981). Steric course of reaction catalyzed by the enoyl acyl-carrier-protein reductase of *Escherichia coli*. *European Journal of Biochemistry*, **116**, 581–6.

Slabas, A.R., Sidebottom, C.M., Hellyer, A., Kessell, R.M.J. & Tombs, M.P. (1986). Induction, purification and characterization of NADH-specific enoyl acyl carrier protein reductase from developing seeds of oil seed rape (*Brassica napus*). *Biochimica et Biophysica Acta*, **877**, 271–80.

Slabas, A.R., Hellyer, A., Sidebottom, C., Bambridge, H., Cottingham, I.R., Kessell, R., Smith, C.G., Sheldon, P., Kekwick, R.G.O., de Silva, J., Windust, J., James, C.M., Hughes, S.G. & Safford, R. (1987). Molecular structure of plant fatty acid synthesis enzymes. In *Plant Molecular Biology*, ed. D. von Wettstein & N.H. Chua, vol. 140, NATO Series A, pp. 265–77.

Slabas, A.R., Cottingham, I.R., Austin, A., Hellyer, A., Safford, R. & Smith, C.G. (1990). Immunological detection of NADH-specific enoyl-ACP reductase from rape seed (*Brassica napus*): induction relationship of α and β polypeptides, mRNA translation and interaction with ACP. *Biochimica et Biophysica Acta*, **1039**, 181–8.

Turnowsky, F., Fuchs, K., Jeschek, C. & Högenauer, G. (1989). *envM* genes of *Salmonella typhimurium* and *Escherichia coli*. *Journal of Bacteriology*, **171**, 6555–65.

Varughese, K.I., Skinner, M.M., Whiteley, J.M., Matthews, D.A. & Xuong, N.H. (1992). Crystal structure of rat liver dihydropteridine reductase. *Proceedings of the National Academy of Sciences, USA*, **89**, 6080–4.

Wagner, U.G., Bergler, H., Fuchsbichler, S., Turnowsky, F., Högenauer, G. & Kratky, C. (1994). Crystallization and preliminary X-ray diffraction studies of the enoyl-ACP reductase from *Escherichia coli*. *Journal of Molecular Biology*, **243**, 126–7.

Zhong, S., Jordan, F., Kettner, C. & Polgar, L. (1991). Observation of tightly bound ^{11}B nuclear magnetic resonance signals on serine proteases. Direct solution evidence for tetrahedral geometry around the boron in the putative transition-state analogues. *Journal of the American Chemical Society*, **113**, 9429–35.

PART II: Fatty acid modifications

D.J. MURPHY and P. PIFFANELLI

5 Fatty acid desaturases: structure, mechanism and regulation

Introduction

Fatty acid desaturases are responsible for the insertion of double bonds (normally in the Z or *cis* conformation) into alkyl chains, following the abstraction of two hydrogen atoms. These enzymes are almost universally found in microbial, plant and animal cells, where they play a key role in a wide range of physiological processes. Desaturases regulate the fluidity of membrane lipids and hence influence the interface between cells and their external environment. The importance of desaturases in plant cells is underlined by the dramatic phenotype of the *fab2* mutant of *Arabidopsis* (Lightner *et al.*, 1994). The mutant plants are defective in a single gene encoding a stearoyl-ACP desaturase and consequently accumulate increased levels of the C_{18} saturate, stearic acid, in their membrane lipids. Such plants have gross anatomical irregularities and a dwarf phenotype which can only be rescued by growth at abnormally high temperatures of 36 °C. These characteristics strongly suggest that the mutant phenotype is caused by alterations in membrane structure resulting from a reduction in desaturase activity.

Desaturases are also key enzymes in the biosynthesis of a variety of signalling molecules ranging from highly volatile insect pheromones to several classes of animal hormones and plant growth regulators including prostaglandins, leukotrienes and jasmonic acid derivatives. The physical properties and nutritional value of many animal and plant storage lipids are determined by desaturases. Such storage lipids include mammalian milk, which is relatively saturated, and seed and fruit oils which are often highly unsaturated. Mammals, such as humans, lack the enzyme oleate desaturase and therefore plant-derived polyunsaturates, such as linoleic acid, are essential dietary components. Monounsaturates, which are prominent in the so-called 'Mediterranean diet', are also believed to have an important nutritional role.

Products derived from plant unsaturated fatty acids, including foods, polymers, lubricants, cosmetics and pharmaceuticals, have a value of many billions of $US per year. This means that the manipulation of desaturases, e.g. by genetic engineering, is attractive to biotechnologists, food producers and retailers and to the chemicals industry (Murphy, 1996). The past few years have seen considerable advances in our understanding of the mechanism, structure and regulation of desaturases – particularly in plants. In this chapter, we present a brief outline of some aspects of our current understanding of plant desaturases and their potential for future manipulation.

Substrate specificity

The vast majority of unsaturated fatty acids are synthesised by desaturases acting on preformed long-chain acyl esters, with O_2 and a reduced compound (such as ferredoxin or cytochrome b_5) as cofactors. An alternative anaerobic pathway for unsaturated fatty acid synthesis exists in a limited number of bacteria, and notably the Eubacteriales (Gurr & Harwood, 1991), but will not be considered here. The oxygen-requiring desaturases of animals are normally membrane-bound (typically the endoplasmic reticulum (ER)) and use long-chain acyl-CoA esters as substrates. Plant desaturases fall into two clear groups; soluble and membrane-bound, as shown in Fig. 1. Membrane-bound desaturases are located in both ER and chloroplast membranes and have been shown to use a variety of complex lipid substrates including phosphatidylcholine, phosphatidylglycerol and monogalactosyl diacylglycerol. The soluble plant desaturases are localised in the chloroplast stroma and tend to be specific for saturated acyl-ACP substrates.

Plant desaturases as a whole can act on fatty acids ranging in length from at least C_{12} to C_{24} and can insert double bonds in a variety of positions in the acyl chain. Nevertheless, each individual desaturase tends to be rather specific in the chain length and double bond insertion position of its substrate(s). The specificities of three of the major classes of plant desaturases, as revealed by radiochemical labelling studies, have been reviewed by Heinz (1993) and the main conclusions are shown in Fig. 2. One class of desaturases which includes the soluble Δ_9 and the ER membrane-bound Δ_{12} enzymes will insert a double bond at

Fig. 1. The major classes of fatty acid desaturases in plants. All of the soluble desaturases use acyl-ACP substrates and are located in the stroma of plastids. The membrane-bound desaturases use complex lipid substrates, e.g. phospho- or glyco-lipids, and are located on both ER and plastidial membranes. *Note that all of the desaturases insert Z double bonds, with the exception of the plastidial Δ_3 palmitate desaturase, which inserts an E double bond.

SOLUBLE FATTY ACID DESATURASES

enzyme	normal substrate	immediate product	ultimate product
Δ9 STEAROYL-ACP DESATURASE	18:0	$18{:}1_{\Delta9}$	OLEIC ACID
Δ4 PALMITOYL-ACP DESATURASE	16:0	$16{:}1_{\Delta4}$	PETROSELINIC ACID
Δ6 PALMITOYL-ACP DESATURASE	16:0	$16{:}1_{\Delta6}$	HEXADECENOIC ACID
Δ9 MYRISTOYL-ACP DESATURASE	14:0	$14{:}1_{\Delta9}$	ω-5 ANACARDIC ACIDS
Δ9 PALMITOYL-ACP DESATURASE	16:0	$16{:}1_{\Delta9}$	PALMITOLEIC ACID VACCENIC ACID

MEMBRANE-BOUND DESATURASES

	enzyme	substrate	products
PLASTID-LOCALISED	Δ3 PALMITOYL DESATURASE	16:0	$16{:}1_{\Delta3E}$*
	ω-9 PALMITOYL DESATURASE	16:0	$16{:}1_{\Delta7}$
	ω-6 OLEATE DESATURASE	$16{:}1_{\Delta7}$ $18{:}1_{\Delta9}$	$16{:}2_{\Delta7,10}$ $18{:}2_{\Delta9,12}$
	ω-3 LINOLEATE DESATURASE	$16{:}1_{\Delta7,10}$ $18{:}2_{\Delta9,12}$	$16{:}2_{\Delta7,10,13}$ $18{:}3_{\Delta9,12,15}$
ER-LOCALISED	Δ9 STEAROYL DESATURASE	18:0	$18{:}1_{\Delta9}$
	Δ12 OLEATE DESATURASE	$18{:}1_{\Delta9}$	$18{:}2_{\Delta9,12}$
	Δ12 OLEATE DESATURASE-CYTOCHROME b_5	$18{:}1_{\Delta9}$	$18{:}2_{\Delta9,12}$
	ω-3 LINOLEATE DESATURASE	$18{:}2_{\Delta9,12}$	$18{:}3_{\Delta9,12,15}$
	Δ6 LINOLEATE DESATURASE	$18{:}2_{\Delta9,12}$	$18{:}3_{\Delta6,9,12}$

(Fig. 1)

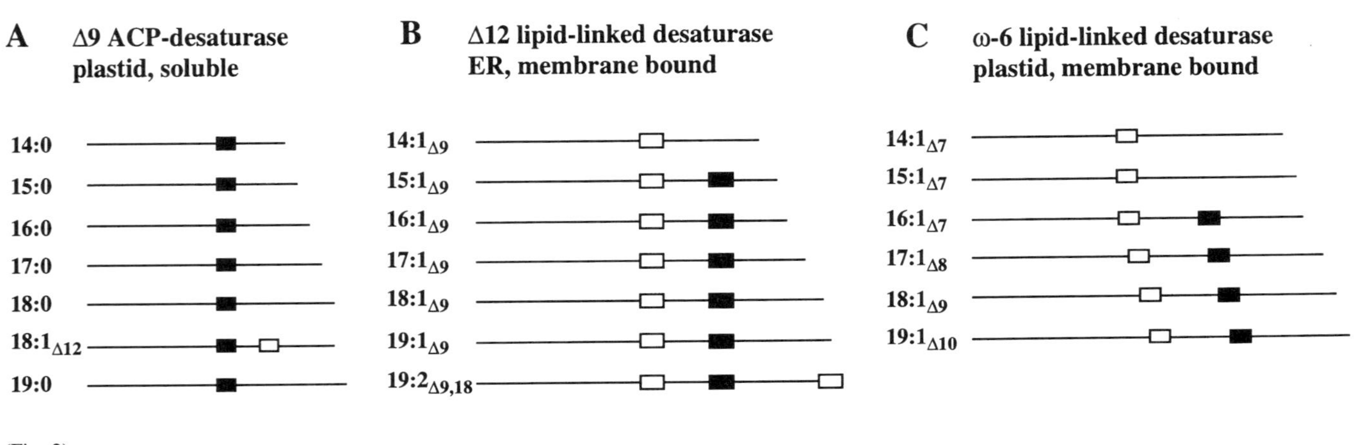
A
Δ9 ACP-desaturase
plastid, soluble
14:0
15:0
16:0
17:0
18:0
18:1Δ12
19:0
B
Δ12 lipid-linked desaturase
ER, membrane bound
14:1Δ9
15:1Δ9
16:1Δ9
17:1Δ9
18:1Δ9
19:1Δ9
19:2Δ9,18
C
ω-6 lipid-linked desaturase
plastid, membrane bound
14:1Δ7
15:1Δ7
16:1Δ7
17:1Δ8
18:1Δ9
19:1Δ10

(Fig. 2)

a specific position relative to the carboxyl (α) terminus of the fatty acid. The other class of desaturases, including the chloroplast membrane-bound ω_6 and the ER and chloroplast membrane-bound ω_3 enzymes locate the double bond position relative to the methyl (ω) terminus of the fatty acid. These differences imply that the substrates may be presented to the enzyme in different ways in the two classes of desaturase.

Structure

Soluble desaturases

A great advance in our understanding of desaturases and related enzymes was the recent publication of the first desaturase crystal structure (Lindqvist *et al.*, 1996). The three-dimensional structure of a recombinant form of the castor bean Δ_9 stearoyl-ACP desaturase has been determined to a resolution of 2.4Å. Each 37 kDa subunit in this homodimeric protein is made up of a barrel-shaped assembly consisting of nine anti-parallel α-helices in a bundle, with two further capping α-helices as shown in Fig. 3.

Each desaturase subunit contains a diiron centre with one iron interacting with the side chains of E_{196} and H_{232} and the other iron with the side chains of E_{105} and H_{146}, while the side chains of E_{143} and E_{229} interact with both iron atoms (Lindqvist *et al.*, 1996). This diiron centre is the active site of the desaturase, wherein the chemistry associated with double bond insertion must occur. Since the diiron centre is buried in the interior of the protein, there must be (a) an electron transfer channel between the iron atoms and the protein surface and (b) a deep hydrophobic cleft to allow the acyl-ACP substrate to interact with the

Fig. 2. Specificities of plant desaturases in relation to chain length, position of newly introduced double bond (■), and location of existing double bond (□). A, The soluble plastidial stearoyl-ACP desaturase always inserts the double bond in the Δ_9 position regardless of substrate chain length. B, The membrane-bound lipid-linked ER oleate desaturase always inserts the double bond in the Δ_{12} position. C, The plastid membrane-bound oleate desaturase inserts the double bond in the ω_6 position, i.e. in contrast to the other two desaturases it locates the double bond with respect to the methyl (ω) end of the acyl chain, rather than the carboxyl (α) end. The data are based on labelling studies with *Chlorella* cells and *Ricinus communis* seed slices, as reviewed by Heinz (1993).

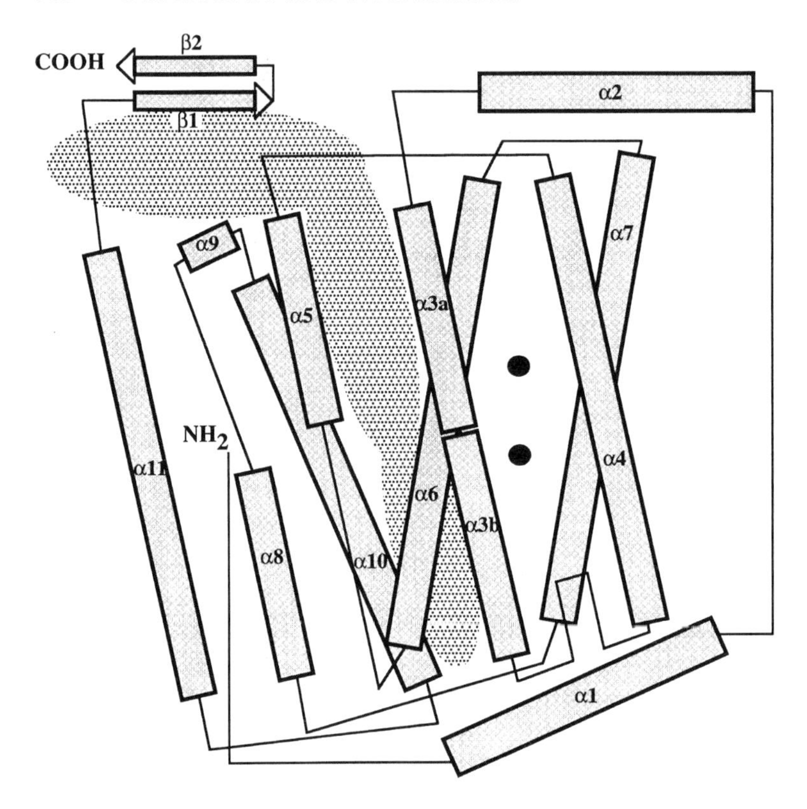

Fig. 3. Topological model of one subunit of the stearoyl-ACP desaturase homodimer. The protein consists of eleven α-helices, nine of which form an antiparallel helix bundle. The location of the diiron centre, which is associated with ligands from four of the α-helices in the bundle, is indicated by the two black spheres. The putative hydrophobic substrate cleft is indicated by shading. Figure adapted from Lindqvist *et al.* (1996).

diiron centre. There are two possible electron transfer channels, one of which extends to the surface via the series of aromatic residues, W_{139}, W_{135}, Y_{236}, F_{189} and W_{132}. This reaches the surface close to the protruding loop between helices α3b and α4 which, with the cap part of helix α1, could then provide a site for binding of the electron donor and electron transfer. The other possibility is for a channel from the surface to the diiron centre via residues W_{62}, D_{228} and H_{146}, which is analogous to the channel suggested for the related diiron protein, ribonucleotide reductase (Nordlund & Eklund, 1993). This channel would emerge from the protein at the flat surface formed by helices α1, α6, α10 and α11. A postulated substrate binding site is provided by a long and narrow cleft, mainly lined with non-polar residues, which extends from the surface deep into the protein, as shown in Fig. 3. This would allow the bulky ACP moiety to reside on the surface, while the acyl chain extended into the interior of the protein. The shape of the substrate cleft would favour a slight kink in the acyl chain in the vicinity of the Δ_9 carbon, which is also close to the diiron centre. This would facilitate the formation of the Z-configuration of the double bond in the oleoyl-ACP product.

Although this is the first crystal structure of a fatty acid desaturase, the structures of two related diiron proteins have previously been resolved, i.e. the R_2 subunit of the *E. coli* ribonucleotide reductase (Nordlund, Sjöberg & Eklund, 1990) and the R_2 subunit of methane monooxygenase from *M. capsulatus* (Rosenzweig *et al.*, 1993). The three-dimensional structures of both of these proteins can be almost perfectly superimposed on that of the plant Δ_9 desaturase. This reveals a close similarity between the three proteins, which belies their lack of conserved amino acids, apart from the iron ligands, as shown in Fig. 4.

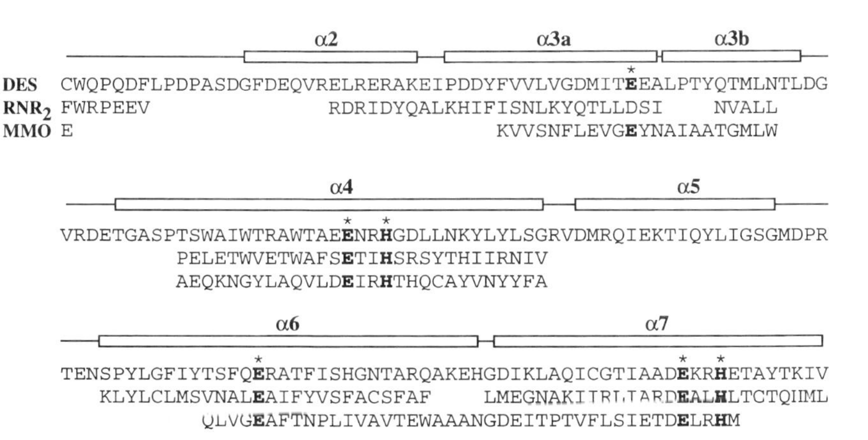

Fig. 4. Amino acid sequence alignment of structurally equivalent residues in: DES, stearoyl-ACP desaturases; RNR_2, the R_2 protein of ribonucleotide reductase; MMO, methane monooxygenase. The top line indicates the positions of α-helices 2–7 in the stearoyl-ACP desaturase. The ligands to the diiron clusters are shown by asterisks. Note the lack of overall primary sequence homology between these three proteins, despite their close structural and functional similarities. Figure adapted from Lindqvist *et al.* (1996).

The structural similarities between these proteins strongly suggests that they are evolutionarily related (see pp. 119–22) and is also consistent with similarities in their O_2-activation chemistry, as discussed later.

Soluble Δ_9 stearoyl-ACP desaturases are found in all plant cells and are essential for the biosynthesis of unsaturated membrane lipids. Several related soluble desaturases have recently been described which are found specifically in lipid-rich tissues, such as seeds and trichomes. The close relationship between the soluble Δ_9 stearate desaturases and their other soluble homologues is shown in Fig. 5. Only six of the dozens of available Δ_9 stearate desaturase sequences are displayed, and these form a distinct group with amino acid identity levels typically in the region of 80–90%. The other soluble desaturases have identities with the Δ_9 stearate desaturases of 60–70%. All of these soluble desaturases are clearly distinct from the membrane-bound Δ_9 desaturases, with which they share very little sequence similarity. Like the soluble stearate desaturases, the Δ_4 palmitate desaturase from coriander (Cahoon & Ohlrogge, 1994), the Δ_6 palmitate desaturase from *Thunbergia alata* (Cahoon *et al.*, 1994), and the Δ_4 palmitate/myristate desaturase from milkweed *Asclepias syriaca* (Cahoon, Coughlan & Shanklin, 1997), all use acyl-ACP substrates and are localised in the stroma fraction of plastids in developing seeds. Another related soluble desaturase, is the Δ_9 myristoyl-ACP desaturase, which is required for the formation of monounsaturates in the glandular trichomes of *Pelargonium hortorum* (Schultz *et al.*, 1996). It is likely that these tissue-specific soluble desaturases have evolved from an ancestral Δ_9 stearoyl-ACP desaturase via relatively minor changes in their amino acid compositions and 3D structures which have led to the observed changes in substrate and double bond insertion specificities (see pp. 119–22).

The 3D structural characterisation of the soluble Δ_9 stearate desaturase and the isolation of genes encoding very similar (at the primary sequence level) Δ_4, Δ_6 and Δ_9 palmitate and myristate desaturases allows for site-directed mutagenesis and domain-swapping studies to elucidate exactly how such similar enzymes have such different chain length and double bond insertion specificities. Preliminary studies have demonstrated that the alteration of as few as five residues of the *Thunbergia* Δ_6 palmitate desaturase can modify its activity to that of a predominantly Δ_9 stearate desaturase (Shanklin *et al.*, 1996; Cahoon & Shanklin, 1997). This approach holds great promise for the elucidation of desaturase specificities and will also be important for future efforts to engineer rationally designed enzymes to produce particular valuable unsaturated fatty acids.

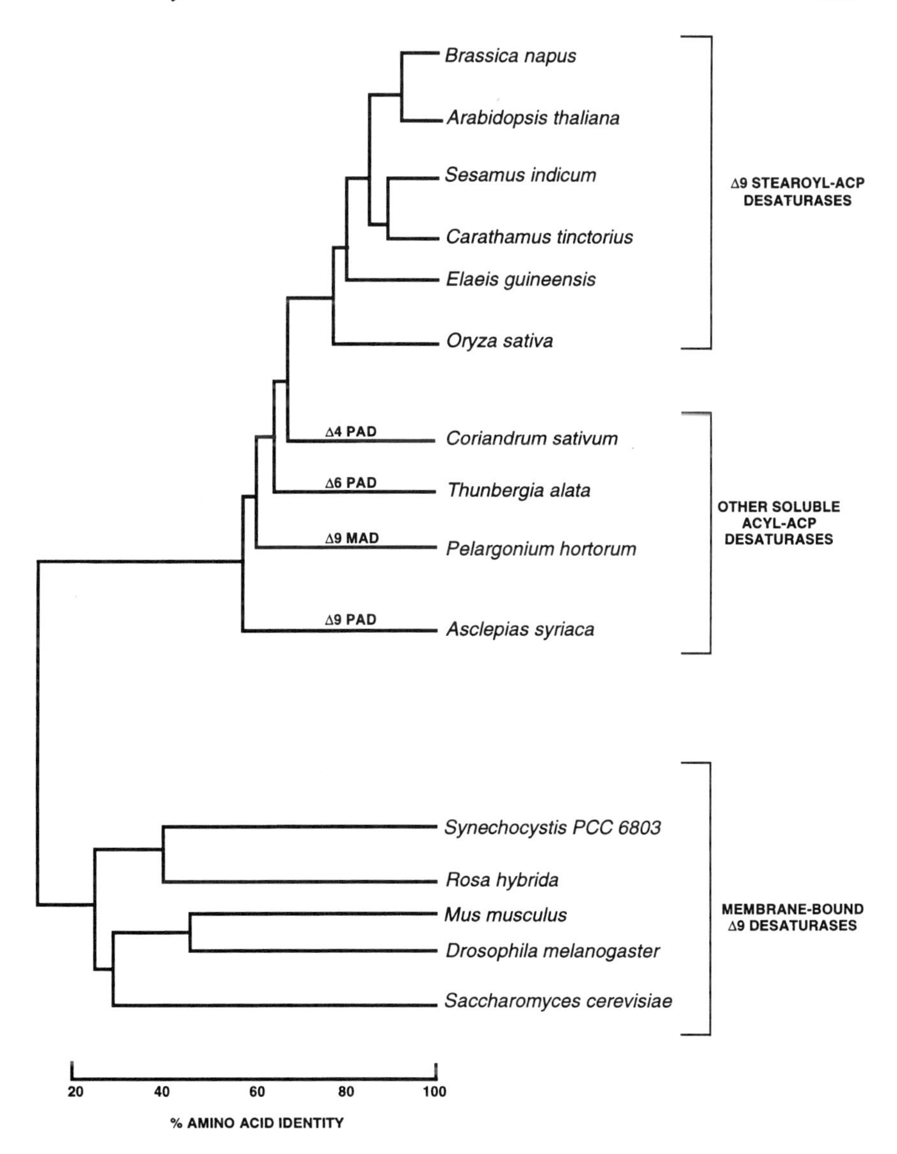

Fig. 5. Phylogenetic relationships of the 'Δ_9 desaturase family'. The family comprises all of the >30 sequenced soluble Δ_9 stearoyl-ACP desaturases from plants (of which only six examples are shown here), the closely related tissue-specific acyl-ACP desaturases from plants, and the relatively unrelated membrane-bound Δ_9 stearoyl-CoA desaturases from plants, animals, yeast and cyanobacteria. PAD, palmitoyl-ACP desaturase; MAD, myristoyl-ACP desaturase.

Membrane-bound desaturases

The majority of plant and animal desaturases, and their accessory electron transport components, are membrane bound. This property has made it much more challenging to isolate and characterise these proteins when compared with the soluble desaturases described above. The important breakthrough that led to the sequencing of most of the plant and cyanobacterial membrane desaturases was the use of molecular genetics. Of particular value has been the availability of mutagenised lines of *Arabidopsis* in conjunction with map-based cloning or T-DNA tagging. This has now resulted in the isolation and sequencing of genes encoding the ER membrane-bound Δ_{12} oleate (Okuley *et al.*, 1994) and ω_3 linoleate (Arondel *et al.*, 1992) desaturases and the plastid membrane-bound ω_3 linoleate desaturase (Hitz *et al.*, 1994). One notable

Fig. 6. Structures of membrane-bound desaturases and related proteins. A, General model of the topology of the desaturase sequences shown in B. All of these desaturases are predicted to have two sets of membrane-anchoring transbilayer α-helical domains. The active sites of the enzymes, containing the iron-binding ligands Ia, Ib and II, are probably located on the cytoplasmic face of the ER or the stromal face of the plastid envelope membranes, where they are available to interact with their electron donors, such as cytochrome b_5 or ferredoxin. B, Domain structures of the membrane-bound desaturases, hydroxylases and monooxygenases[a]. Shaded boxes indicate hydrophobic domains containing more than 40 residues and capable of spanning the bilayer twice. (The locations of the three His-containing regions Ia, Ib and II, that are the putative iron-binding ligands, are represented by solid boxes.) The sequences are aligned relative to the conserved His residues in region Ia. The hydroxylase and monooxygenase have an additional N-terminal region capable of spanning the membrane twice. Figure adapted from Shanklin *et al.* (1994). [a]Bnω_3, *Brassica napus* ER-bound ω_3 linoleate desaturase (Arondel *et al.*, 1992); RcΔ_{12} OH *Ricinus communis* ER-bound oleate hydroxylase (van der Loo *et al.*, 1995); At Δ_{12}, *Arabidopsis thaliana* ER-bound Δ_{12} oleate desaturase (Okuley *et al.*, 1994); SsΔ_{12}, *Synechocystis* sp Δ_{12} oleate desaturase (Wada, Gombos & Murata, 1990); SsΔ_6, *Synechocystis* sp Δ_6 linoleate desaturase (Reddy *et al.*, 1993); RnΔ_9, *Rattus norvegicus* Δ_9, stearate desaturase (Thiede, Ozols & Strittmatter, 1986); ScΔ_9, *Saccharomyces cerevisiae* Δ_9 stearate desaturase (Stukey, McDonough & Martin, 1990); Po alkB, *Pseudomonas oleovorans* alkaline hydroxylase (Kok *et al.*, 1989); Pp xylM *Pseudomonas putida* xylene monooxygenase (Suzuki *et al.*, 1991).

exception was the use of a biochemical approach for the purification of the ω_6 oleate desaturase from plastidial envelope membranes of spinach (Schmidt *et al.*, 1994).

Comparison of available sequence information reveals that there is a high degree of similarity between the same class of membrane-bound desaturases in different plant species, but much less similarity between different classes of desaturases, even in the same species. For example, all of the ω_3 desaturases are >85% identical (Yadav *et al.*, 1993) whereas there is very little sequence identity between the ω_3, ω_6 and Δ_{12} membrane-bound desaturases (Shanklin, Whittle & Fox, 1994). Nevertheless, there are important structural motifs that are shared by all membrane-bound desaturases and other related enzymes as shown in Fig. 6. The most strictly conserved feature is the presence of eight

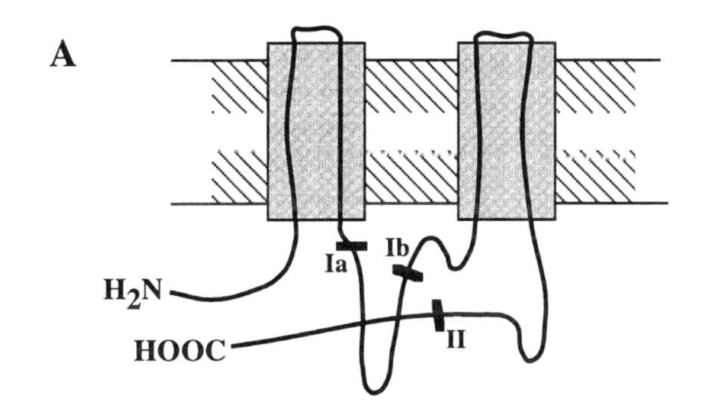

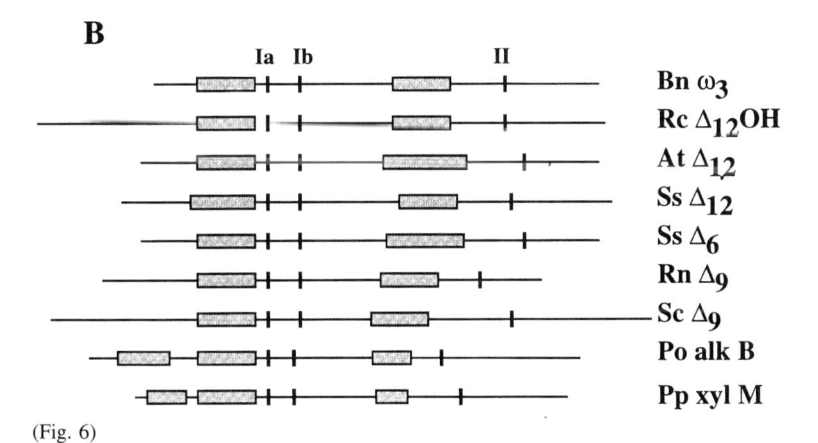

(Fig. 6)

histidines in three separate clusters with the following consensus sequence:

$$[HX_{(3\ \mathrm{or}\ 4)}\ H]\ X_{(20\text{–}50)}\ [HX_{\ (2\ \mathrm{or}\ 3)}\ HH]\ X_{\ (100\text{–}200)}\ [HX_{\ (2\ \mathrm{or}\ 3)}\ HH]$$

This motif was also found in two bacterial alkane hydroxylases and more recently in the Δ_{12} oleate hydroxylases from castor bean and *Lesquerella fendleri* (van der Loo *et al.*, 1995; Broun, Hawker & Somerville, 1997) and the Δ_6 linoleate desaturase from borage (Beremand *et al.*, 1997). The individual conversion, by site-directed mutagenesis, of each of these eight conserved histidine residues to alanine abolished the catalytic activity of a recombinant membrane-bound desaturase (Shanklin *et al.*, 1994). In contrast, the mutation of three non-conserved histidines located elsewhere in the protein had no effect. This demonstrates that each of the conserved histidines is essential for desaturase function and is consistent with their role as iron ligands. The eight histidine residues are sufficient to act as ligands for two iron atoms and may therefore be functionally equivalent to the conserved motif $[EX_2\ H]\ X_{\ (80\text{–}120)}\ [EX\ _2H]$, which is involved in binding the diiron centre of the soluble desaturases and related proteins (see Fig. 4). More recently, it has been reported that the site-directed mutagenesis of as few as six residues of a *Lesquerella* Δ_{12} oleate hydroxylase was sufficient to convert the enzyme substantially into a Δ_{12} oleate desaturase (Broun, Hawker & Somerville, 1997). This highlights the close similarity of the two enzyme types and implies that these hydroxylases may be derived from modified desaturases (see section pp. 119–22).

Electron donors to desaturases

There are at least two major electron donor systems that are believed to be utilised by plant desaturases, depending on the intracellular location. The plastidial soluble and membrane-bound desaturases are normally assumed to use ferredoxin and NAD(P)H/ferredoxin oxidoreductase, while the ER membrane-bound desaturases probably use cytochrome b_5 and NADH/cytochrome b_5 oxidoreductase (Fig. 7). There is evidence that at least some desaturases can use either of these two electron donor systems. For example, a cyanobacterial Δ_6 linoleate desaturase that probably normally uses ferredoxin was able to use the cytochrome b_5 donor system when it was produced in transgenic tobacco plants (Reddy & Thomas, 1996). It has also been observed that the Δ_9 stearoyl-ACP desaturase from safflower seeds can be reduced *in vitro* either by ferredoxin or by flavodoxin (albeit more efficiently with ferredoxin) (Jaworski & Stumpf, 1974). This implies that the interaction

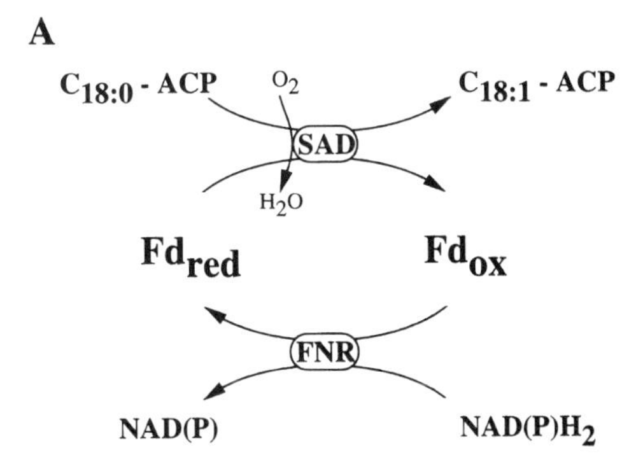

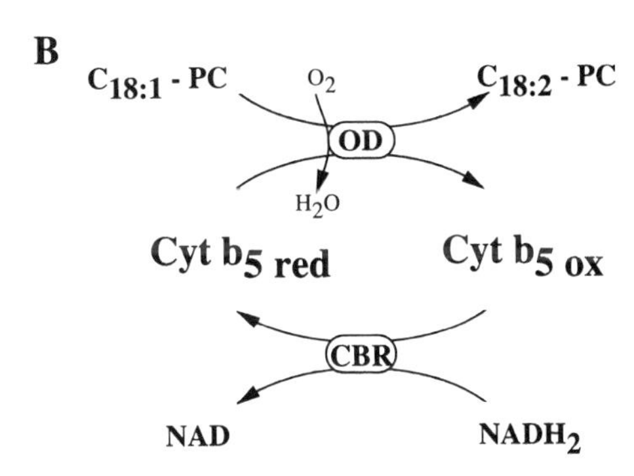

Fig. 7. Electron donor systems for fatty acid desaturases. A, The ferredoxin (Fd), NAD(P)H/ferredoxin oxidoreductase (FNR) system is used by all cyanobacterial and plastidial desaturases, whether soluble or membrane-bound. Examples include plastidial Δ_9 stearoyl-ACP desaturase (SAD), and the ω_6 and ω_3 linoleate desaturases which often use galactolipid substrates. B, The cytochrome b_5 (Cytb$_5$), NADH/cytochrome b_5 oxidoreductase (CBR) system is used by the ER-membrane-bound desaturases. Examples include the Δ_{12} oleate (OD) and ω_3 linoleate desaturases, which predominantly use phosphatidylcholine (PC) substrates.

between the terminal electron donor and the desaturase is relatively non-specific and it is possible that other electron donors, apart from ferredoxin and cytochrome b_5, may also be capable of reducing desaturases. The identity of the true *in vivo* electron donor(s) in the case of plastidial desaturases therefore remains to be resolved.

Numerous gene sequences encoding putative plant cytochrome b_5 and b_5 reductases have been reported in the databases. Most of the derived protein sequences are very similar to the homologous animal and fungal proteins, all of which are membrane bound. An interesting variant is a cytochrome b_5-containing fusion protein of 52 kDa from sunflower, which also contains a domain exhibiting the classical eight-histidine motif found in all membrane-bound desaturases (Sperling, Schmidt & Heinz, 1995). This protein therefore contains a cytochrome b_5 domain fused to the N-terminus of a putative membrane-bound desaturase. More recently, a Δ_6 linoleate desaturase–cytochrome b_5 fusion protein has been reported in borage (Sayanova *et al.*, 1997). Interestingly, a yeast protein has also been described in which a cytochrome b_5-like domain is located at the C-terminus of a membrane-bound Δ_9 stearoyl-ACP desaturase (Mitchell & Martin, 1995). In the case of the yeast fusion protein, truncation or disruption of the cytochrome b_5-like domain resulted in unsaturated fatty acid auxotrophy, suggesting that this domain is essential for desaturase activity. This process of the fusion of two or more metabolically related proteins to create multifunctional polypeptides is relatively common in eukaryotes and other examples relating to lipid metabolism include acetyl-CoA carboxylase and the type I fatty acid synthetases of animals and fungi. The existence of desaturase–cytochrome b_5 fusion proteins is good circumstantial evidence that cytochrome b_5 is the major, and maybe the exclusive, electron donor to many of the ER membrane-bound desaturases.

Mechanism

The recent availability of purified recombinant Δ_9 stearoyl-ACP desaturase has allowed for more detailed characterisation of the chemical mechanism of double bond insertion. This desaturase converts stearoyl-ACP to oleoyl-ACP in the presence of O_2, NAD(P)H, NAD(P)H/ ferredoxin oxidoreductase and ferredoxin as shown in Fig. 7. Using optical and Mössbauer spectroscopy, it was shown that the oxidised enzyme contains one pair of antiferromagnetically coupled Fe^{3+} sites per subunit (Fox *et al.*, 1993). Incubation with dithionate or stearate/O_2 reduced the site to a diferrous state. The data are consistent with the presence of a diiron-oxo cluster similar to the oxo- or hydroxo-bridged diiron clusters in hemerythrin, ribonucleotide reductases and bacterial hydrocarbon hydroxylases. This conclusion has been strengthened by evidence from resonance Raman spectroscopy which detects vibrational modes in the Δ_9 desaturase that are characteristic of diiron-oxo clusters (Fox *et al.*, 1994). Few kinetic data are currently available for desaturases, and speculations concerning their catalytic mechanism have therefore been based on analogy with related enzymes, principally methane monooxygenase hydroxylase (MMOH), and its proposed mechanism (for recent examples, see Kurtz, 1997 and Wallar & Lipscomb, 1996).

A speculative model for desaturase reaction cycle, based largely on kinetic data from MMOH, is shown in Fig. 8. Like other class II diiron enzymes, the Δ_9 desaturase has a diferric resting state with a catalytic cycle probably involving a thermodynamically favoured reduction directly to the diferrous state (Paulsen *et al.*, 1994; Ai *et al.*, 1997). It is likely that the reactions then proceed via reduction of O_2 to generate a bound peroxo intermediate (Des^P). It is then proposed that this intermediate isomerises to produce the diamond-shaped diferryl intermediate, Des^Q, which would have enough energy to abstract two hydrogens from a fatty acid and oxidise them to water resulting in the formation of a double bond in the final fatty acid product. It should be noted that the respective structures of these intermediates represent only one of several possibilities based on recent kinetic data on both Δ_9 stearate desaturases (Ai *et al.*, 1997) and other diiron oxo proteins (Kurtz, 1997). Some of the alternative mechanisms are reviewed by Wallar and Lipscomb (1996).

Although progress has begun to be made recently in the elucidation of the reaction mechanism of the soluble desaturases, much remains to be done. In particular, more detailed kinetic studies are needed to resolve the structures of the various iron-oxo intermediates. At present, these structures and the proposed catalytic mechanism of both desaturases and related diiron enzymes remain hypothetical. Nevertheless, the availability of high resolution crystal structures, relatively facile over-expression systems (Hoffman *et al.*, 1995) and closely related homologous enzymes should allow a much more detailed understanding of desaturase reaction chemistry in the near future. The description of what is probably the very similar chemistry involved in membrane-bound fatty acid desaturation will be a more formidable challenge. However, even here there has been encouraging progress with a preliminary report of the solubilisation and purification of a membrane-bound alkane hydroxylase that shares the conserved eight-histidine motif with the membrane-bound desaturases (Shanklin *et al.*, 1996). This protein contains three iron atoms per molecule, of which two may comprise a diiron centre. If this enzyme is still catalytically active in the micellar state in which it is purified, it may be amenable to the kind of spectroscopic techniques which have been so informative in studying the reaction mechanism of the soluble desaturases.

A

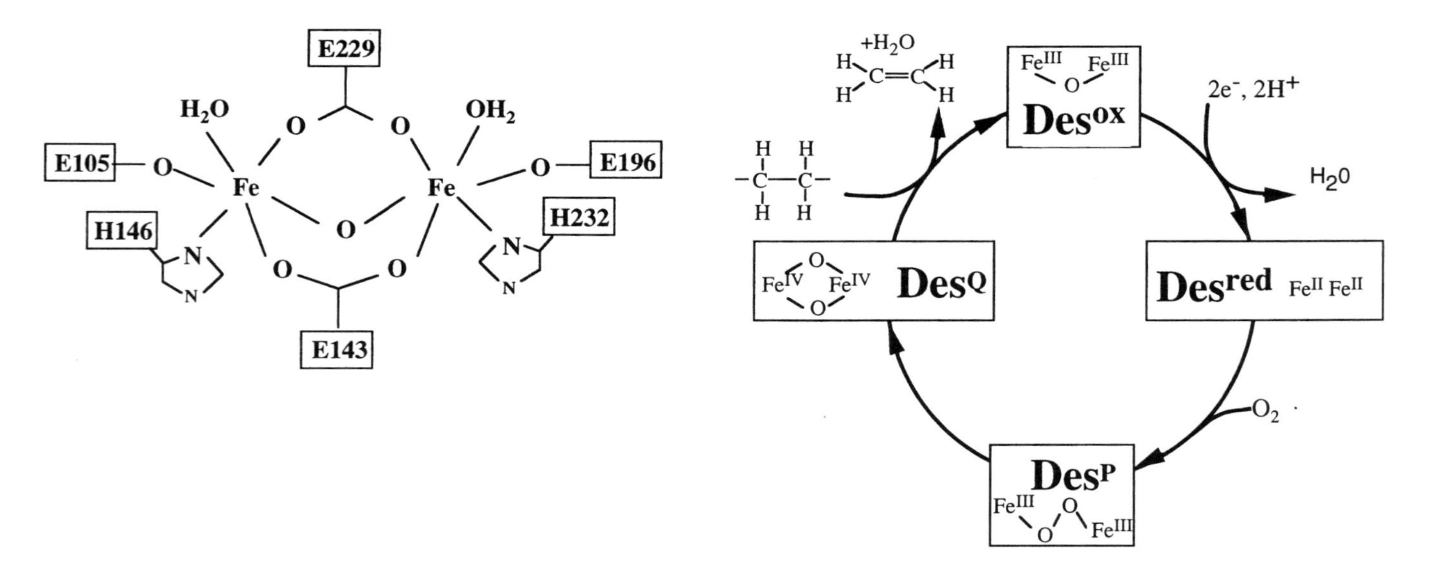

E229
H2O
OH2
E105
O
Fe
E196
H146
H232
N
E143
+H2O
H
C
DesOX
FeIII
2e-, 2H+
H20
DesQ
FeIV
Desred
FeII FeII
O2
DesP

(Fig. 8.)

Regulation

Fatty acid desaturases in all organisms are subject to several different types of regulation, depending on their localisation and function as shown in Tables 1 and 2. Those desaturases involved in membrane lipid biosynthesis, e.g. Δ_9 stearoyl-ACP desaturase and Δ_{12} oleate desaturase, have important 'housekeeping' functions and are therefore constitutively regulated. However, there may be requirements for the additional biosynthesis of specific membrane lipids at certain stages of development. For example, the greening of cotyledons or the expansion of young leaves requires a huge increase in the polyunsaturated fatty acids that made up the bulk of the photosynthetic thylakoid membrane lipids. Hence the basic constitutive control of desaturase gene expression is often supplemented by finely tuned developmental and environmental regulation of the same desaturase genes, as discussed below.

Plant development

The requirement for additional desaturase activities during certain developmental processes can be met either by the up-regulation of existing constitutively expressed desaturase genes and/or by the induction of other desaturase genes that are specific to the particular stages(s) of development. There are examples of both types of desaturase regulation in plants. One of the best characterised examples of a desaturase gene that is both constitutively and developmentally

Fig. 8. Catalytic mechanism of desaturation. A, Proposed arrangement of ligands for the diiron centre in the oxidised form of Δ_9 stearoyl-ACP desaturase. This structure is based on X-ray crystallography and on Mössbauer and resonance Raman spectroscopy. B, Putative reaction cycle of desaturation showing reduction of the resting diferric state (Des^{ox}) to the diferrous form (Des^{red}) followed by oxygen binding to create a peroxo intermediate, Des^{P} which isomerises to form a high-energy diferryl species, Des^{Q}. This species can then remove two hydrogens from a fatty acid to form water plus an unsaturated fatty acid. Note that designation of the peroxo and diferryl intermediates as Des^{P} and Des^{Q}, respectively, is an arbitary nomenclature based on the anthology with methane monooxygenase (see Wallar & Lipscomb, 1996). This scheme is based on preliminary kinetic data from a Δ_9 stearate desaturase (Ai *et al.*, 1997) and more detailed data from related class II diiron proteins (Que & Dong, 1996).

Table 1. *Pattern of expression of plant fatty acid desaturases*

Enzyme	Type of regulation
Plastidial Δ_9 Stearoyl-ACP desaturase	Housekeeping up-regulated in lipid-accumulating tissues (e.g. developing embryos, anthers)
ER-localised $\Delta_{12}(\omega_6)$ oleate desaturase I	Housekeeping
ER-localised $\Delta_{12}(\omega_6)$ oleate desaturase II	Seed-specific [soybean, cotton]
ER-localised ω_3 linoleate desaturase I	Housekeeping mainly expressed in non-photosynthetic tissues (e.g. roots) up-regulated in storage lipid-accumulating tissues (e.g. embryos, pollen grains)
Plastidial ω_6 oleate desaturase	Housekeeping
Plastidial ω_3 linoleate desaturase I	Housekeeping mainly expressed in photosynthetic tissues (e.g. leaf, stem)
Plastidial ω_3 linoleate desaturase II	Low-temperature inducible induced by a shift to suboptimal growth temperatures
ER-localised Δ_6 linoleate desaturase	Seed-specific up-regulated in storage lipid-accumulating seed tissues of *Borago officinalis*
Plastidial Δ_9 myristoyl-ACP desaturase	Trichome-specific induced in glandular trichomes of pest-resistant *Pelargonium hortorum* plants
Plastidial Δ_4 palmitoyl-ACP desaturase	Seed-specific up-regulated in storage lipid-accumulating endosperm tissues of *Coriandrum sativum*
Plastidial Δ_6 palmitoyl-ACP desaturase	Seed-specific up-regulated in storage lipid-accumulating endosperm tissues of *Thunbergia alata*
Plastidial Δ_9 palmitoyl-ACP desaturase	Seed-specific up-regulated in storage lipid-accumulating endosperm tissues of *Asclepias syriaca*

Table 2. *Developmental and environmental factors influencing fatty acid desaturase activity in plants*

Stimulus	Enzyme	Effect
Senescence	ER Δ_9 stearate desaturase	Induced mRNA accumulation in senescing rose petals[a]
Hormones	**Abscisic acid**	
	ER ω_3 linoleate desaturate	Increased mRNA levels in rapeseed microspore-derived embryos[b]
	Δ_9 stearoyl-ACP desaturase	Increased mRNA levels in rapeseed zygotic embryos[c]
	Auxin	
	ER ω_3 linoleate desaturate	Increased mRNA levels in mung bean hypocotyls[d]
Temperature	ER ω_3 linoleate desaturate	Decreased mRNA levels at chilling temperatures in rice roots[e]
	Plastid ω_3 linoleate desaturase	Induced mRNA accumulation in *Arabidopsis* leaves[f]
Light	Plastid ω_3 linoleate desaturase	Increased mRNA accumulation during greening of aetiolated *Arabidopsis* and wheat leaves[g,h]
Wounding	Plastid ω_3 linoleate desaturase	Increased mRNA accumulation in wounded tobacco leaves[i]
Water deficit	Δ_9 stearoyl-ACP desaturase ER Δ_{12} oleate desaturase	Decreased mRNA levels in acclimation of potato cells to low water potential[j]
Pathogen attack	Plastid ω_3 linoleate desaturase	Rapid and localised increase in mRNA levels around fungal penetration sites in parsley[k]
Pathogen attack	ER-Δ_{12} oleate desaturase	Up-regulated in *Arabidopsis* cellscultured with a fungal elicitor[l]

[a]; Fukuchi-Mizutani *et al.*, 1995, [b]; Zou *et al.*, 1995, [c]; Slocombe *et al.*, 1994, [d]; Yamamoto, 1994, [e]; Kodama *et al.*, 1997, [f]; Gibson *et al.*, 1994; [g]; Horiguchi *et al.*, 1996, [h]; Nishiuchi *et al.*, 1995; [i]; Hamada *et al.*, 1996, [j]; Leone *et al.*, 1996, [k]; Kirsch *et al.*, 1997, [l]; Trezzini *et al.*, 1993.

regulated is found in the case of the soluble Δ_9 stearoyl-ACP desaturase. This enzyme is responsible for 'housekeeping' functions relating to general cell membrane synthesis/turnover but is also involved in the synthesis of storage lipids (in seeds and pollen), structural lipids (in the tapetum) and thylakoid membrane lipids (in expanding leaves). Detailed analysis using gene-specific probes and transgenic plants showed that the same stearoyl-ACP desaturase genes were both constitutively and developmentally regulated in all of the above tissues (Slocombe *et al.*, 1994; Piffanelli *et al.*, 1997; Piffanelli, 1997). The regulation of this class of genes was found to be mainly exerted at the transcriptional level in both photosynthetic and storage lipid-accumulating tissues. However, analysis of Δ_9 stearoyl-ACP desaturase mRNA and protein levels in the *Arabidopsis chs1* chilling-sensitive mutant, which shows reduced accumulation of newly synthesised desaturase polypeptides upon transfer of *chs1* plants from 23 °C to 13 °C, suggests that the regulation of this class of desaturase genes is also exerted at the post-transcriptional level (Schneider, Nielsen & Somerville, 1995).

Other evidence that housekeeping-type desaturases, such as the soluble Δ_9 and ER-bound Δ_{12} enzymes, have central roles in plant growth and development comes from hormonal studies. For example, the involvement of auxin in the up-regulation of mRNA levels of the ER-bound Δ_{12} desaturase suggests an intriguing causal link between the correct level of membrane desaturation and cell elongation (Yamamoto, 1994; Lightner *et al.*, 1994). Analysis of double desaturase mutants of *Arabidopsis* has also revealed a complexity of superimposed regulatory mechanisms of Δ_{12} and ω_3 fatty acid desaturases acting at both transcriptional and post- transcriptional levels (McConn *et al.*, 1994). The phenotypes of other (non-desaturase) mutants of *Arabidopsis* (*act1, ela, rod1*) suggest that expression of the plastidial and ER-localised desaturase genes during plant development is coordinated in order to maintain the continuous bidirectional exchange of acyl lipids which occurs between plastids and the ER membranes (Kunst, Browse & Somerville, 1989). It has recently been suggested that an important role for α-linolenic acid in plant development may be related to pollen development. Triple desaturase mutants of *Arabidopsis*, which almost completely lacked any α-linolenate, were male-sterile but normal pollen maturation could be restored by exogenous α-linolenate or jasmonate (McConn & Browse, 1996). It appears that jasmonate performs at least two key roles in flower development; firstly in the maturation of viable pollen grains and secondly in enabling successful dehiscence of the anthers to occur. Therefore, the expression of desaturase genes is essential for sexual plant reproduction.

A contrasting example of specific developmental regulation via the expression of additional desaturase genes is found during the senescence of rose petals. Senescence is normally associated with lipid degradation rather than biosynthesis. However, there is evidence that polyunsaturated fatty acids are more susceptible to catabolism than are more saturated species (Brown, Chambers & Thomspon, 1991). Therefore, there may be a requirement for the transient induction of desaturase activities at the onset of senescence in order to increase the pool of polyunsaturates available for recycling via catabolism. One of the major genes that is specifically induced during senescence in rose petals encodes a protein resembling the membrane-bound Δ_9 stearoyl-CoA desaturases that were hitherto only found in animals and fungi (Fukuchi-Mizutami *et al.*, 1995). This enzyme is dramatically different from all previously described plant stearate desaturases, which are soluble plastidial enzymes that use an acyl-ACP substrate (see pp. 99–103). The existence of a distinct class of stearate desaturase may be due to its specialised role in a particular developmental process, i.e. senescence. It will be interesting to confirm the existence of stearoyl-CoA desaturases in other plant tissues and species. This has been implied recently by the deposition in the databases of *Arabidopsis* EST clones with significant homology to genes encoding mammalian stearoyl-CoA desaturases (Piffanelli, 1997).

Environmental factors

Desaturases are subject to two main types of regulation by changes in the environment of the cell or organism. First, the availability or lack of substrate can radically alter both desaturase enzyme activity and gene expression, particularly in animals and fungi. For example, the Δ_9 desaturase is strongly induced by feeding after a period of fasting in rats (Cook, 1991) or the addition of exogenous stearate to yeast cells (Choi *et al.*, 1996). This type of regulation is less important in autotrophic organisms such as plants.

The second major environmental effect on desaturases is caused by temperature and this is common to many plants, animals and microbes. For example, the Δ_9 desaturase gene of poikilothermic animals like fish is rapidly induced by cooling to 10 °C (Tiku *et al.*, 1996). Cyanobacteria, which resemble the chloroplasts of higher plants in their membrane structure and glycerolipid composition, also have a desaturase-mediated response to changes in temperature (Murata & Wada, 1995). In plants, there is a general inverse relationship between polyunsaturation of fatty acids and growth temperature, both in membrane and storage lipids. The overall reason for the increase in desaturase activities in organisms subjected to low temperatures is probably related to the need to maintain membrane fluidity, which is directly affected by the extent of unsaturation of the acyl lipid components of the bilayer. The analysis of the *fad* single, double or triple mutants of *Arabidopsis* suggests that polyunsaturation, and particularly the high levels of α-linolenic acid normally found in photosynthetic membranes, may not be required for plant growth and development at optimal growing temperatures. For example, a triple desaturase mutant containing only negligible amounts of α-linolenic acid was barely affected in photosynthetic membrane function at 22 °C (McConn & Browse, 1996). However, the single desaturase mutants from which these plants were derived showed gradually more severe abnormalities, compared with the wild-type plants, as the growth temperature was reduced to 12 °C and then 6 °C (Miquel *et al.*, 1993). These observations reinforce the importance of desaturases in acclimation of organisms to low temperatures.

The mechanism of low-temperature adaptation in plants probably involves both transcriptional and/or post-translational regulation of the desaturases involved in polyunsaturate formation. A direct effect of temperature on desaturase enzyme activity has been shown in developing soybean seeds (Cheesebrough, 1989). On the other hand, a cold-inducible plastidial ω_3 desaturase gene has been isolated from *Arabidopsis* (Gibson *et al.*, 1994) and there are several other reports that are consistent with the presence of cold-inducible ω_3 and Δ_{12} desaturase genes in soybeans (Kinney, 1994; Rennie & Tanner, 1989). However, there are other reports of the isolation of *Arabidopsis* and soybean Δ_{12} desaturase genes that are not regulated by low temperature (Okuley *et al.*, 1994; Heppard *et al.*, 1996). Since many desaturases are encoded by multigene families, it is possible that some plant species may have both cold-inducible and non-cold-inducible forms of the same class of desaturase enzyme and/or gene.

It is clear that desaturases play an important role in determining the extent of cold-tolerance of particular plant genotypes. This has been demonstrated most convincingly in transgenic plants that contain additional desaturase genes. The transfer of genes encoding either a cyanobacterial Δ_9 desaturase (Ishizaki-Nishizawa *et al.*, 1996) or an *Arabidopsis* plastidial ω_3 desaturase (Kodama *et al.*, 1994, 1995) both conferred a significant increase in chilling tolerance in the leaves of transgenic tobacco plants, compared with their wild-type counterparts.

A final example of an environmental response involving desaturases is that of water stress. Water deficits are a severe stress for cellular membrane systems of plants. Homeoviscous adaptation is an important mechanism to preserve the correct state of membrane fluidity. This is achieved by regulating the degree of unsaturation of the fatty acids (Quinn, Joo & Vigh, 1989; Tiku *et al.*, 1996). The acclimation of potato cells to lower water potential was shown to involve a net decrease of unsaturation of membrane phospholipids such as phosphatidylcholine and phosphatidylethanolamine. The reduced levels of fatty acid unsaturation was partially associated with decreased steady-state levels of the Δ_9 and Δ_{12} desaturase genes (Leone *et al.*, 1996), although other levels of regulation were also invoked to explain the extensive changes in membrane lipid observed upon PEG acclimation of potato cells in culture.

Pest and disease resistance

There is a growing body of evidence concerning the importance of fatty acid desaturases in the responses of plants to attack by various pest and disease organisms. One recently described example concerns the trichomes that are present on the surfaces of many leaves and which often play a role in defence against invertebrate herbivores. Tall glandular trichomes are found on the leaves of varieties of the garden geranium, *Pelargonium hortorum*, that are resistant to damage by small arthropod pests. This pest resistance has been found to be due to a single Mendelian locus encoding a Δ_9 palmitoyl-ACP desaturase (Schultz *et al.*, 1996). The desaturase is expressed uniquely in the glandular trichomes and is involved in accumulation of ω_5 ana cardic acids which form a sticky and ultimately toxic substance that traps potential arthropod pests.

Desaturases also play a key role in plant responses to fungal pathogens as demonstrated by recent evidence from parsley, *Petroselinum crispum*. When parsley cells or leaves were challenged with an elicitor from the fungal pathogen, *Phytopthora sojae*, there was a rapid and highly localised induction of the plastidial ω_3 linoleate desaturase mRNA at fungal infection sites (Kirsch *et al.*, 1997). This desaturase is responsible for the synthesis of α-linolenic acid which is a precursor of the well-known plant signalling molecule, jasmonic acid. The fungal elicitor is known to cause a rapid and massive increase in jasmonate production concurrent with the 'oxidative burst' (release of hydrogen peroxide) which occurs before the induction of plant defence-related genes (Kirsch *et al.*, 1997). It was also found that the expression of a putative ER-localised Δ_{12} fatty acid desaturase was strongly induced in tomato plants infected with the citrus exocortis viroid (Gadea *et al.*, 1996) and the expression of the Δ_{15} desaturase gene was shown to be strongly induced in *Arabidopsis* cells cultured with a fungal elicitor (Trezzini, Horrichs & Somssich, 1993). Moreover, the gene encoding a plastidial isoform of the ω_3 linoleate desaturase was found to be significantly and relatively rapidly (12 hours) up-regulated following wounding of tobacco leaves (Hamada *et al.*, 1996). These results demonstrate that the induction of desaturase gene expression is one of the key components of the response of plants to infection or trauma. The resulting elevated levels of polyunsaturates, and particularly α-linolenic acid, can either act as substrates for jasmonate synthesis and hence trigger a cascade of physiological responses both locally and systemically (Doares *et al.*, 1995) and/or they can be involved in the generation of free radicals and lipid hydroperoxides during the oxidative burst (see above).

Lipid-storing tissues

Several kinds of plant desaturase are found uniquely in lipid-storing tissues, such as the embryos and endosperm of certain seeds and the oil-rich mesocarp of some fruits. For example, most plant tissues contain an ω_3 linoleate desaturase that produces α-linolenic acid, but the seeds of species including borage and evening primrose also contain a Δ_6 linoleate desaturase that produces γ-linolenic acid. This fatty acid, which is marketed as a component of evening primrose oil or starflower (borage) oil, is claimed to have useful therapeutic qualities and is now being produced in larger quantities via transgenic rapeseed plants containing an introduced Δ_6 linoleate desaturase gene (Beremand *et al.*, 1997). The seeds of coriander and most other Umbelliferae contain a Δ_4 palmitate desaturase that is responsible for the accumulation of petroselinic acid. This fatty acid makes up as much as 80% of the total seed oil but is found nowhere else in the plant (Kleiman & Spencer, 1982). Petroselinic acid can be oxidatively cleaved to yield two industrially useful products, lauric and adipic acids, which can be used respectively in the manufacture of detergents and polymers. Acetylinic double bounds are inserted into fatty acids by seed specific desaturases in numerous plant species including *Crepis alpina* (Banas *et al.*, 1997; Lee *et al.*, 1997). Fatty acids containing conjugated double bonds are produced as the principal components of tung (*Aleurites fordii*) and marigold (*Calendula officinalis*) oils and high levels of epoxy fatty acids are found in seeds of *Vernonia* and *Euphorbia* spp. (Smith, 1971; Gunstone, Harwood & Padley, 1994). These conjugated and epoxy fatty acids are used widely in the chemicals industry, e.g. in the manufacture of paints, varnishes and other drying agents and for the production of certain cosmetics. Although the biological significance of these tissue-specific desaturases is unclear, the economic value of their end products has made them of great interest for biotechnological manipulation (Murphy, 1996).

Evolution of desaturases

During the past few years, a great deal of new information has become available concerning the composition and structure of a large number of desaturases and related enzymes. This enables us to speculate on possible evolutionary relationships between all of these proteins as shown in Fig. 9. From both sequence and structural analysis it is clear that there are two major groups of desaturases, i.e. the soluble acylACP desaturases and the membrane-bound acyl-CoA or acyl-lipid desaturases.

Soluble desaturases are only found in plants and are all plastidial proteins of which the major example is the Δ_9 stearoyl-ACP desaturase. This is an essential housekeeping enzyme and is found in all cells. It is likely that the highly related seed-specific Δ_4 and Δ_6 palmitoyl desaturases and trichome-specific Δ_9 myristoyl desaturases found in some plants (see pp. 97, 102–3) have evolved from the Δ_9 stearoyl-ACP desaturase. Indeed, a relatively subtle modification of the seed-specific Δ_6 desaturase by site-directed mutagenesis results in its conversion back to a mainly Δ_9 desaturase (see p. 102). All of these soluble plant desaturases are structurally related to a number of bacterial diiron proteins with similar, but not identical, reaction mechanisms (Fox *et al.*, 1994). Together, these proteins can be grouped together to form the class II diiron-oxo proteins and it is possible that they share a common evolutionary lineage.

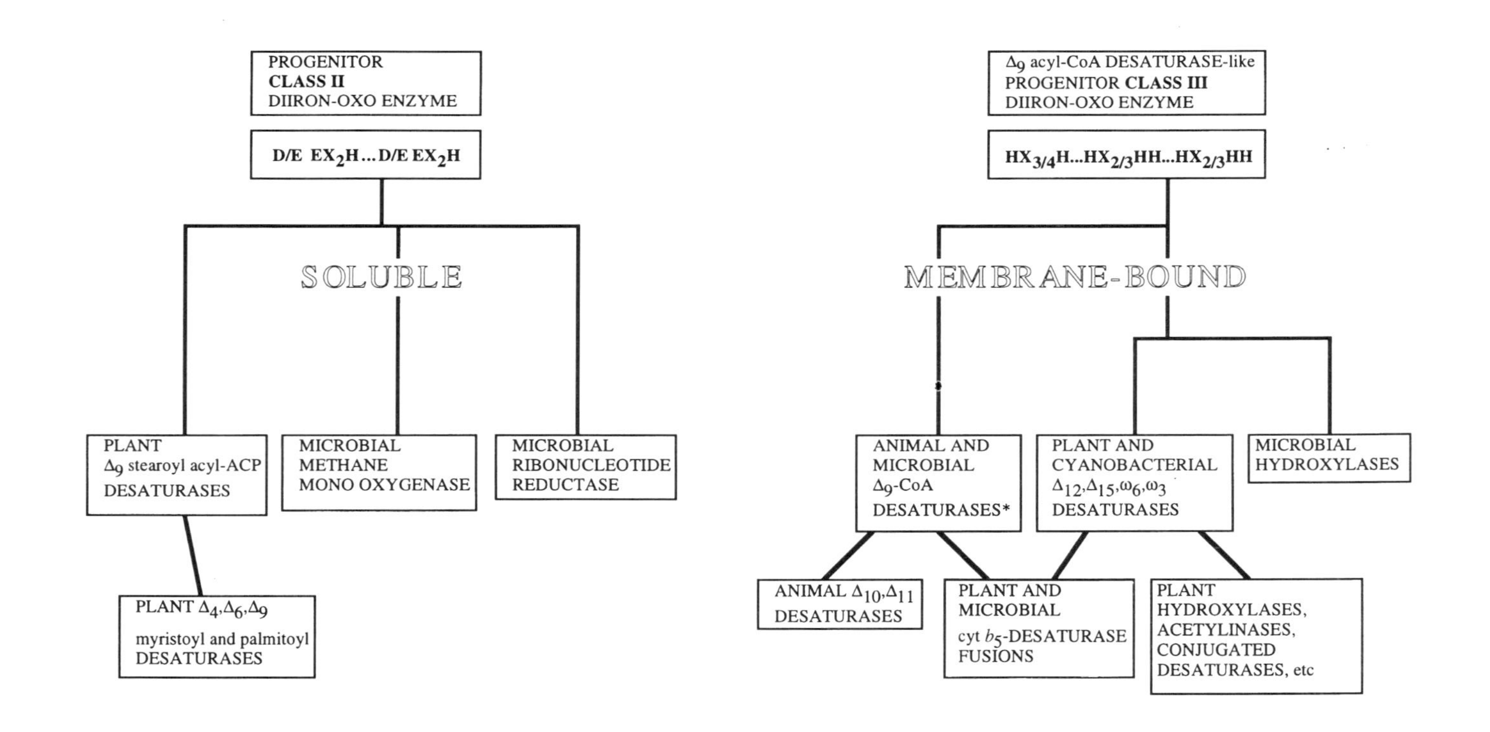
PROGENITOR
CLASS II
DIIRON-OXO ENZYME
D/E EX_2H...D/E EX_2H
SOLUBLE
PLANT
Δ_9 stearoyl acyl-ACP
DESATURASES
MICROBIAL
METHANE
MONO OXYGENASE
MICROBIAL
RIBONUCLEOTIDE
REDUCTASE
PLANT $\Delta_4,\Delta_6,\Delta_9$
myristoyl and palmitoyl
DESATURASES
Δ_9 acyl-CoA DESATURASE-like
PROGENITOR CLASS III
DIIRON-OXO ENZYME
$HX_{3/4}H...HX_{2/3}HH...HX_{2/3}HH$
MEMBRANE-BOUND
ANIMAL AND
MICROBIAL
Δ_9-CoA
DESATURASES*
PLANT AND
CYANOBACTERIAL
$\Delta_{12},\Delta_{15},\omega_6,\omega_3$
DESATURASES
MICROBIAL
HYDROXYLASES
ANIMAL Δ_{10},Δ_{11}
DESATURASES
PLANT AND
MICROBIAL
cyt b_5-DESATURASE
FUSIONS
PLANT
HYDROXYLASES,
ACETYLINASES,
CONJUGATED
DESATURASES, etc

(Fig. 9.)

The membrane-bound desaturases in plants, whether plastidial or ER located, are clearly related as shown in Fig. 6. These plant desaturases are also related to the animal and microbial membrane-bound desaturases and the microbial membrane-bound hydroxylases. According to a phylogenetic tree derived by Shanklin *et al.* (1994), all of these proteins probably evolved from a prototype most closely resembling the mammalian Δ_9 stearoyl-CoA desaturase. This then gave rise to one cluster corresponding to the microbial hydroxylases and another cluster including all of the plant and cyanobacterial membrane desaturases. In view of the emerging similarities between the ER Δ_{12} desaturases and the castor and *Lesquerella* seed-specific ER Δ_{12} hydroxylases, it is also proposed that, in some plants, novel enzymes such as hydroxylases, acetylinases, epoxidases and conjugated desaturases are likely to have arisen from modified ER Δ_{12} and ω_3 desaturases. It has been suggested that all of these membrane-bound desaturases and related enzymes can be grouped together to form a class III of diiron-oxo proteins (Shanklin *et al.*, 1994).

It is possible that the class II (soluble) and class III (membrane-bound) diiron-oxo proteins are evolutionarily related. Despite their obvious structural differences, they share many features that suggest that a common mechanism is used for the desaturase reaction. These include the catalytic requirement for iron, the inhibition by metal chelators, the stereospecificity of the desaturase reaction, and the kinetic isotope effects observed for C-H bond cleavage (Fox *et al.*, 1994). Nevertheless, the fundamental differences in the nature of the iron-coordination site, and the lack of any other sequence similarities, make it difficult to hypothesise a common ancestry between the class II and class III diiron-oxo proteins at present. It is possible that these

Fig. 9. Evolution of desaturases and related proteins. Although desaturases and related diiron-oxo proteins have similar catalytic mechanisms, they fall into two distinct groups based on their structure and ligand chemistry. The class II diiron-oxo proteins are all soluble and have the consensus iron coordination site: [D/E EX_2H] – [D/E EX_2H]. All of the plant acyl-ACP desaturases are included in this group. The class III diiron-oxo proteins are all membrane-bound and their consensus diiron ligand site is: [$HX_{3\text{ or }4}H$] . . . [$HX_{2\text{ or }3}H$] . . . [$HX_{2\text{ or }3}H$]. This group includes all of the plant ER and plastidial lipid-linked desaturases and the related seed-specific hydroxylases. Note that both class II and class III iron-oxo proteins include several enzymes which are not desaturases.

protein classes are an example of convergent evolution, where structures arise independently to carry out similar functions, i.e. they are analogous rather than homologous.

Future prospects

The explosion of interest in, and information about, plant desaturases over the past few years opens up a whole series of exciting prospects for their manipulation that may have far-reaching consequences for agriculture. For example, the enhancement of low-temperature tolerance via desaturase gene insertion allows the extension of the climatic range of some of our most important crops to be envisaged. Much of the World's most productive farmland is located in temperate regions that are at, or near to, the limits of cultivation of chilling-sensitive crops such as rice, soybeans, maize and sunflowers. The extension of their cultivation by several hundred kilometres could have a hugely beneficial effect on food yields and the overall economy of the regions involved. Fatty acid desaturases play a number of key roles in the development of plants and their responses to the external environment, including injury and disease. More detailed knowledge of the function of pathogen- or trauma-induced desaturases may shed light on the mechanisms, which have been evolved by some plants to combat such stresses.

There is much interest in the biotechnological manipulation of desaturases to produce useful edible and industrial products (Murphy, 1996). In some cases it is possible to manipulate desaturases, whether by conventional breeding or transgenic methods, to modify only the fatty acids in the seed storage lipids (Heppard *et al.*, 1996; Yadav, 1995). However, there are other cases, such as the high-oleate mutants of soybean and other plants, where the valuable seed phenotype is associated with poor agronomic performance of the crop, probably due to effects on membrane lipids (Heppard *et al.*, 1996). Another example is the high-stearic variety of transgenic rapeseed plants containing an antisense Δ_9 stearoyl-ACP desaturase; although the increased stearate levels are mainly found in the storage lipids, many of the transgenic plants exhibited over three times the normal amount of stearate in their membrane lipids, and this was associated with low germination rates and a lack of vigour in the seedlings (Thompson & Li, 1996). It is therefore important that our understanding of the role of desaturases in plant development is increased in order to effect their manipulation, whether by conventional or transgenic methods, in a more rational and successful manner. Ultimately, engineering of plant desaturases can be considered to produce large quantities of any given fatty acid of chain length C_8 to C_{24}

or more and with any desired degree and positional specificity of unsaturation, hydroxylation or epoxidation in the oil crop of choice. Such crops would then serve as environmentally friendly and literally 'green' factories for the production of valuable renewable oils for use by the pharmaceutical, food or chemicals industries.

Acknowledgements

We are grateful to Dr J Ross for assistance with the figures and constructive comments on the manuscript and to Dr S Bornemann for useful discussions on desaturase mechanisms. The support of the Biotechnology and Biological Sciences Research Council competitive strategic grant to the John Innes Centre is also acknowledged.

References

Ai, J., Broadwater, J.A., Loehr, T.M., Sanders-Loehr, J. & Fox, B.G. (1997). Azide adducts of stearoyl-ACP desaturase: a model for bridging by dioxygen in the binuclear iron active site. *Journal of Biological and Inorganic Chemistry*, **2**, 37–45.

Arondel, V., Lemieux, B., Hwang, I., Gibson, S., Goodman, H.M. & Somerville, C.R. (1992). Map-based cloning of a gene controlling omega-3 fatty acid desaturation in *Arabidopsis*. *Science*, **258**, 1353.

Banas, A., Bafor, M., Wiberg, E., Lenman, M., Stahl, U. & Stymne, S. (1997). Biosynthesis of an acetylinic fatty acid in microsomal preparations from developing seeds of *Crespis alpina*. In *Physiology, Biochemistry and Molecular Biology of Plant Lipids* ed. J.P. Williams, M.U. Khan & N.W. Lem, pp. 57–9. Dordrecht: Kluwer.

Beremand, P.D., Nunberg, A.N., Reddy, A.S. & Thomas, T.L. (1997). Introduction of γ-linolenic acid by transgenic plants expressing cyanobacterial or plant Δ_6 desaturase genes. In *Physiology, Biochemistry and Molecular Biology of Plant Lipids*, ed. J.P. Williams, M.U. Khan & N.W. Lem, pp. 351–4. Dordrecht: Kluwer.

Broun, P., Hawker, N. & Somerville, C.R. (1997). Expression of castor and *L. fendleri* oleate 12-hydroxylases in transgenic plants. In *Physiology, Biochemistry and Molecular Biology of Plant Lipids*, ed. J.P. Williams, M.U. Khan & N.W. Lem, pp. 342–4. Dordrecht: Kluwer.

Brown, J.H., Chambers, J.A. & Thompson, J.E. (1991). Acyl chain and headgroup regulation of phospholipid catabolism in senescing carnation flowers. *Plant Physiology*, **95**, 909–16.

Cahoon, E. & Ohlrogge, J.B. (1994). Metabolic evidence for the involvement of a Δ^4-palmitoyl-acyl carrier protein desaturase in petroselinic acid synthesis in coriander endosperm and transgenic tobacco cells. *Plant Physiology*, **104**, 827–37.

Cahoon, E.B. & Shanklin, J. (1997). Approaches to the design of acyl-ACP desaturases with altered fatty acid chain-length and double bond positional specificities. In *Physiology, Biochemistry and Molecular Biology of Plant Lipids*, ed. J.P. Williams, M.U. Khan & N.W. Lem, pp. 374–6. Dordrecht: Kluwer.

Cahoon E.B., Cranmer, A.M., Shanklin, J. & Ohlrogge, J.B. (1994). Δ_6 hexadecenoic acid is synthesized by the activity of a soluble Δ^6 palmitoyl-acyl carrier protein desaturase in *Thunbergia alata* endosperm. *Journal of Biological Chemistry*, **44**, 27 519–26.

Cahoon, E.B., Coughlan, S.J. & Shanklin, J. (1997). Characterisation of a structurally and functionally diverged acyl-acyl carrier protein desaturase from milkweed seed. *Plant Molecular Biology*, **33**, 1105–10.

Cheesebrough, T.M. (1989). Changes in the enzymes for fatty acid synthesis and desaturation during acclimation of developing soybean seeds to altered growth temperature. *Plant Physiology*, **90**, 760–4.

Choi, J-Y., Stukey, J., Hwang, S-Y. & Martin, C.E. (1996). Regulatory elements that control transcription activation and unsaturated fatty acid-mediated repression of the *Saccharomyces cerevisiae OLE1gene*. *Journal of Biological Chemistry*, **271**, 3581–9.

Cook, H.W. (1991). Fatty acid desaturation and chain elongation in eukaryotes. In *Biochemistry of Lipids, Lipoproteins and Membranes*, ed. D.E. Vance & J. Vance, pp. 141–69. New York: Elsevier.

Doares, S.H., Syrovets, T., Weiler, E.W. & Ryan, C.A. (1995). Oligogalacturonides and chitosan activate plant defensive genes through the octadecanoid pathway. *Proceedings of the National Academy of Sciences, USA*, **92**, 4095–8.

Fox, B.G., Shanklin, J., Somerville, C. & Münck, E. (1993). Stearoyl-acyl carrier protein Δ_9 desaturase from *Ricinus communis* is a diiron-oxo protein. *Proceedings of the National Academy of Sciences, USA*, **90**, 2486–90.

Fox, B.G., Shanklin, J., Ai, J., Loehr, T.M. & Sanders-Loehr, J. (1994). Resonance Raman evidence for an Fe-o-Fe center in stearoyl-ACP desaturase. Primary sequence identity with other diiron-oxo proteins. *Biochemistry*, **33**, 12776–86.

Fukuchi-Mizutani, M., Savin, K., Cornish, E., Tanaka, Y., Ashikari, T., Kusumi, T. & Murata, N. (1995). Senescence-induced expression of a homologue of Δ_9 desaturase in rose petals. *Plant Molecular Biology*, **29**, 627–35.

Gadea, J., Mayda, M.E., Conejero, V. & Vera, P. (1996). Characterization of defense- related genes ectopically expressed in viroid-infected tomato plants. *American Phytopathological Society*, **9**, 409–15.

Gibson, S., Arondel, V., Iba, K. & Somerville, C.R. (1994). Cloning of a temperature- regulated gene encoding a chloroplast ω-3 desaturase from *Arabidopsis thaliana*. *Plant Physiology*, **106**, 1515–621.

Gunstone, F.D., Harwood, J.L. & Padley, F.B. (1994). *The Lipid Handbook*. London: Chapman and Hall.

Gurr, M.I. & Harwood, J.L. (1991). *Lipid Biochemistry*. London: Chapman and Hall.

Hamada, T., Nishiuchi, T., Kodama, H., Nishimura, M. & Iba, K. (1996). cDNA cloning of a wounding-inducible gene encoding a plastid ω-3 fatty acid desaturase from tobacco. *Plant Cell Physiology*, **37**, 606–11.

Heinz, E. (1993). Biosynthesis of polyunsaturated fatty acids. In *Lipid Metabolism in Plants*, ed. T.S. Moore, pp. 33–89. Boca Raton: CRC Press.

Heppard, E.P., Kinney, A.J., Stecca, K.L. & Miao, G.H. (1996). Developmental and growth temperature regulation of two different microsomal ω-6 desaturase genes in soybeans. *Plant Physiology*, **110**, 311–19.

Hitz, W.D., Carlson, T.J., Booth, Jr, J.R., Kinney, A.J., Stecca, K.L. & Yadav, N.S. (1994). Cloning of a higher-plant plastid ω-6 fatty acid desaturase cDNA and its expression in a cyanobacterium, *Plant Physiology*, **105**, 635–41.

Hoffman, B.J., Broadwater, J.A., Johnson, P., Harper, J., Fox, B.G. & Kenealy, W.R. (1995). Lactose fed-batch overexpression of recombinant metalloproteins in *Escherichia coli* BL21(DE3): Process control yielding high levels of metal-incorporated, soluble protein. *Protein Expression and Purification*, **6**, 646–54.

Horiguchi, G., Iwakawa, H., Kodama, H., Kawakami, N., Nishimura, M. & Iba, K. (1996). Expression of a gene for plastid ω-3, fatty acid desaturase and changes in lipid and fatty acid compositions in light- and dark-grown wheat leaves. *Physiologia Plantarum*, **96**, 275–83.

Iba, K., Gibson, S., Nishiuchi, T., Fuse, T., Nishimura, M., Arondel, V., Hugly, S. & Somerville, C. (1993). A gene encoding a chloroplast ω-6 fatty acid desaturase complements alterations in fatty acid desaturation and chloroplast copy number of the *fad7* mutant of *Arabidopsis thaliana*. *Journal of Biological Chemistry*, **268**, 24 099–105.

Ishizaki-Nishizawa, O., Fujii, T., Azuma, M., Sekiguchi, K., Murata, N., Ohtani, T. & Toguri, T. (1996). Low-temperature resistance of higher plants is significantly enhanced by a nonspecific cyanobacterial desaturase. *Nature Biotechnology*, **14**, 1003–6.

Jaworski, J.G. & Stumpf, P.K. (1974). Fat metabolism in higher plants. Properties of a soluble stearoyl-acyl carrier protein desaturase from maturing *Carthamus tinctoris*. *Archives of Biochemistry and Biophysics*, **162**, 158–65.

Kinney, A.J. (1994). Genetic modification of the storage lipids of plants. *Current Opinion in Biotechnology*, **5**, 144–51.

Kirsch, C., Takamiya-Wik, M., Reinold, S., Hahlbrock, K. & Somissich I.E. (1997). Rapid, transient, and highly localized induction of plastidial ω-3 fatty acid desaturase mRNA at fungal infection sites in *Petroselinum crispum*. *Proceedings of the National Academy of Sciences, USA*, **94**, 2079–84.

Kleiman R. & Spencer, G.F. (1982). Search for new industrial oils: XV1 umbelliferae- seed oils rich in petroselinic acid. *Journal of the American Oil Chemists Society*, **59**, 29–38.

Kodama, H., Hamada, T., Horiguchi, G., Nishimura, M. & Iba, K. (1994). Genetic enhancement of cold tolerance by expression of a gene for chloroplast ω-3 fatty acid desaturase in transgenic tobacco. *Plant Physiology*, **105**, 601–5.

Kodama, H., Horiguchi, G., Nishiuchi, T., Nishimura, M. & Iba, K. (1995). Fatty acid desaturation during chilling acclimation is one of the factors involved in conferring low-temperature tolerance to young tobacco leaves. *Plant Physiology*, **107**, 1177–85.

Kodama, H., Akagi, H., Kusumi, K., Fujimura, T. & Iba, K. (1997). Structure, chromosomal location and expression of a rice gene encoding the microsome ω-3 fatty acid desaturase. *Plant Molecular Biology*, **33**, 493–502.

Kok, M., Oldenhuis, R., van der Linden, M.P.G., Raatjes, P., Kingma, J., van Lelyveld, P.H. & Witholt, B. (1989). The *Pseudomonas oleovorans* alkane hydroxylase gene. *Journal of Biological Chemistry*, **264**, 5435–41.

Kunst, L., Browse, J. & Somerville, C. (1989). A mutant of *Arabidopsis*, deficient in desaturation of palmitic acid in leaf lipids. *Plant Physiology*, **90**, 943–7.

Kurtz, D.M. (1997). Structural similarity and functional diversity in diiron-oxo proteins. *Journal Biological Inorganic Chemistry*, **2**, 159–67.

Lee, M., Lenman, M., Banas, A., Bafor, M., Sjodahl, S., Dahlqvist, A., Gummeson, P., Schweizer, M., Singh, S., Nilsson, R., Liljenberg, C. & Stymne, S. (1997). Cloning of a cDNA encoding an acetylenic acid forming enzyme and its functional expression in yeast. In *Proceedings of the Symposium on Biochemistry and Molecular Biology of Plant Fatty Acids and Glycerolipids*. Lake Tahoe, USA. Abstract A7.

Leone, A., Costa, A., Grillo, S., Tucci, M., Horvath, I. & Vigh, L. (1996). Acclimation to low water potential determines changes in membrane fatty acid composition and fluidity in potato cells. *Plant Cell and Environment*, **19**, 1103–9.

Lightner, J., James, Jr, D.W., Dooner, H.K. & Browse, J. (1994). Altered body morphology is caused by increased stearate levels in a mutant of *Arabidopsis*. *The Plant Journal*, **6**, 401–12.

Lindqvist, Y., Huang, W., Schneider, G. & Shanklin, J. (1996). Crystal structure of Δ_9 stearoyl-acyl carrier protein desaturase from castor seed and its relationship to other di-iron proteins. *The EMBO Journal*, **15**, 4081–92.

McConn, M. & Browse, J. (1996). The critical requirement for linolenic acid is pollen development, not photosynthesis, in an *Arabidopsis* mutant. *The Plant Cell*, **8**, 403–16.

McConn, M., Hugly, S., Browse, J. & Somerville, C. (1994). A mutation at the *fad8* locus of *Arabidopsis* identifies a second chloroplast ω-3 desaturase. *Plant Physiology*, **106**, 1609–14.

Miquel, M., James, Jr, D., Dooner, H. & Browse, J. (1993). *Arabidopsis* requires polyunsaturated lipids for low-temperature survival. *Proceedings of the National Academy of Sciences, USA*, **90**, 6208–12.

Mitchell, A.G. & Martin, C.E. (1995). A novel cytochrome b_5-like domain is linked to the carboxyl terminus of the *Saccharomyces cerevisiae* Δ_9, fatty acid desaturase. *Journal of Biological Chemistry*, **270**, 29766–772.

Murata, N. & Wada, H. (1995). Acyl-lipid desaturases and their importance in the tolerance and acclimatization to cold of cyanobacteria. *Biochemical Journal*, **308**, 1–8.

Murphy, D.J. (1996). Engineering oil production in rapeseed and other oil crops. *Trends in Biotechnology*, **14,** 206–13.

Nishiuchi, T., Nakamura, T., Abe, T., Kodama, H., Nishimura, M. & Iba, K. (1995). Tissue-specific and light-responsive regulation of the promoter region of the *Arabidopsis thaliana* chloroplast ω-3 fatty acid desaturase gene (FAD7). *Plant Molecular Biology*, **29**, 599–609.

Nordlund, P. & Eklund, H. (1993). Structure and function of the *Escherichia coli* ribonucleotide reductase protein R2. *Journal of Molecular Biology*, **232**, 123–64.

Nordlund, P., Sjöberg, B-M. & Eklund, H. (1990). Three dimensional structure of the free radical protein of ribonucleotide reductase. *Nature*, **345**, 592–8.

Okuley, J., Lightner, J., Feldmann, K., Yadav, N., Lark, E. & Browse, J. (1994). *Arabidopsis FAD2* gene encodes the enzyme that is essential for polyunsaturated lipid synthesis. *The Plant Cell*, **6**, 147–58.

Paulsen, K.E., Liu, Y., Fox, B.G., Lipscombe, J.D., Münck, E. & Stankovich, M.T. (1994). Oxidation-reduction potentials of the methane monooxygenase hydroxylase component from *Methylosinus* trichosporium OB3b. *Biochemistry*, **33**, 713–22.

Piffanelli, P. (1997). Molecular and biochemical studies of stearoyl-ACP desaturase in gametophytic and sporophytic generations of plants. PhD thesis, University of East Anglia, Norwich, UK.

Piffanelli, P., Ross, J.H.E. & Murphy D.J. (1997). Intra- and extracellular lipid composition and associated gene expression patterns during pollen development in *Brassica napus*. *The Plant Journal*, **11**, 549–62.

Que, L. & Dong, Y.H. (1996). Modeling the oxygen activation chemistry of methane monooxygengase and ribonucleotide reductase. *Accounts of Chemical Research*, **29**, 190–6.

Quinn, P.J., Joo, F. & Vigh, L. (1989). Role of unsaturated lipids in membrane structure and stability. *Progress of Biophysiology and Molecular Biology,* **53**, 71–103.

Reddy, A.S. & Thomas, T.L. (1996). Expression of a cyanobacterial Δ_6-desaturase gene results in γ-linolenic acid production in transgenic plants. *Nature*, **14**, 630–42.

Reddy, A.S., Nuccio, M.L., Gross, L.M. & Thomas, T.L. (1993). Isolation of a Δ_6-desaturase gene from the cyanobacterium *Synechocystis* sp. strain PCC 6803 by gain-of-function expression in *Anabaena* sp. strain PCC 7120. *Plant Molecular Biology*, **22**, 293–300.

Rennie, B.D. & Tanner, J.W. (1989). Fatty acid composition of oil from soybean seeds grown at extreme temperatures. *Journal of the American Oil Chemists Society*, **86**, 1622–4.

Rosenzweig, A.C., Frederick, C.A., Lippard, S.J. & Norlund, P. (1993). Crystal structure of a bacterial non-haem iron hydroxylase that catalyses the biological oxidation of methane. *Nature*, **366**, 537–40.

Sayanova, O., Smith, M.A., Lapinskas, P., Stobart, A.K., Dobson, G., Christie, W.W., Shewry, P.R. & Napier, J. (1997). Expression of a borage desaturase cDNA containing an N-terminal cytochrome b5 domain results in the accumulation of high levels of Δ6 desaturated fatty acids in transgenic tobacco. *Proceedings of the National Academy of Science, USA*, **94**, 4211–16.

Schmidt, H., Dresselhaus, T., Buck, F. & Heinz, E. (1994). Purification and PCR-based cDNA cloning of a plastidial n-6 desaturase. *Plant Molecular Biology*, **26**, 631–42.

Schneider, J.C., Nielsen, E. & Somerville, C. (1995). A chilling-sensitive mutant of *Arabidopsis* is deficient in chloroplast protein accumulation at low temperature. *Plant, Cell and Environment*, **18**, 23–32.

Schultz, D.J., Cahoon, E.B., Shanklin, J., Craig, R., Cox-Foster, D.L., Mumma, R.O. & Medford, J.I. (1996). Expression of a delta-9 14 : 0-acyl carrier protein fatty acid desaturase gene is necessary for the production of omega-5 anacardic acids found in pest-resistant geranium (*Pelargonium hortorum*). *Proceedings of the National Academy of Sciences, USA*, **93**, 8771–5.

Shanklin, J., Whittle, E. & Fox, B.G. (1994). Eight histidine residues are catalytically essential in a membrane-associated iron enzyme, stearoyl-CoA desaturase, and are conserved in alkane hydroxylase and xylene monooxygenase. *Biochemistry*, **33**, 12787–94.

Shanklin, J., Cahoon, E.B., Whittle, E., Lindqvist, Y., Huang, W., Schneider, G. & Schmidt, H. (1996). Structure–function studies on desaturases and related hydrocarbon hydroxyloses. In *Physiology, Biochemistry and Molecular Biology of Plant Lipids*, ed. J.P. Williams, M.U. Khan & N.W. Lem, pp. 6–10. Dordrecht: Kluwer.

Slocombe, S.P., Piffanelli, P., Fairbairn, D., Bowra, S., Hatzopoulos, P., Tsiantis, M. & Murphy, D.J. (1994). Temporal and tissue-specific regulation of a *Brassica napus* stearoyl-acyl carrier protein desaturase gene. *Plant Physiology*, **104**, 1167–76.

Smith, C.R. (1971). Occurrence of unusual fatty acids in plants. *Progress in the Chemistry of Fats and Other Lipids*, **11**, 137–77.

Sperling, P., Schmidt, H. & Heinz, E. (1995). A cytochrome-b_5-containing fusion protein similar to plant acyl lipid desaturases. *European Journal of Biochemistry*, **232**, 798–805.

Stukey, J.E., McDonough, V.M. & Martin, C.E. (1990). The *OLE1* gene of *Saccharomyces cerevisiae* encodes the Δ_9 fatty acid desaturase and can be functionally replaced by the rat stearoyl-CoA desaturase gene. *The Journal of Biological Chemistry*, **265**, 20 144–9.

Suzuki, M., Hayakawa, T., Shaw, J.P., Rekik, M. & Harayama, S. (1991). Primary structure of xylene monooxygenase: similarities to and differences from the alkane hydroxylation system. *Journal of Bacteriology*, **173**, 1690–5.

Thiede, M.A., Ozols, J. & Strittmatter, P. (1986). Construction and sequence of cDNA for rat liver stearoyl coenzyme A desaturase. *The Journal of Biological Chemistry*, **261**, 13 230–5.

Tiku, P.E., Gracey, A.Y., Macartney, A.I., Beynon, R.J. & Cossins, A.R. (1996). Cold-induced expression of Δ_9-desaturase in carp by transcriptional and posttranslational mechanisms. *Science*, **271**, 815–18.

Thompson, G.A. & Li, G. (1996). Altered fatty acid composition of membrane lipids in seeds and seedling tissues of high-saturate canolas. In *Physiology, Biochemistry and Molecular Biology of Plant Lipids*, ed. J.P. Williams, M.U. Khan & N.W. Lem, pp. 351–54. Dordrecht: Kluwer.

Trezzini, G.F., Horrichs, A. & Somssich, I.E. (1993). Isolation of putative defense-related genes from *Arabidopsis thaliana* and expression in fungal elicitor-treated cells. *Plant Molecular Biology*, **21**, 385–9.

van der Loo, F.J., Broun, P., Turner, S. & Somerville, C. (1995). An oleate 12-hydroxylase from *Ricinus communis* L. is a fatty acyl desaturase homolog. *Proceedings of the National Academy of Sciences, USA*, **92**, 6743–7.

Wada, H., Gombos, Z. & Murata, N. (1990). Enhancement of chilling tolerance of a cyanobacterium by genetic manipulation of fatty acid desaturation. *Nature*, **347**, 200–3.

Wallar, B.J. & Lipscomb, J.D. (1996). Dioxygen activation by enzymes containing binuclear nonheme iron clusters. *Chemical Reviews*, **96**, 2625–57.

Yadav, N.S. (1995). Genetic modification of soybean oil quality. In *Soybean Biotechnology*, ed. D.P.S. Verma & R. Shoemaker. Wallingford: CAB International.

Yadav, N.S., Wierzbicki, A., Aegerter, M., Caster, C.S., Pérez-Grau, L., Kinney, A.J., Hitz, W.D., Booth, J.R., Schweiger, B., Stecca, K.L., Allen, S.M., Blackwell, M., Reiter, R.S., Carlson, T.J., Russell, S.H., Feldmann, K.A., Pierce, J. & Browse, J. (1993). Cloning of higher plant ω-3 fatty acid desaturases. *Plant Physiology*, **103**, 467–76.

Yamamoto, K.T. (1994). Further characterization of auxin-regulated mRNAs in hypocotyl sections of mung bean [*Vigna radiata* (L.) Wilczek]: sequence homology to genes for fatty-acid desaturases and atypical late-embryogenesis-abundant protein, and the mode of expression of the mRNAs. *Planta*, **192**, 359–64.

Zou, J., Abrams, G.D., Barton, D.L., Taylor, D.C., Pomeroy, M.K. & Abrams, S.R. (1995). Induction of lipid and oleosin biosynthesis by (+)-abscisic acid and its metabolites in microspore-derived embryos of *Brassica napus* L. cv Reston. *Plant Physiology*, **108**, 563–71.

J. BROWSE, J. SPYCHALLA, J. OKULEY
and J. LIGHTNER

6 Altering the fatty acid composition of vegetable oils

Introduction

Vegetable oils constitute one of the world's most important plant commodities, with current annual production in excess of 65 million metric tons with a total value of $US 25 billion (Murphy, 1994). Production has increased steadily since 1970 at an average annual rate of 4% – about twice the rate of growth in world population (Stymne & Stobart, 1984). The major use of plant oils is for human consumption, but a significant proportion find use in manufacturing industries, particularly in the production of surfactants, paints, plasticisers and specialty lubricants. For both food and industrial applications, it is the fatty acid composition of the oil which determines its usefulness and therefore, its commercial value. The relationship between fatty acid composition and oil characteristics is complex, since it depends not only on the overall fatty acid composition but also on the combinations of fatty acids in the different molecular species of triacylglycerol and on which of the three distinct positions of the glycerol is occupied by a particular fatty acid.

To a large extent, the increases in oil production over the last 15–20 years have been fuelled by the release of improved varieties and efficiencies of cultivation for a relatively few species – soybean, oil palm, canola (low-erucic rapeseed) and sunflower. In the case of soybean, demand for vegetable protein has also been a major factor, since protein is the economically important product from this crop. As a result, expansion of oil production has continued even though these major vegetable oils exhibit fatty acid compositions which make them less than ideal for human nutrition and the requirements of the food industry. For example, both soybean and canola oils presently contain levels of linolenic acid which threaten the shelf-life of products made from them (Weiss, 1983), while palm, coconut and, to a lesser extent, other oils contain high levels of saturated fatty acids which are undesirable because they may contribute to the development of atherosclerosis. Modifying the fatty acid compositions of food oil crops is therefore an attractive goal. At the same time, there is an increasing realisation that the molecular tools of biotechnology will allow the production of industrially valuable oils in transgenic crop plants. This will provide for efficient biological synthesis of specialised chemicals and at the same time permit diversification of agricultural production into a new generation of custom-designed crops. The economic benefits to the agricultural industry involve both the greater value of the new crops and an easing of the over-production that has characterised many commodity crops over the last several decades. As a result, there is unprecedented interest among plant biotechnology companies in modifying oil composition by mutation breeding or by the use of cloned genes to alter the products of seed lipid metabolism.

In this chapter, we shall first describe the significance and biosynthesis of oil components and then outline the ways in which genetics and other approaches have allowed the cloning of fatty acid desaturases.

Goals of oilseed modification

Seed oils are composed almost entirely of triacylglycerols in which fatty acids are esterified to each of the three hydroxyl groups of glycerol. The use of triacylglycerols as a seed reserve maximises the quantity of stored energy within a limited volume, because the fatty acids are a highly reduced form of carbon (Miquel & Browse, 1994). A large variety of different fatty acid structures are found in nature (Gunstone, Harwood & Padley, 1994; Hilditch & Williams, 1964; Murphy, 1994; van de Loo, Fox & Somerville, 1993), but just five account for 90% of the commercial vegetable oil produced: palmitic (16 : 0), stearic (18 : 0), oleic (18 : 1), linoleic (18 : 2), and α-linolenic (18 : 3) acids. These fatty acids are the ones also found most commonly in membrane lipids. As mentioned above, the factors governing the physical properties of a particular oil are complex, but in broad terms, the important characteristics of these major fatty acid constituents are as follows.

Palmitic, stearic and other saturated fatty acids

These are solid at room temperature, in contrast to the unsaturated fatty acids which remain liquid. Because saturated fatty acids have no double bonds in the acyl chain, they remain stable to oxidation at elevated temperatures. They are important components in margarines and chocolate formulations, but for most food applications, reduced levels of saturated fatty acids are desired. (Short chain fatty acids (4 : 0, 6 : 0,

8 : 0) are exceptions to the saturated fatty acid paradigm because they are taken up and metabolised by animals in a similar fashion to carbohydrates. Seed oils contain only very low levels of short chain fatty acids.)

Oleic acid

This has one double bond, but is still relatively stable at high temperatures, and oils with high levels of oleic acid are suitable for cooking and other processes where heating is required. Recently, increased consumption of high oleic oils has been recommended, because oleic acid appears to lower blood levels of low density lipoproteins without affecting levels of high density lipoproteins.

Linoleic acid

This is the major 'polyunsaturated' fatty acid in foods and is an essential nutrient for humans, since it is the precursor for arachidonic acid and prostaglandin synthesis. It is a desirable component for many food applications, but it has limited stability when heated.

α-Linolenic acid

This is also an important component of the human diet. It is used to synthesise the ω3 family of long-chain fatty acids and the prostaglandins derived from these. However, the three double bonds are susceptible to oxidation, so that oils containing high levels of 18 : 3 deteriorate rapidly on exposure to air, especially at high temperatures. Partial hydrogenation of such oils is often necessary before they can be used in food products.

Trans isomers

Trans isomers of unsaturated fatty acids are produced during partial hydrogenation of vegetable oils. The catalyst used to bring about addition of H_2 across the double bonds also converts the natural *cis* double bonds to the *trans* isomers (Gunstone *et al.*, 1994; Weiss, 1983). An all-*trans* unsaturated fatty acid behaves biophysically (and thus physiologically) more like a saturated fatty acid. There are also some concerns about whether *trans* fatty acids are metabolised effectively in the human body.

The considerations summarised here provide the basis for understanding some of the choices involved in deciding on the best oil composition for a particular purpose. Many attempts to alter the fatty acid compositions of oils aim to reduce the proportions of 16 : 0, 18 : 0, 18 : 2 and 18 : 3 in favor of 18 : 1. However, production of margarine and other solid vegetable fats currently involves catalytic hydrogenation of unsaturated oils. The cost of processing and the fact that hydrogenation also converts the natural *cis* double bonds to the *trans* isomers (Gunstone *et al.*, 1994; Murphy, 1994) have provided incentives to produce oils with increased levels of saturated fatty acids (Knutzon *et al.*, 1992) or a high melting-point monounsaturated fatty acid, petroselinic acid Δ6–18 : 1 (Cahoon, Shanklin & Ohlrogge, 1992). Attempts to engineer the synthesis of POS (palmitate, oleate, stearate) triacylglycerols as a substitute for cocoa butter (Ohlrogge, 1994; Stymne & Stobart, 1987) represent one example of a specialised, high-value food use as a target of biotechnology.

Vegetable oils are also important for a range of industrial applications. Many industrial uses of plant oils depend on particular, less common fatty acids. Medium-chain acids (particularly lauric acid, 12 : 0) are used in detergents, while the hydroxylated ricinoleic acid (12-OH 18 : 1) and long-chain eicosenoic (20 : 1) and erucic (22 : 1) acids are components of specialty lubricants and coatings (Gunstone *et al.*, 1994; Murphy, 1994; van de Loo *et al.*, 1993). Epoxy fatty acids are used in plastics manufacture (Ohlrogge, 1994). Producing these unusual fatty acids in established crop species would overcome the poor agronomic performance and other undesirable characteristics that are typically found in the plant species that normally accumulate them in the seed oil (Hilditch & Williams, 1964). In many cases identifying and cloning the gene responsible for the synthesis of a particular fatty acid is not difficult. However, transformation of the gene into a new plant host has often resulted in disappointingly low yields of the desired product. This is a key problem because an extremely important but frequently overlooked criterion for successful introduction of a new source of an industrial oil is the need to ensure very high levels of the desired fatty acid. In general, extraction and purification of a specialised fatty acid will not be economic unless this component accounts for at least 80–90% of total fatty acids in the oil. This means that we need to understand the biochemistry and regulation of oilseed metabolism in detail so that we can preclude or eliminate competing reactions that lead to the synthesis of unwanted fatty acid products. Recent studies have also demonstrated the importance of ensuring that any novel fatty acid product is suitably recognised as a substrate by the enzymes necessary for its further metabolism and assembly into triacylglycerols (Cahoon *et al.*, 1992; Eccleston *et al.*, 1996). These enzymes may include (among others) ketoacyl-ACP synthases, thioesterases, lipases, acyl-CoA ligases, cholinephosphotransferase and acyltransferases (Cao & Huang, 1987; Laurent & Huang, 1992; Miquel & Browse, 1994; Ohlrogge, 1994; Stahl, Banas & Stymne, 1995; Sun, Cao & Huang, 1988).

The biochemistry of seed lipid synthesis

Most of the major oilseed crops (including soybean, canola, sunflower and safflower) are characterised by oils containing predominantly 18-carbon unsaturated fatty acids that are also the major fatty acids of plant membrane lipids. The pathways for triacylglycerol (TAG) synthesis are summarised in Fig. 1. Although this scheme is drawn for *Arabidopsis*, it incorporates information from other oilseeds and has provided an excellent framework for discussing lipid metabolism in many different species (Browse & Somerville, 1991; Miquel & Browse, 1994). Indeed, *Arabidopsis* seed lipids contain substantial proportions of both unsaturated 18-carbon fatty acids (32% 18 : 2; 20% 18 : 3) and long-chain fatty acids (17% 20 : 1) derived from 18 : 1. This suggests that *Arabidopsis* is a good model of both 18 : 2/18 : 3-rich seeds (such as those listed above) and for those species (such as high-erucic *Brassica* varieties) containing longer fatty acids.

Fatty acid synthesis occurs almost exclusively in the plastids ([1] in Fig. 1) (Stumpf, 1980; Browse & Somerville, 1991; Somerville & Browse, 1991) to produce 16 : 0-ACP. Some 16 : 0 is released by a thioesterase of the *FatB* family (Jones *et al.*, 1995) but most is elongated to 18 : 0-ACP [2] and efficiently desaturated by the 18 : 0-ACP desaturase [3] before 18 : 1 is released by a *FatA* thioesterase. Very little 18 : 0 is released from the plastid and 16 : 0 and 18 : 1 are the main products that are converted to CoA thioesters to become the primary substrates for subsequent reactions in other cellular compartments. Newly produced 18 : 1-CoA and 16 : 1-CoA can be used for the synthesis of phosphatidylcholine (PC) by reactions of the Kennedy pathway [7–10] (Stymne & Stobart, 1987). PC is the major structural lipid of the endoplasmic reticulum, but it is also the substrate for the sequential desaturation of 18 : 1 to 18 : 2 and 18 : 3 by the *FAD2* and *FAD3* gene products [11, 12] (Browse *et al.*, 1993; Miquel & Browse, 1992; Stymne & Appelqvist, 1978). The synthesis of PC from diacylglycerol (DAG) by the enzyme CDP-choline : DAG cholinephosphotransferase (the gene is called AAPT1 in soybean (Dewey *et al.*, 1994)) is freely reversible (Slack *et al.*, 1983) as shown by [13], so that in many oilseeds PC is a direct precursor of highly unsaturated species of DAG used for TAG synthesis [16] (Slack, Roughan & Balasingham, 1978). However, the

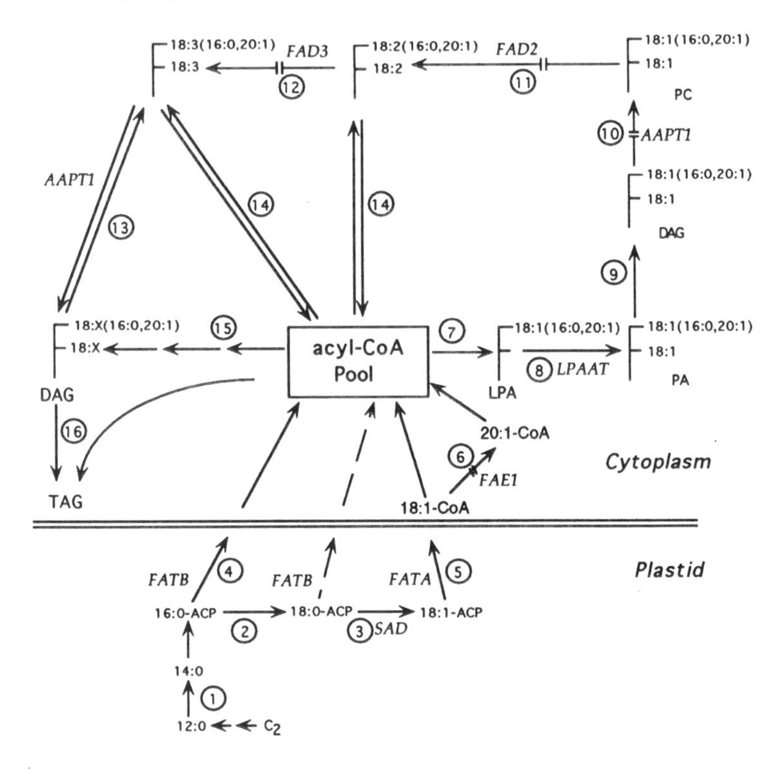

Fig. 1. An abbreviated scheme for the reactions of triacylglycerol synthesis in seeds of *Arabidopsis*. For reactions where a structural gene has been cloned, the gene symbol is shown. The enzymatic steps indicated by the numbers are: **1**, fatty acid synthesis; **2**, 16 : 0 elongation; **3**, stearoyl-ACP desaturase; **4**, palmitoyl-ACP thioesterase; **5**, oleoyl-ACP thioesterase; **6**, 18 : 1 elongation; **7,** acylCoA : glycerol-3-phosphate acyltransferase; **8,** acyl-CoA : lysophosphatidic acid acyltransferase; **9**, phosphatidic acid phosphatase; **10** and **13**, CDP choline : diacylglycerol cholinephosphotransferase (amino alcohol phosphotransferase); **11**, oleoyl-phosphatidylcholine desaturase, FAD2; **12**, linoleoyl-phosphatidylcholine desaturase, FAD3; **14**, acyl-CoA : sn-1 acyllysophosphatidylcholine acyltransferase; **15**, as in 7, 8, and 9 except indicating the use of any fatty acids from the acyl-CoA pool; **16**, acyl-CoA : sn-1,2 diacylglycerol acyltransferase.

acyl-CoA pool does not contain only 16 : 0 and 18 : 1. Exchange of 18 : 1 from CoA with the fatty acid at position *sn-2* of PC [14] provides inputs of 18 : 2 and 18 : 3 (Stymne & Stobart, 1984). Elongation of 18 : 1-CoA to 20 : 1-CoA and 22 : 1-CoA also occurs in *Arabidopsis* as it does in other cruciferous oilseeds (Stumpf & Pollard, 1983). Obviously, the synthesis of DAG [15] (and also PC) may also involve these components of the acyl-CoA pool, as does the final acylation of DAG to form TAG [16] by the enzyme acyl-CoA : DAG acyltransferase (Cao & Huang, 1987; Stobart, Stymne & Hoglund, 1986; Stymne & Stobart, 1984). Thus, the pool of DAG used for TAG synthesis is fed by both phosphatidic acid phosphatase and cholinephosphotransferase. Very recently, evidence has been presented for a reversible dismutase reaction (DAG + DAG $\longleftrightarrow$ TAG + MAG) being involved in TAG synthesis and possibly TAG turnover in developing oilseeds (Mancha *et al.*, 1995). It is not yet clear how important this reaction is in oil synthesis, but the rates measured *in vitro* appear to be of a similar order to the rates of DAG acyltransferase activity.

The present formulation of the pathways shown in Fig. 1 has benefited greatly from studies of *Arabidopsis* mutants. In particular, cloning of the *FAD3* (Arondel *et al.*, 1992) and *FAD2* (Okuley *et al.*, 1994) desaturases and of the *FAE1* condensing enzyme (James *et al.*, 1994) was possible through the use of mutants defective in these steps.

The fatty acyl desaturases are diiron-oxo proteins

The formation of carbon–carbon double bonds in the hydrocarbon chains of fatty acids is an energetically demanding reaction. This reaction is catalysed by a non-haem iron structure in the desaturases now known to be an Fe-O-Fe (diiron-oxo) centre. The non-haem nature of the proteins has been known for a long time because fatty acid desaturases are inhibited by cyanide but not by carbon monoxide. The first major breakthrough in understanding the mechanistic biochemistry of desaturation followed the availability of gram quantities is the soluble Δ9–18 : 0-ACP desaturase obtained through expression of a castor bean cDNA in *E. coli*. Mössbauer spectroscopy and other studies of the desaturase revealed characteristics in common with known diiron-oxo proteins including ribonucleotide reductase and methane monooxygenase (Fox *et al.*, 1993). The crystal structure of the iron-containing component of ribonucleotide reductase was known (Nordlund, Sjöberg & Eklund, 1990) and this class of enzymes were characterised by a duplicated primary sequence motif, EXXH, that forms part of the coordination sphere for the Fe-O-Fe cluster.

The identification of two EXXH boxes in the predicted sequence of the 18 : 0-ACP desaturase provided additional support for the conclusion even though the two proteins overall show no significant sequence homology. Subsequent studies using resonance Raman spectroscopy (Fox *et al.*, 1994) confirmed the diiron-oxo nature of the active site. More recently, the crystal structure of the desaturase has been determined (Shanklin *et al.*, 1995) and the three-dimensional similarities to the ribonucleotide reductase structure are now obvious.

There are a number of related soluble acyl-ACP desaturases that perform desaturation at different positions in the acyl chain or on acyl groups of different lengths. A cDNA encoding a Δ6–16 : 0-ACP desaturase has been isolated from *Thunbergia alata* (Cahoon & Ohlrogge, 1994). Production and expression of chimeric cDNAs composed of parts of the *Thunbergia* and castor bean genes have already provided evidence about the specific amino acid residues involved in determining the positional specificity of desaturation (Cahoon & Shanklin, 1995). Mapping of site-specific amino acid changes onto the Δ9–18 : 0-ACP desaturase crystal structure and the eventual solving of the Δ6–16 : 0-ACP desaturase structure promise to provide important theoretical and practical insights on these enzymes.

The soluble acyl-ACP desaturases are unusual since all other fatty acid desaturases from animals, yeast, cyanobacteria and plants are integral membrane proteins. The plant and cyanobacterial enzymes desaturate fatty acids on glycerolipid substrates while the yeast and animal desaturases act on acyl-CoAs. Investigation of the plant enzymes by traditional biochemical approaches has been limited because solubilising and purifying them has proven very difficult. To date, the only glycerolipid desaturase purified to homogeneity is the 16 : 1/18 : 1 desaturase from spinach chloroplasts (Schmidt *et al.*, 1994). Instead, much of our understanding of the mechanisms and regulation of glycerolipid desaturation has come from investigations of seven classes of *Arabidopsis* mutants, each one deficient in a specific desaturation step (Browse & Somerville, 1994). The *FAD4*, *FAD5*, *FAD6*, *FAD7* and *FAD8* gene products are chloroplast enzymes that have only small effects on seed lipid composition. Two endoplasmic reticulum desaturases, the *FAD2* and *FAD3* gene products are responsible for desaturation of 18 : 1 to 18 : 2 and of 18 : 2 to 18 : 3, respectively, in seeds and other tissues of *Arabidopsis* (Browse *et al.*, 1993). In other oilseed species, the equivalent genes often exist as seed-specific isoforms (Ohlrogge, Browse & Somerville, 1991).

As discussed below, the availability of mutants at the *fad2* and *fad3* loci allowed the wild-type genes to be cloned in the absence of purified desaturase proteins. Thus, *FAD3* was cloned by chromosome walking (Arondel *et al.*, 1992) while a gene tagging approach was used to clone *FAD2* (Okuley *et al.*, 1994). Sequence analysis of these desaturases revealed that they do not contain conserved EXXH motifs like those found in the soluble diiron-oxo enzymes. Instead, all the membrane desaturases from plants, animals and microbes show a common set of histidine-rich boxes. The composition and location within the protein of each of these boxes is highly conserved (Fig. 2) suggesting that the eight histidine residues are important components of the enzyme active site. Histidine residues are common iron ligands in metalloproteins and site-directed mutagenesis has now been used to determine that changing any one of the equivalent histidines in the animal 18 : 0-CoA desaturase eliminates enzyme activity (Shanklin, Whittle & Fox, 1994). This circumstantial evidence and the limited spectroscopic information available for the 18 : 0-CoA desaturase (Strittmatter *et al.*, 1974) suggest that the membrane desaturases are also diiron-oxo proteins although the coordination sphere for the Fe-O-Fe cluster (and overall three-dimensional structure) must be very different from those reported for other diiron-oxo proteins (Shanklin *et al.*, 1994).

Traditional cloning strategies in eukaryotes and prokaryotes

Almost all of the approaches available to molecular biologists have been used to clone one or more of the fatty acid desaturases. Information on these is summarised in Table 1. Not all of them will be discussed in this chapter but more details can be found in the references cited.

In eukaryotes, gene cloning has traditionally begun with biochemistry. A purified protein is needed to provide the sequence (for PCR approaches) or antibodies needed to begin library screening. As discussed above, the membrane desaturases present a challenge in this respect. The spinach chloroplast 16 : 1/18 : 1 desaturase (a *FAD6* homologue)

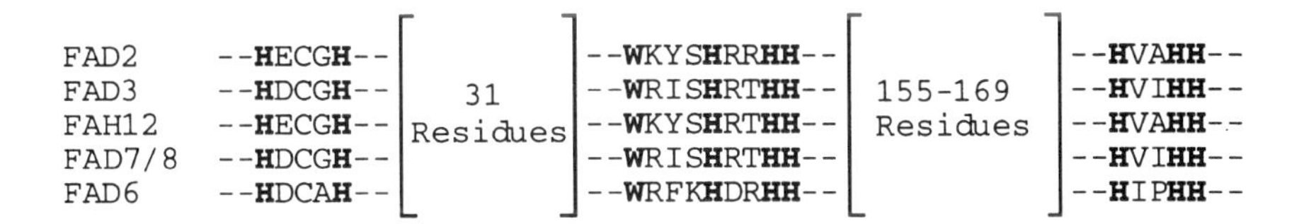

Fig. 2. Conserved histidine residues in membrane-bound desaturases. The histidines are believed to form the coordination sphere for the diiron-oxo group at the active site.

Table 1. *The approaches that have been used to clone fatty acid desaturases from cyanobacteria and plants*

Approach		Genes cloned	Reference
1.	Cloning by complementation	*desA*	Wada *et al.*, 1990
2.	Serendipity	*desC*	Sakamoto *et al.*, 1994
3.	Heterologous gain of function	*desD*	Reddy *et al.*, 1993
4.	Protein purification	*SAD*, *FAD6*	Shanklin & Somerville 1991; Schmidt *et al.*, 1994
5.	Map-based cloning	*FAD3*	Arondel *et al.*, 1992
6.	Homology (probes and PCR)	acyl-ACP desaturases *desB*, *desG*, *FAD6* *FAD7*, *FAD8*, *des6*	Cahoon *et al.*, 1994; Sakamoto *et al.*, 1995; Ishizaki-Nishizawa *et al.*, 1996; Hitz *et al.*, 1994; Iba *et al.*, 1993; Gibson *et al.*, 1994; Sayanova *et al.*, 1997
7.	Gene tagging	*FAD2*	Okuley *et al.*, 1994
8.	Random sequencing	*FAH12*, des6	van der Loo *et al.*, 1995; Thomas *et al.*, 1995
9.	Database searches	*fat-1*, *fat-2*	Spychalla *et al.*, 1997

was cloned in this way (Schmidt *et al.*, 1994) but only after other approaches had already been successful in providing an *Arabidopsis* cDNA (Hitz *et al.*, 1994).

In bacterial systems, genes are often cloned by complementation of a known mutant. This approach is possible because bacteria are easy to transform and every gene in a genomic expression library can thus be tested for its ability to correct a defined (and preferably selectable) mutant phenotype. The observation that *fad12* mutants of *Synechocystis sp.* PCC 6803 grow more slowly than wild type at 22 °C allowed Norio Murata's group to clone the *desA* gene. This breakthrough, and other tools available in prokaryotic genetics, has since led to the isolation of five distinct fatty acid desaturases from cyanobacteria – *desA* to *desG* in Table 1 (Murata & Wada, 1995).

Map-based cloning of *FAD3*

The first plant gene to be cloned by chromosome walking (map-based cloning) was *FAD3* which encodes the endoplasmic reticulum 18 : 2

desaturase (Arondel *et al.*, 1992). A defined mutant phenotype and old-fashioned genetics are essential for chromosome walking because progress towards the objective is estimated by measuring recombination between the most recently identified DNA marker – for example an RFLP (restriction fragment length polymorphism) or CAPS (cleaved amplified polymorphic sequence) – and the mutant locus in the F_2 population derived from an appropriate cross. As one approaches the target, crossover events become increasingly rare so that it is difficult to complete the walk. This endgame problem was solved by Arondel *et al.* (1992) using a clever trick. Based on the easy part of the walk, they had identified three YAC clones that were likely to span the *fad3* locus (Fig. 3). The most promising of these, EW7D11 was likely to contain 12–25 genes. Based on previous work with oilseeds, the authors reasoned that the *FAD3* transcript would be moderately abundant in developing seeds. On the other hand, it was quite likely that most or all of the other genes on EW7D11 would not be strongly expressed in seed tissue. When the entire EW7D11 YAC was radiolabelled and used to probe a cDNA library derived from developing seeds of *Brassica napus* , one family of cDNAs accounted for half of the clones identified. As predicted, these turned out to be clones of *FAD3*.

The *FAD3* gene was also cloned by T-DNA tagging (Yadav *et al.*, 1993). The expectation was that all the *FAD* genes would be similar

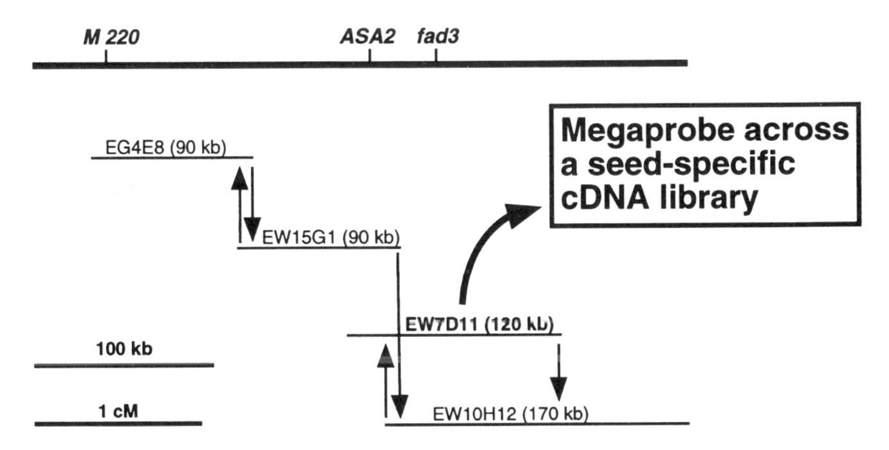

Fig. 3. Strategy used for map-based cloning of *FAD3*. Identifying YACs spanning the *fad3* locus represents the easy part of a chromosome walking project. The most promising YAC, EW7D11, was used to probe a cDNA library from developing *Brassica napus* seeds in which the most abundant family of cross-hybridising sequences were *FAD3* cDNAs. (Adapted from Arondel *et al.*, 1992.)

enough to allow cloning of the complete collection by using *FAD3* as a probe or by designing PCR primers based on regions of homology between *FAD3* and the cyanobacterial desaturases. Indeed, clones for three chloroplast desaturases, *FAD6*, *FAD7* and *FAD8*, were quickly obtained by these approaches. However, extensive efforts in several laboratories using available gene sequences failed to isolate *FAD2* or the chloroplast *FAD4* and *FAD5* genes. These last two have still not been cloned.

T-DNA tagging of *FAD2*

One of many tools available to the *Arabidopsis* community is a collection of lines, developed by Dr Ken Feldmann (DuPont Co. and University of Arizona), in which mutations have been caused by random insertion of a T-DNA into the genome. In 1990, it was decided to determine whether there were any fatty acid mutants represented in the T-DNA population. At that time, several genes had already been obtained from *Arabidopsis* by T-DNA tagging. These included *GLABOROUS1* (Herman & Marks, 1989; Marks & Feldmann, 1989) and *AGAMOUS* (Yanofsky *et al.*, 1990). Work on many more genes was also under way.

However, the tagged mutants so far identified had been obtained from searches that involved simple visual screens and that generally targeted relatively broad classes of mutants. In terms of searching for a particular tagged locus such as *fad2*, the chances of success can be calculated from the knowledge that the T-DNA lines contain an average of 1.4 inserts and the assumption that the inserts are randomly located throughout the genome. If a figure of 100 000 kb is used for the size of the *Arabidopsis* genome then for a target gene of 2 kb (including essential 5′ and 3′ sequences), there is approximately a 3% chance of finding a tagged mutant for every 1000 T-DNA lines that are screened.

Because the T-DNA containing lines are segregating for the insert (and thus for any resultant mutation), it was necessary for us to sample several individuals from each line. To simplify the procedure, single leaves were harvested from ten plants from one line and pooled for analysis. Based on knowledge of the chemically induced mutants, it was expected to be able to successfully identify any of the mutants in a segregating population. Fig. 4 shows a typical wild type gas chromatogram from the T-DNA population, together with the chromatogram obtained from a pooled sample of one line, number 658, that showed a small increase in 18 : 1 relative to the wild type. Subsequent analysis of individual plants from line 658 showed that it was segregating 3 : 1 for increased 18 : 1. However, the overall fatty acid composition of the

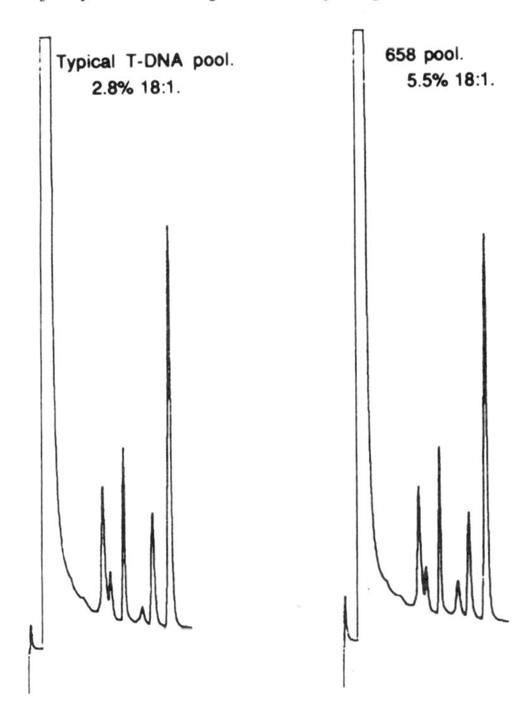

Fig. 4. Gas chromatograph separations of fatty acid methyl esters derived from leaves of wild-type *Arabidopsis* (left) and the T-DNA containing line 658 (right). Leaves from ten plants were pooled for analysis.

homozygous mutant segregant suggested that the mutation had produced a leaky allele of *fad2*. Such leaky alleles can be caused by T-DNA insertion in the 5′ (or 3′) non-coding region of the gene. A portion of plant DNA adjacent to the left border of the T-DNA was cloned by plasmid recovery and was then used to identify a full-length cDNA that showed low but significant homology (about 40% identity at the deduced amino acid level) to the *Arabidopsis FAD3* gene.

The identity of the new gene was confirmed by transforming the cDNA into *fad2* mutant roots and showing complementation of the fatty acid phenotype (Okuley *et al.*, 1994). Sequence analysis of the *fad2* locus in the T-DNA mutant showed that insertion of the T-DNA had occurred at the 3′ end of the transcribed sequence, upstream of a large intron and the ATG start codon of the *FAD2* open reading frame (Fig. 5). This structure can explain the leaky phenotype of this mutant if it

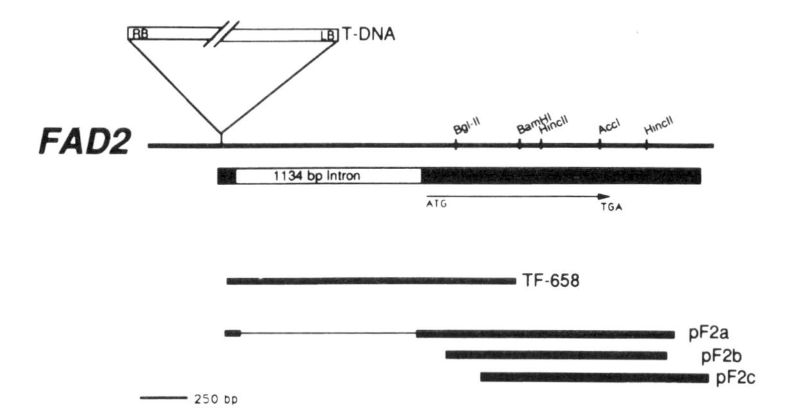

Fig. 5. Structure of the *fad2* locus in the T-DNA mutant line. The lengths and locations of the 1.6 kb flanking sequence obtained by plasmid rescue (TF-658) and three cDNAs are shown. (Adapted from Okuley *et al.*, 1994.)

is assumed that there is a cryptic promotor in the left border of the T-DNA or in the 1134 bp intron.

Random sequencing of cDNAs

Many of the unusual fatty acids found in nature have important industrial uses. However, the plants producing them are most often poorly suited for extensive commercial production. As an alternative, the isolation of key genes directing the synthesis of a particular fatty acid can provide the means to genetically engineer agronomically suitable oilseed crops to produce the desired oil more easily and cheaply. For example, ricinoleic acid (18 : 1-OH) produced by the castor oil plant (*Ricinus communis*) is an extremely versatile natural product with industrial applications that include the synthesis of nylon-11, lubricants, hydraulic fluids, plastics, cosmetics and other materials. However, the castor plant contains ricin, an extremely toxic lectin, as well as other poisons and allergens. In addition, agronomic problems result in poor yield compared with other crops. These factors mean that castor is a minor crop grown mainly in non-industrialised countries.

The successful cloning of a gene that encodes the hydroxylase responsible for 18 : 1-OH synthesis from 18 : 1 relied on a detailed understanding of the biochemistry involved. It has been known for some time that the conversion of 18 : 1 to 18 : 1-OH probably involves a single enzyme that inserts a hydroxyl group at the same position ($\Delta 12$)

as the double bond that is introduced by the FAD2 desaturase (Moreau & Stumpf, 1981; Bafor *et al.*, 1991). Consideration of the probable reaction cycle of the desaturases suggested that 18 : 1-OH might be produced by a stalled desaturase (van de Loo *et al.*, 1993). Partial sequencing of randomly chosen cDNAs from a library derived from developing castor bean endosperm identified a relatively abundant gene (three of 468 cDNAs) with homology to the glycerolipid desaturases of other oilseeds (van de Loo *et al.*, 1995). Expression of the castor cDNA in transgenic tobacco plants led to the synthesis of 18 : 1-OH in seed tissues. The *FAH12* sequence shows striking homology to both the *FAD2* gene of *Arabidopsis* (which encodes the 18 : 1 phosphatidylcholine Δ12 desaturase) and to the castor FAD2 homologue (approximately 67% amino acid identity in both cases).

The successful use of random sequencing projects to clone genes involved in the synthesis of unusual fatty acids does not require the high level of homology found between the *FAH12* and *FAD2* genes. Because all the membrane-bound desaturases described from animals, microbes and plants contain three, highly conserved histidine-rich sequences, new members of the gene family can be identified easily even when overall homology is low. For example, the production of γ-linolenate (Δ6, 9, 12–18 : 3) in seeds of borage, blackcurrant and evening primrose is known to involve a Δ6 desaturase acting on 18 : 2 at the *sn-2* position of PC (Stymne & Stobart, 1986). To clone the glycerolipid Δ6 desaturase gene, Thomas *et al.* (1995) partially sequenced 600 cDNAs from a subtracted library prepared from developing borage seeds. One class of cDNAs showed low (*ca.* 30% identity at the amino acid level) overall homology to other desaturases but did contain the characteristic histidine boxes with spacings similar to those in the Δ6 desaturase of *Synechocystis* (Reddy *et al.*, 1993). Expression of the borage cDNA in transgenic tobacco and carrot suspension culture protoplasts led to the production of γ-linolenate and Δ6, 9, 12, 15–18 : 4.

Constructing a cDNA library and partially sequencing many clones may appear to be a daunting prospect. However, in practice it is a straightforward and relatively inexpensive project for any well-equipped molecular biology laboratory. Identification of the *FAH12* and Δ6 desaturase genes was facilitated by their similarity to previously characterised desaturases but it is likely that the strategy could be successful even without this aid. Available information suggests that an enzyme that is involved in the synthesis of the major fatty acid of an oilseed will be translated from a moderately abundant transcript (0.5–1% of the cDNAs sequenced). Furthermore, in the majority of cases, this transcript will be specific for the seed and will not be found in other tissues of the plant. This feature will assist differential screening strategies. Most very abundant cDNAs (e.g. storage proteins, rRNAs) and the majority of moderately abundant cDNAs can have functions ascribed to them on the basis of homologies to genes of known function from other organisms. Given these parameters, it is likely that the search for a particular gene – the gene that encodes the enzyme responsible for vernolic acid synthesis (Bafor *et al.*, 1993), for example, could be quickly reduced to 10–20 different cDNAs. Identifying the correct gene from this number would be quite easily accomplished using RNA gel blot analysis and heterologous expression in tobacco cells or another appropriate system.

Many of the enzymes responsible for the synthesis of unusual fatty acids are integral membrane proteins and will probably be difficult to purify. For this reason, direct approaches, such as the one described here, may offer the most efficient way to clone and identify the desired genes.

Database searches and *Caenorhabditis elegans* genes

In the search for agriculturally relevant desaturases, we do not need to be limited to plant genes. For example, animal and yeast genes for 18 : 0-CoA desaturases have been expressed in plants (Polashok, Chin & Martin, 1992; Grayburn, Collins & Hildebrand, 1992). The authors set out to use the *Arabidopsis* desaturase genes and the extensive information now available from the *Caenorhabditis elegans* genome project (Waterston & Sulston, 1995) to identify and clone genes encoding fatty acid desaturases in animals. *C. elegans* is a good model in this respect because it elaborates a wide range of polyunsaturated fatty acids including arachidonic and eicosapentaenoic acids, from very simple precursors available in its diet (Satouchi *et al.*, 1993).

A search of the *C. elegans* DNA databases, using the sequences of *Arabidopsis* genes, identified several putative desaturases. The first of these to be cloned was *fat-1* (Spychalla, Kinney & Browse, 1997). The predicted protein encoded by a *fat-1* cDNA showed 32–35% identity with both FAD2 and FAD3 of *Arabidopsis*. When expressed in transgenic plants, *fat-1* resulted in a 90% increase in the proportion of α-linolenic acid in root lipids, thus demonstrating that the cDNA encodes a desaturase that acts on 18 : 2. Wild-type *Arabidopsis* incorporated ω-6 fatty acids (Δ8,11,14–20 : 3 and Δ5,8,11,14–20 : 4) into membrane lipids but did not desaturate them. By contrast, *fat-1* transgenic plants efficiently desaturated both of these fatty acids to the corresponding ω-3 products (Fig. 6). These findings indicate that the *C. elegans fat-1*

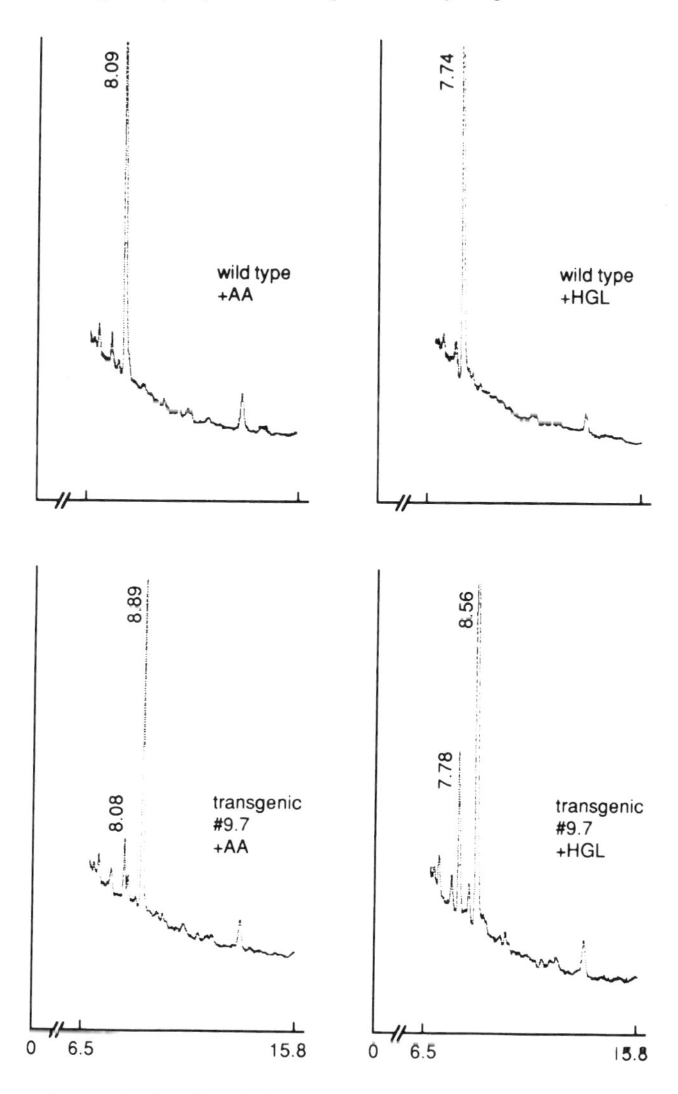

Fig. 6. Partial gas chromatograph traces of the component fatty acids of phosphatidylcholine extracted from leaves of wild-type (top) or *fat-1* transgenic (bottom) *Arabidopsis* sprayed with arachidonic acid (AA, 20 : 4 ω-6). Only the *fat-1* transgenic plants produce significant amounts of the ω-3 desaturated products.

gene encodes the first animal representative of a class of glycerolipid desaturates that have previously been characterised in plants and cyanobacteria. The FAT-1 protein is an ω-3 fatty acyl desaturase that recognises a range of 18- and 20-carbon ω-6 substrates.

The potential value of the FAT-1 desaturase is that it efficiently converts arachidonic acid to eicosapentaenoic acid. This function may be useful for producing specialised vegetable oils.

Conclusion

The membrane-bound fatty acid desaturases belong to a large family of enzymes that carry out energetically demanding reactions on hydrocarbon substrates. The sequences of the desaturase proteins are very diverse but they have a common set of histidine boxes that include residues that are believed to bind the diiron-oxo active site. Projects to clone the desaturases have used almost every approach that is available in modern molecular genetics.

References

Arondel, V., Lemieux, B., Hwang, I., Gibson, S., Goodman, H.M. & Somerville, C.R. (1992). Map-based cloning of a gene controlling omega-3 fatty acid desaturation in *Arabidopsis*. *Science*, **258**, 1353–5.

Bafor, M., Smith, M.A., Jonsson, L., Stobart, K. & Stymne, S. (1991). Ricinoleic acid biosynthesis and triacylglycerol assembly in microsomal preparations from developing castor bean (*Ricinus communis*) endosperm. *Biochemical Journal*, **280**, 507–14.

Bafor, M., Smith, M.A., Jonsson, L., Stobart, K. & Stymne, S. (1993). Biosynthesis of vernoleate (*cis*-12-epoxyoctadeca-*cis*-9-enoate) in microsomal preparations from developing endosperm of *Euphorbia lagascae*. *Archives of Biochemistry and Biophysics*, **303**, 145–51.

Browse, J. & Somerville, C. (1991). Glycerolipid synthesis: biochemistry and regulation. *Annual Review of Plant Physiology and Plant Molecular Biology*, **42**, 467–506.

Browse, J. & Somerville, C. (1994). Glycerolipids. In *Arabidopsis*, ed. E. Meyerowitz & C. Somerville, pp. 881–912. New York: Cold Spring Harbor Press.

Browse, J., McConn, M., James, D. & Miquel, M. (1993). Mutants of *Arabidopsis* deficient in the synthesis of α-linolenate. Biochemical and genetic characterization of the endoplasmic reticulum linoleoyl desaturase. *The Journal of Biological Chemistry*, **268**, 16 345–51.

Cahoon, E.B. & Ohlrogge, J.B. (1994). Metabolic evidence for the involvement of a Δ^4-palmitoyl-acyl carrier protein desaturase in petroselinic acid synthesis in coriander endosperm and transgenic tobacco cells. *Plant Physiology*, **104**, 827–37.

Cahoon, E.B. & Shanklin, J. (1995). Engineering of the chain-length and double bond positional specificities of acyl-ACP desaturases. In *National Plant Lipid Cooperative Symposium on Biochemistry and Molecular Biology of Plant Fatty Acids and Glycerolipids*, pp. 0–22. National Plant Lipid Cooperative.

Cahoon, E.B., Shanklin, J. & Ohlrogge, J.B. (1992). Expression of a coriander desaturase results in petroselinic acid production in transgenic tobacco. *Proceedings of the National Academy of Sciences, USA*, **89**, 11 184–8.

Cahoon, E.B., Cranmer, A.M., Shanklin, J. & Ohlrogge, J.B. (1994). Δ^6 Hexadecenoic acid is synthesized by the activity of soluble Δ_6 palmitoyl-acyl carrier protein desaturase in *Thunbergia alata* endosperm. *Journal of Biological Chemistry*, **269**, 27 519–26.

Cao, Y-Z. & Huang, A.H.C. (1987). Acyl coenzyme a preference of diacylglycerol acyltransferase from maturing seeds of *Cuphea*, maize, rapeseed and canola. *Plant Physiology*, **84**, 762–9.

Dewey, R.E., Wilson, R.F., Novitzky, W.P. & Goode, J.H. (1994). The *AAPTI* gene of soybean complements a cholinephosphotransferase-deficient mutant of yeast. *Plant Cell*, **6**, 1495–507.

Eccleston, V.S., Cranmer, A.M., Voelker, T.A. & Ohlrogge, J.B. (1996). Medium-chain fatty acid biosynthesis and utilization in *Brassica napus* plants expressing lauroyl-acyl carrier protein thioesterase. *Planta*, **198**, 46–53.

Fox, B.G., Shanklin, J., Somerville, C. & Münck, E. (1993). Stearoyl-acyl carrier protein Δ^9 desaturase from *Ricinius communis* is a diioron-oxo protein. *Proceedings of the National Academy of Sciences, USA*, **90**, 2486–90.

Fox, B.G., Shanklin, J., Ai, J., Loehr, T.M. & Sanders-Loehr, J. (1994). Resonance raman evidence for an Fe-O-Fe center in stearoyl-ACP desaturase. Primary sequence identity with other diiron-oxo proteins. *Biochemistry*, **33**, 12 776–86.

Gibson, S., Arondel, V., Iba, K. & Somerville, C. (1994). Cloning of a temperature-regulated gene encoding a chloroplast ω-3 desaturase from *Arabidopsis thaliana. Plant Physiology*, **106**, 1615–21.

Grayburn, W.S., Collins, G.B. & Hildebrand, D.F. (1992). Fatty acid alteration by a delta 9 desaturase in transgenic tobacco tissue. *Biotechnology*, **10**, 675–8.

Gunstone, F.D., Harwood, J.L. & Padley, F.B. (1994). *The Lipid Handbook*. London: Chapman & Hall.

Herman, P.L. & Marks, M.D. (1989). Trichome development in *Arabidopsis thaliana* II. Isolation and complementation of the *GLABOROUS1* gene. *The Plant Cell*, **1**, 1051–5.

Hilditch, T.P. & Williams, P.N. (1964). *The Chemical Constituents of Natural Fats*. London: Chapman & Hall.

Hitz, W.D., Carlson, T.J., Booth, J.R., Kinney, A.J., Stecca, K.L. & Yadav, N.S. (1994). Cloning of a higher plant plastid ω6 fatty acid desaturase cDNA and its expression in a cyanobacterium. *Plant Physiology*, **105**, 635–41.

Iba, R., Gibson, S., Nishiuchi, T., Puse, T., Nishimua, M., Arondel, V., Hughly, S. & Somerville, S. (1993). A gene encoding a chloroplast ω-3 fatty acid desaturase complements alterations in fatty acid denaturation and chloroplast copy number of the *fad7* mutant of *Arabidopsis thallana. Journal of Biological Chemistry*, **268**, 24 099–105.

Ishizaki-Nishizawa, O., Fujil, T., Aruma, M., Sckiguchi, K., Murnen, N., Ohrani, T. & Toguri, T. (1996) Low-temperature resistance of higher plants is significantly enchanced by a nonspecific cyanobacterial desaturase. *Nature Biotechnology*, **14**, 1003–6.

James, Jr, D.W., Lim, E., Keller, J., Plooy, I., Ralston, E. & Dooner, H.K. (1994). Directed tagging of the *Arabidopsis* fatty acid elongation 1 (*FAE1*) gene with the maize transposon activator. *The Plant Cell*, **7**, 309–19.

Jones, A., Aubrey, J., Davies, H.M. & Voelker, T.A. (1995). Palmitoyl-acyl carrier protein (ACP) thioesterase and the evolutionary origin of plant acyl-ACP thioesterases. *The Plant Cell*, **7**, 359–71.

Knutzon, D.S., Thomson, G.A., Radke, S.E. & Johnson, W.B. (1992). Modification of *Brassica* seed oil by antisense expression of a steroyl-acyl carrier protein desaturase gene. *Proceedings of the National Academy of Sciences, USA*, **89**, 2624–8.

Laurent, P. & Huang, A.H.C. (1992). Organ- and development-specific acyl coenzyme A: lysophosphatidate acyltransferase in palm and meadowfoam. *Plant Physiology*, **99**, 1711–5.

Mancha, M., Stobart, K., Lenman, M., Stahl, U. & Stymne, S. (1995). Triacylglycerol turnover in microsomal membranes from developing oil seeds. In *National Plant Lipid Cooperative Symposium*, pp. P-226. National Plant Lipid Cooperative.

Marks, M.D. & Feldmann, K.A. (1989). Trichome development in *Arabidopsis thaliana* I. T-DNA tagging of the *GLABOROUS1* gene. *The Plant Cell*, **1**, 1043–53.

Miquel, M. & Browse, J. (1992). *Arabidopsis* mutants deficient in polyunsaturated fatty acid synthesis. Biochemical and genetic characterization of a plant oleoyl-phosphatidylcholine desaturase. *The Journal of Biological Chemistry*, **267**, 1502–9.

Miquel, M. & Browse, J. (1994). Lipid biosynthesis in developing seeds. In *Seed Development and Germination*, ed. G. Galili, J. Kigel & M. Negli, pp. 169–93. New York: Marcel Dekker.

Moreau, R. & Stumpf, P. (1981). Recent studies of the enzymatic synthesis of ricinoleate by developing castor bean. *Plant Physiology*, **67**, 672–6.

Murata, N. & Wada, H. (1995). Acyl-lipid desaturases and their importance in the tolerance and acclimatization to cold of cyanobacteria. *Biochemical Journal*, **308**, 1–7.

Murphy, D.J. (1994). *Designer Oil Crops.* Weinheim: VCH.

Nordlund, P., Sjöberg, B.-M. & Eklund, H. (1990). Three-dimensional structure of the free radical protein of ribonucleotide reductase. *Nature*, **345**, 593–8.

Ohlrogge, J.B. (1994). Design of new plant products: engineering of fatty acid metabolism. *Plant Physiology*, **104**, 821–6.

Ohlrogge, J.B., Browse, J. & Somerville, C.R. (1991). The genetics of plant lipids. *Biochimica et Biophysica Acta*, **1082**, 1–26.

Okuley, J., Lightner, J., Feldmann, K., Yadav, N., Lark, E. & Browse, J. (1994). The *Arabidopsis FAD2* gene encodes the enzyme that is essential for polyunsaturated lipid synthesis. *The Plant Cell*, **6**, 147–58.

Polashok, J.J., Chin, C-K. & Martin, C.E. (1992). Expression of the yeast Δ-9 fatty acid desaturase in *Nicotiana tabacum. Plant Physiology*, **100**, 894–901.

Reddy, A.S., Nuccio, M.L., Gross, L.M. & Thomas, T.L. (1993). Isolation of a delta-6-desaturase gene from the cyanobacterium *Synechocystis* sp. strain PCC-6803 by gain-of-function expression in *Anabaena* sp. strain PCC-7120. *Plant Molecular Biology*, **27**, 293–300.

Sakamoto, T., Wada, H., Nishida, I., Ohmori, M. & Murata, N. (1994). Δ9 acyl-lipid desaturases of cyanobacteria. *Journal of Biological Chemistry*, **269**, 25575–80.

Satouchi, K., Hirano, K., Sakaguchi, M., Takehara, H. & Matsuura, F. (1993). Phospholipids from the free-living nematode *Caenorhabditis elegans. Lipids*, **28**, 837–40.

Sayanova, O., Smith, M.A., Lapinskas, P., Stobart, A.K., Dobson, G., Christie, W.W., Shewry, P.R. & Napier, J.A. (1997). Expression of a borage desaturase cDNA containing an N-terminal cytochrome b_5 domain results in the accumulation of high levels of Δ^6-desaturated fatty acids in transgenic tobacco. *Proceedings of the National Academy of Sciences, USA*, **94**, 4211–16.

Schmidt, H., Dresselhaus, T., Buck, F. & Heinz, E. (1994) Purification and PCR-based cDNA cloning of a plastidial n-6 desaturase. *Plant Molecular Biology*, **26**, 631–42.

Shanklin, J. & Somerville, C. (1991). Stearoyl-acyl-carrier-protein desaturase from higher plant is structurally unrelated to the animal fungal homologs. *Proceedings of the National Academy of Sciences, USA*, **88**, 2510–14.

Shanklin, J., Whittle, E. & Fox, B.G. (1994). Eight histidine residues are catalytically essential in a membrane-associated iron enzyme, stearoyl-CoA desaturase, and are conserved in alkane hydroxylase and xylene monooxygenase. *Biochemistry*, **33**, 12 787–94.

Shanklin, J., Lindqvist, L., Huang, W., Fox, B. & Schneider, G.

(1995). Structure–function studies on desaturase and related hydrocarbon hydroxylases. In *National Plant Lipid Cooperative Symposium*, pp. 0–21. National Plant Lipid Cooperative.

Slack, C.R., Roughan, P.G. & Balasingham, N. (1978). Labeling of glycerolipids in the cotyledons of developing oilseeds by [1-^{14}C]acetate and [2-^{3}H]glycerol. *Biochemical Journal*, **179**, 421–33.

Slack, C.R., Campbell, L.C., Browse, J.A. & Roughan, P.G. (1983). Some evidence for the reversibility of choline phosphotransferase-catalysed reaction in developing linseed cotyledons *in vivo*. *Biochimica et Biophysica Acta*, **754**, 10–20.

Somerville, C. & Browse, J. (1991). Plant lipids: metabolism mutants and membranes. *Science*, **252**, 80–7.

Spychalla, J.P., Kinney, A.J. & Browse, J. (1997). Identification of an animal *omega*-3 fatty acid desaturase by heterologous expression in *Arabidopsis*. *Proceedings of the National Academy of Sciences, USA*, **94**, 1142–7.

Stahl, U., Banas, A. & Stymne, S. (1995). Plant microsomal phospholipid acyl hydrolases have selectivities for uncommon fatty acids. *Plant Physiology*, **107**, 953–62.

Stobart, A.K., Stymne, S. & Hoglund, S. (1986). Safflower microsomes catalyse oil accumulation *in vitro*: a model system. *Planta*, **169**, 33–7.

Strittmatter, P., Spatz, L., Corcoran, D., Rogers, M.J., Setlow, B. & Redline, R. (1974). Purification and properties of rat liver microsomal stearyl coenzyme a desaturase. *Proceedings of the National Academy of Sciences, USA*, **71**, 4565–9.

Stumpf, P.K. (1980). Biosynthesis of saturated and unsaturated fatty acids. In *The Biochemistry of Plants*, ed. P.K. Stumpf & E.E. Conn. New York: Academic Press.

Stumpf, P.K. & Pollard, M.R. (1983). Pathways of fatty acid biosynthesis in higher plants with particular reference to developing rapeseed. In *High and Low Erucic Acid Rapeseed Oils*, ed. J.K.G. Kramer, F.D. Sauer & W.J. Pigden. Toronto: Academic Press.

Stymne, S. & Appelqvist, A. (1978). The biosynthesis of linoleate from oleoyl-CoA via oleoyl phosphatidylcholine in microsomes of developing safflower seeds. *European Journal of Biochemistry*, **90**, 223–9.

Stymne, S. & Stobart, A.K. (1984). Evidence for the reversibility of the acyl-CoA: Lysophosphatidylcholine acyltransferase in the microsomes of developing safflower cotyledons and rat liver. *Biochemical Journal*, **223**, 305–14.

Stymne, S. & Stobart, A.K. (1986). Biosynthesis of γ-linolenic acid in cotyledons on microsomal preparations of the developing seeds of common borage (*Borago officinalis*). *Biochemical Journal*, **240**, 385–93.

Stymne, S. & Stobart, A.K. (1987). Triacylglycerol biosynthesis. In

The Biochemistry of Plants, ed. P.K. Stumpf & E.E. Conn, vol. 9, pp. 175–214. New York: Academic Press.

Sun, C., Cao, Y-Z. & Huang, A.H.C. (1988). Acyl coenzyme A preference of the glycerol phosphate pathway in the microsomes from the maturing seeds of palm, maize, and rapeseed. *Plant Physiology*, **88**, 56–60.

Thomas, T.L., Nunberg, A., Reddy, A.S., Nuccio, M.L. & Beremand, P. (1995). Cloning and expression of Δ6-desaturase genes in transgenic plants. In *National Plant Lipid Cooperative Symposium*, South Lake Tahoe, CA, June 1–4, 1995, National Plant Lipid Cooperative, 0–25.

van de Loo, F.J., Fox, B.C. & Somerville, C.R. (1993). Unusual fatty acids. In *Lipid Metabolism in Plants*, ed. T.S. Moore, pp. 91–126. Boca Raton, FL: CRC Press.

van de Loo, F.J., Broun, P., Turner, S. & Somerville, C. (1995). An oleate 12-hydroxylase from *Ricinus communis* is a fatty acyl desaturase homolog. *Proceedings of the National Academy of Sciences, USA*, **92**, 6743–7.

Wada, H., Gombos, Z. & Murata, N. (1990). Enhancement of chilling tolerance of a cyanobacterium by genic manipulation of fatty acid desaturation. *Nature*, **347**, 200–3.

Waterston, R. & Sulston, J. (1995). The genome of *Caenorhabditis elegans*. *Proceedings of the National Academy of Sciences, USA*, **92**, 10 836–40.

Weiss, T.J. (1983). *Food Oils and Their Uses*. Westport, CT: AVI Publishing.

Yadav, N.S., Wierzbicki, A., Aegerter, M., Caster, C.S., Pérez-Grau, L., Kinney, A.J., Hitz, W.D., Booth, Jr, R.J., Schweiger, B., Stecca, K.L., Allen, S.M., Blackwell, M., Reiter, R.S., Carlson, T.J., Russell, S.H., Feldmann, K.A., Pierce, J. & Browse, J. (1993). Cloning of higher plant ς-3 fatty acid desaturases. *Plant Physiology*, **103**, 467–76.

Yanofsky, M., Ma, H., Bowman, J., Drews, G., Feldmann, K.A. & Meyerowitz, E.M. (1990). The protein encoded by the *Arabidopsis* homeotic gene *Agamous* resembles transcription factors. *Nature*, **346**, 35–9.

P.J. QUINN

7 Engineering frost resistance in plants by genetic manipulation

Introduction

The production of plants that can tolerate or exhibit improved tolerance to freezing conditions has obvious and considerable commercial potential. Much effort has therefore been devoted to understanding the molecular mechanisms of freezing-induced injury to plants. It is generally agreed that the ability of plants to tolerate freezing is a multifactorial process. Moreover, the ability of particular plants to survive frost conditions is clearly under overt genetic control because prior exposure of many plant species to low temperatures above 0 °C induces physiological and biochemical changes that serve to adapt them to survive subsequent freezing conditions. This process, referred to as acclimation, is associated with changes in cellular constituents such as sugars, amino acids and similar compatible solutes, the appearance of different isoforms of proteins, and changes in the lipid composition of the cell membranes. These alterations are said to be responsible for improving survival of the plant by protecting vulnerable constituents from the harmful effects of ice formation or aiding in recovery of functional organelles during the thawing process.

Different stress factors including exposure to high or low temperatures, drought or salinity, often result in similar physiological and biochemical responses which are ultimately traceable to changes in expression of particular genes (Rhy *et al.*, 1995). It may be cogently argued that this supports the notion that different stress conditions cause similar instabilities in cellular constituents. Nevertheless, differences in temporal and/or organ responses are observed, depending on the particular stress conditions imposed. It has been suggested that this phenotypic variation of plants in response to different environmental factors may be the result of differential expression of constitutent members of multigene families (Smith, 1990).

Acclimation is known to be associated with the up-regulation of a number of plant genes as evidenced by an increase in steady-state levels of mRNA. A number of these genes have been isolated from both monocotyledon and dicotyledon species (for review see Hughes &

Dunn, 1996). From a genetic perspective, freezing tolerance is a complex multigenic process and with different genes exerting a dominant influence in different species. Some of these genes have now been cloned, but the precise function of many of the protein products encoded by these genes remains uncertain. Nevertheless, studies of recombinant proteins together with detailed expression data are beginning to build up a picture of control at both transcriptional and post-translational levels. One approach that is proving useful for the molecular analysis of cold acclimation is the study of low temperature responsive promoters using reporter gene constructs. Other approaches include the isolation of non-acclimating mutants and the production of transgenic plants.

The consequence of acclimation is that changes in cellular constituents impose stability on proteins or organelles that are otherwise irreversibly denatured or damaged by exposure to the more extreme conditions. These adaptive changes, however, are not permanent so that when plants are restored to more clement conditions, the process of acclimation is reversed and sensitivity to stress conditions returns.

In this chapter the molecular mechanisms that have been invoked to explain freezing-induced injury to plant membranes will be considered. Evidence, which suggests how the changes associated with acclimation may mitigate the effects of exposure to freezing conditions, will be reviewed. Finally, an account will be provided of the possibilities for improving frost tolerance of plants by genetic methods.

Stability of membranes subjected to low temperatures

Loss of functional activity of cell membranes is believed to be one of the primary mechanisms of cell death resulting from exposure of susceptible plants to frost conditions. Unlike biopolymers such as proteins, nucleic acids and polysaccharides, the constituents of biological membranes are not linked together by covalent bonds. Instead, the assembly of polar lipids and proteins is stabilised by entropic forces, resulting in the orientation of the the molecules into a two-dimensional array with the creation of a hydrophobic domain within the structure from which water is excluded. The polar lipids are the most conspicuous amphipaths and, because of their relatively low critical micelle concentration, they form characteristic assemblies when dispersed in aqueous systems. In biological membranes they are said to form a fluid bilayer matrix which serves to orient and support the different membrane proteins.

The composition of lipid molecular species, even of membranes that perform relatively simple functions is highly complex, and individual molecular species may number hundreds in particular biomembranes. Although the bilayer arrangement appears to be the dominant phase of lipids present in most biological membranes (Blaurock, 1982), this is not necessarily the phase preferred by individual molecular species of lipid isolated from biological membranes. If polar lipids extracted from biological membranes are separated into molecular species and then are dispersed individually in excess water or dilute salt solutions, for example, they are found to form one of a number of well-characterised phases (Luzzati, 1968). These phases include, with increasing temperature, bilayer phases in which the hydrocarbon chains are arranged in either a crystal lattice, a gel phase or a disordered liquid-crystal configuration. In addition, a number of non-bilayer phases, such as hexagonal-II or cubic phases are also found to exist at temperatures approximating that of the growth temperature of the organism from which the lipid was extracted (Lindblom & Rilfors, 1989; Seddon & Templer, 1993). Indeed, in the case of the major polar lipid of the photosynthetic membrane of higher plants, monogalactosyldiacylglycerol, the isolated lipid exists in a hexagonal-II structure at temperatures less than −30 °C (Shipley, Green & Nichols, 1973). Since the matrix of biomembranes is said to be represented by a fluid bilayer of polar lipids, the tendency of a proportion of these lipids to form different phases under physiological conditions and thus separate into distinct structural domains (Tenchov, 1985; Quinn, 1987) must be subverted by the interaction of such lipids with other membrane components.

The reason why the membrane lipid matrix is rendered inherently unstable by the presence of non-bilayer forming lipid constituents is thought to be because of the exigencies of certain physiological functions. Fusion between membranes, which is necessary for cellular growth, differentiation and reproduction, for example, must involve destabilisation and formation of non-bilayer lipid phases (Ellens *et al.*, 1989; Siegel *et al.*, 1989). In other situations, non-bilayer lipids may serve to seal the interface between the bilayer lipid matrix and the intrinsic membrane proteins and prevent leakage of solutes (Quinn & Williams, 1983). Finally, shifting poikilothermic cells or organisms to different or even hostile environmental conditions brings about phase changes in the membrane bilayer matrix. Such phase changes may be reversible and the distribution of membrane components are restored when the cells are returned to physiological environments. When the conditions are sufficiently severe, the phase changes may be irreversible and cell death ensues.

Evidence for irreversible phase separations associated with exposure of biological membranes to low temperatures is shown in Fig. 1. When a suspension of chloroplasts is frozen to liquid nitrogen temperatures and thawed before thermally quenching from 20 °C, phase separations of non-bilayer lipids presumably oriented into tubular inverted micelles can be observed within the chloroplast structure. The sequence of events which may be responsible for phase separation of the non-bilayer forming lipids is illustrated in the cartoon presented in Fig. 2. This shows that, at physiological temperatures, the non-bilayer forming lipids are constrained in a bilayer arrangement by their interaction with the intrinsic membrane proteins. Upon cooling, the first transition that the membrane lipids experience is a non-bilayer to lamellar liquid-crystal transition. The non-bilayer forming lipids would no longer preferentially interact with the intrinsic membrane proteins but would be interchangable with the fluid bilayer forming lipids. This phase change is only likely to alter solute permeability across the membrane, but would be reversible when the temperature is increased again to the growth temperature. Further cooling induces a phase transition from lamellar liquid-crystal to lamellar gel phase. Because of differences in hydration of the polar groups of bilayer and non-bilayer-forming lipids, the temperature of this transition is higher for non-bilayer-forming lipids, with equivalent acyl chain composition (cf Fig. 3). The consequence is that non-bilayer-forming lipids tend to phase-separate into a domain of gel

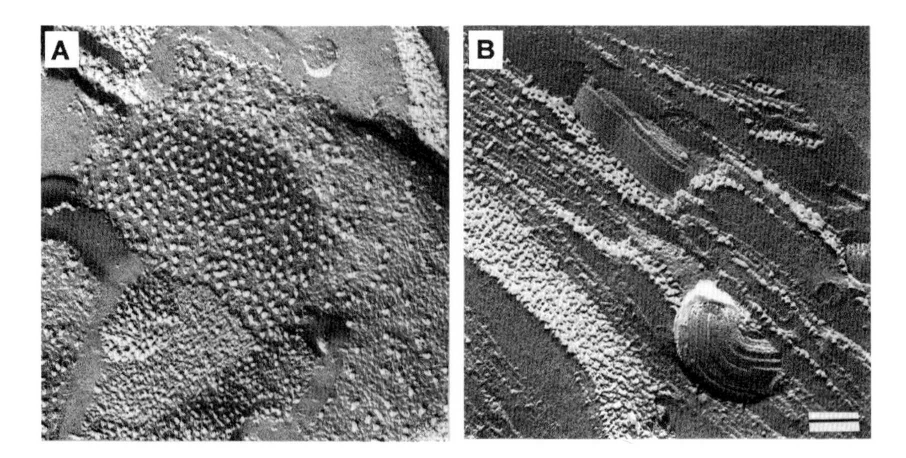

Fig. 1. Non-bilayer lipid phase separations in higher plant chloroplasts. A, chloroplasts thermally quenched from 20 °C; B, chloroplasts frozen to −196 °C and thawed to 20 °C prior to thermal quenching from 20 °C. The magnification bar corresponds to 100 nm.

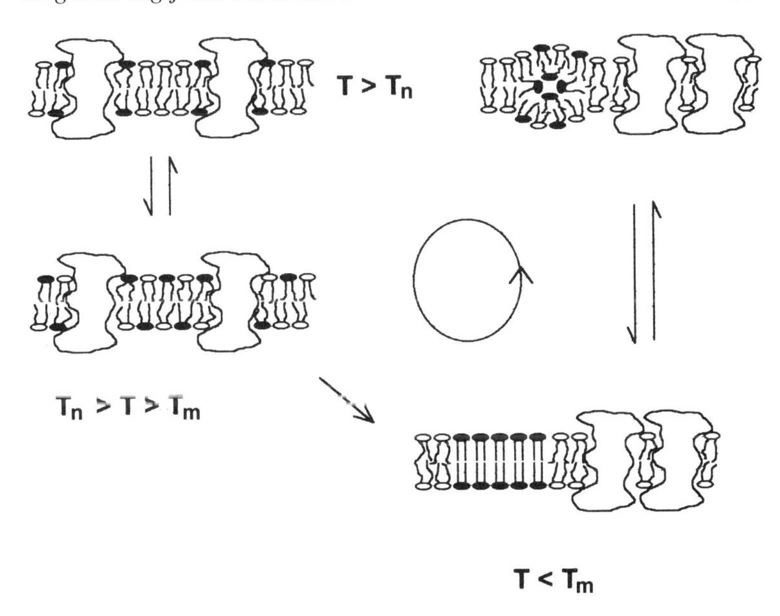

Fig. 2. A phase separation model of irreversible structural changes in biological membranes during cooling and rewarming. T_n, temperature of transition non-bilayer–bilayer phase; T_m, temperature of transition gel–liquid–crystal bilayer. Redrawn from Quinn (1985).

phase lipid from which the intrinsic membrane proteins are excluded together with the bilayer-forming lipids. The formation of domains of phase-separated lipids in bilayer gel structures can be observed in chilling-sensitive organisms cooled to low temperatures above 0 °C (Furtado *et al.*, 1979). When the temperature is subsequently increased to the physiological range, the non-bilayer lipid is no longer constrained into bilayer arrangement by interaction with the protein and adopts a non-bilayer structure.

The changes in membrane lipid composition associated with acclimation of rye and oat has been correlated with ultrastructural lesions in membranes identified by freeze–fracture electron microscopy (Steponkus, Lynch & Uemura, 1990; Steponkus, Uemura & Webb, 1993). In non-acclimated plants, phase separation of tubular inverted micelles, which may represent a hexagonal-II structure, are formed as a consequence of freeze-induced fusion of the plasma membrane with membranes of subcellular organelles (Webb & Steponkus, 1993). Acclimation results in an increase in the ratio of phospholipids to cerebrosides of the plasma membrane, which is associated with a decrease in the threshold temperature at which so-called fracture-jump lesions, which are believed to result from fusion between the plasma membrane and endomembranes (primarily the chloroplast envelope), are observed (Fujikawa, 1990; Fujikawa & Steponkus, 1990). Phase separation of non-bilayer lipid into recognisable hexagonal-II structures does not occur in acclimated leaves and protoplasts.

The existence of such lipid in a hexagonal-II type arrangement, although resulting in a major perturbation of structure, is unlikely to be the primary cause of fatal lesions. The most serious situation exists when the non-bilayer lipid phase separates into inverted micelles sandwiched within the bilayer. These structures, which are likely to precede large-scale lipid phase separations, are believed to be responsible for membrane fusion and possibly for rendering the membranes permeable to soluble macromolecules.

Factors governing lipid phase separations in membranes

As we have seen, the induction of lipid phase separations in cell membranes can lead to irreversible changes and ultimately to cell death. The potential for lipids to form separate domains resides in the fact that the lipids are highly diverse in their chemical structure, and a proportion of them do not form a stable fluid bilayer in excess water. The phase behaviour and structural properties of a membrane are therefore largely influenced by the overall composition of lipids represented in the membrane. Membrane lipids that form bilayer or non-bilayer structures when dispersed in excess water or dilute salt solutions at physiological temperatures are listed in Table 1. The ratio of bilayer-forming and non-bilayer-forming lipids varies from one membrane to another, but molecular species of both type of lipid are invariably present.

The role of acidic lipids

Studies of total polar lipid extracts of plant membranes in aqueous media have revealed that charged lipids exert a marked influence on the phase formed (Gounaris *et al.*, 1983). When such extracts are dispersed in distilled water, only small single bilayer vesicles are formed. On addition of inorganic salts or reducing the pH, which serves to screen the charges on the acidic lipids, the vesicles firstly undergo fusion followed by structural rearrangements with increased screening that result in the appearance of three-dimensional arrays of non-bilayer lipid structure.

Similar phase separations of non-bilayer lipid structures are observed

Table 1. *Bilayer-forming and non-bilayer-forming lipids*

Bilayer-forming lipids	Non-bilayer-forming lipids
phosphatidylcholine (PC)	phosphatidylethanolamine (PE)
phosphatidylglycerol (PG)	monohexosyldiacylglycerol
phosphatidylserine (PS)	diphosphatidylglycerol (cardiolipin)/M^{2+}
phosphatidylinositol (PI)	diacylglycerol (DAG)
dihexosyldiacylglycerol	phosphatidic acid (PA)/M^{2+}
sulphoquinovosyldiacylglycerol	monogalactosyldiacylglycerol (MGalDG)
cerebroside	monoglucosyldiacylglycerol (MGlcDG)
sphingolipids	

when intact plant membranes, such as the mitochondria, are treated with high concentrations of manganese (Van Venetie & Verkleij, 1982) or chloroplasts when exposed to high magnesium concentrations (Carter & Staehelin, 1980). Suspension of chloroplasts in medium of pH 4.5–5.0 or subjecting them to phospholipase A_2 digestion to remove the small proportion of phosphatidylglycerol also serves to induce a phase separation of non-bilayer lipid structures (Thomas *et al.*, 1985). In all cases of this type so far reported, the appearance of phase separated non-bilayer structure requires the close proximity of more than one destabilised membrane, presumably because a relatively large amount of lipid is needed to create a domain of sufficient size to be visible by the freeze–fracture method. A similar conclusion has been derived from studies of freeze-stressed protoplasts (Gordon-Kam & Steponkus, 1984).

Lipid unsaturation

One of the characteristic features of membrane lipids is the presence of unsaturated fatty acid chains. These unsaturated residues are introduced into the saturated products of fatty acid synthetases by specific desaturases. With the exception of a unique *trans*Δ^3 hexadecenoic acid acylated to the C-1 of phosphatidylglycerol found as a relatively minor component of some plant membranes, the double bonds are invariably in the *cis* configuration. The first double bond is inserted in the middle of the chain at the Δ^9 position, and subsequent desaturations take place sequentially at Δ^{12} and ω^3 positions. This methylene-interrupted configuration stabilises the structure and reduces susceptibility to oxidation.

The consequences of the presence of double bonds in the hydrocarbon chains on the phase behaviour of lipids in aqueous dispersions is dramatic. This is illustrated in Fig. 3 which compares the lamellar gel to liquid-crystalline (Lα → Lβ) phase transition temperature of a bilayer-forming lipid (phosphatidylcholine) with different numbers of *cis* unsaturated bonds. The transition temperature decreases by more than 50 °C with the introduction of a single double bond and even further when two double bonds are present. The same trend is also observed in non-bilayer-forming lipids (phosphatidylethanolamine) and in this case the effect is reflected in the temperature of transition between bilayer and non-bilayer phases (Lα → Hex-II) as well.

The behaviour of the saturated and unsaturated lipids can be rationalised on the basis of the conformation of the molecules when they pack into lipid structures. Thus the angle between carbon atoms linked by a single covalent bond is 111 °C whereas the angle beteween the single C–C bond and a *cis* double bond is 123°. When the hydrocarbon chain is in a fully extended configuration, the presence of the *cis* bond creates

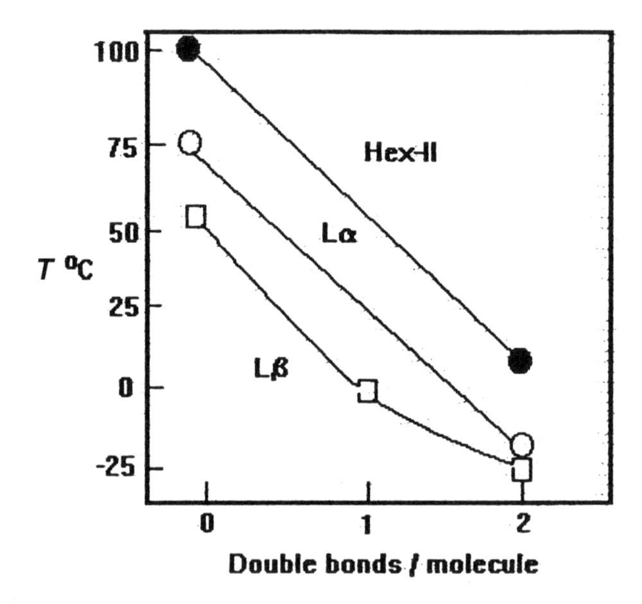

Fig. 3. The effect on the temperature of Lβ phase to α phase transitions of the presence of *cis* double bonds in C-18 derivatives of phosphatidylcholine (□) and phosphatidylethanolamine (●, ○).

a kink of about 30° because of the restricted rotation about the double bond and the failure of a combination of *trans* and *gauche* ± rotamers about the remaining single C–C bonds to compensate for the disjuncture. As a consequence, the molecules are not able to pack together with other fatty acid chains, especially if the double bonds are located in the same position with respect to the functional group of the fatty acid. The location of the first double bond in the centre of the chain causes the maximum disturbance compared to locations towards either end of the chain. This packing constraint is reflected in a slightly increased average cross-sectional area of the hydrocarbon chain. It is noteworthy that *trans*-double bonds have very little effect on the melting point of hydrocarbons, and this is related to the ability of the chains to interact throughout their length by van der Waals attractive forces (Murata & Yamaya, 1984).

Effect of solutes

Some plant species synthesise polyhydroxyalcohols and free amino acids under extreme conditions. Species of lichen have become adapted to extreme drought and can survive dehydration for a few months retaining only 1% or 2% by weight of water. In the case of *Lichina pygmeae*, this is achieved by accumulation of mannosidomannitol; the thermal stability of the lichen correlates with the amount of this polyol (Feige, 1975).

Many salt-tolerant plants accumulate high concentrations of proline as a dominant intracellular osmolite (Treichel *et al.*, 1984). In *Triglochin maritima L.* the content of proline reaches 70% of the free amino acid pool (Stewart & Lee, 1975). Many forest species in the far North synthesise trisaccharides (raffinose and stachiose) at low temperatures. The freezing temperature of the intracellular water decreases in a manner dependent on the concentration of these polyols, allowing survival in conditions of deep undercooling (Newsted, Chibbar & Georges, 1991; McCue & Hanson, 1992; Nolte, Hanson & Gage, 1997). These plants are known to restrict ice formation to the extracellular fluid.

High salinity induces synthesis of glycine betaine in sugar beet and spinach. The activity of the key enzyme in glycine betaine biosynthetic pathway (betaine aldehyde dehydrogenase) increases two- to four-fold in leaves and roots as the NaCl level is raised from zero to 0.5M (George, Becwar & Burke, 1982). This increase in betaine aldehyde dehydrogenase activity was found to be accompanied by an increase in the level of translatable mRNA of the enzyme. Similarly, sucrose synthase was accumulated in spring and winter wheat plants exposed to low temperature (2 °C) (Hirsh, 1987).

Compatible solutes have a relatively small effect (0–3 °C) on temperatures of the various phase transitions of bilayer-forming lipids. Their effect on non-bilayer-forming lipids, however, is more dramatic (Tsvetkova & Quinn, 1994). The effect of sugars, amino acids and polyols is to stabilise the subgel (L_c) and gel (L_β) phases and the non-lamellar (H_{II}) phase at the expense of the lamellar liquid-crystalline (L_α) state. This effect had been noted previously in studies of dipalmitoylphosphatidylethanolamine in aqueous solutions of sugars and is consistent with a tendency to lower the surface potential energy of the system by decreasing the extent of an energetically unfavourable interaction between the lipid headgroups and the surrounding medium.

The effect of compatible solutes is exemplified by the presence of sugars on phase transitions in phosphatidylethanolamines. This is illustrated in Fig. 4. The temperature of the $L_\beta \rightarrow L_\alpha$ and $L_\alpha \rightarrow H_{II}$ transitions in distearoylphosphatidylethanolamine are 74 °C and 100 °C, respectively. Calorimetric scans of the phospholipid dispersed in sucrose and trehalose solutions show that the presence of disaccharides shift the $L_\alpha \rightarrow H_{II}$ transition to lower temperatures and at the same time increase slightly the temperature of the L_β–L_α transition. In 2.4 M sucrose these two transitions merge and convert to a single transition centred at 79 °C (Fig. 4). This transition is expected to reflect a direct $L_\beta \rightarrow H_{II}$ phase transformation.

Polyhydroxy compounds and some di- and tri-substituted amines are known as 'kosmotropic', water-structure-making reagents (Collins & Washabaugh, 1985). Koynova, Tenchov and Quinn (1989) have proposed that these solutes interact indirectly via Hofmeister effects to stabilise the structure of bulk water, thereby increasing interfacial tension, and consequently their presence in the aqueous phase results in a tendency to reduce the area of unfavourable interface between aqueous and lipid phases. This tendency would be expressed in a preferential formation of the more closely packed lipid phases (subgel, gel and inverted hexagonal), with less surface area per polar group, relative to the more expanded phases (lamellar liquid-crystalline).

Hydration

Under freezing conditions, ice is nucleated and ice crystals grow as liquid water is frozen. The removal of liquid water serves to dehydrate solutes and organelles which are not incorporated into the ice. The

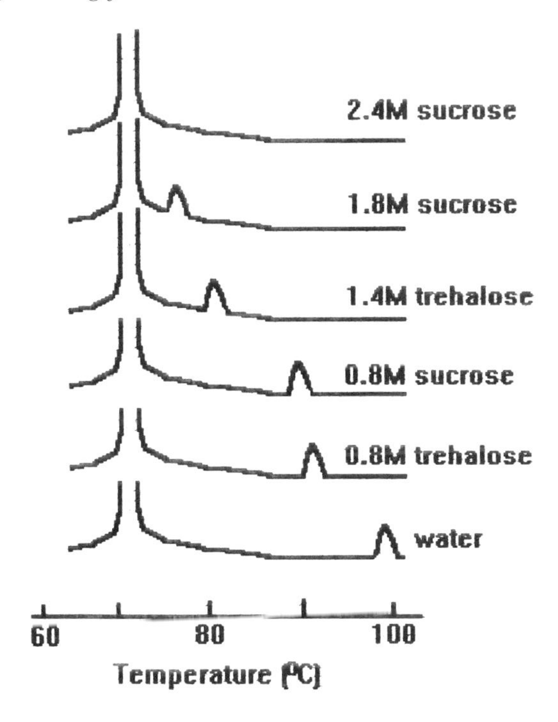

Fig. 4. Differential scanning calorimetric heating thermograms of distearoylphosphatidylethanolamine dispersed in aqueous solutions of trehalose or sucrose. Data derived from Koynova *et al.* (1989).

extent to which the cell can tolerate freeze-induced desiccation due to the crystallisation of water is generally reckoned to be the principal factor that determines survival of the cell under frost conditions (Uemura & Steponkus, 1989).

The phase behaviour of membrane lipids is determined by the activity of water in which they are in contact, i.e. they exhibit lyotropic mesomorphism. Membrane lipids that form non-bilayer structures have polar groups that are, in general, less hydrated than their bilayer-forming counterparts. Paradoxically, the hexagonal-II phase in excess water contains the greatest amount of water per lipid. Water activity, however, plays an important role in the stability of non-bilayer-forming lipids such that non-bilayer phases are favoured in limiting water conditions at equivalent temperatures. This is exemplified by the study of Brentel, Selstam and Lindblom (1985) who examined the phase structure of mixtures of monogalactosyl- and digalactosyldiacylglycerides dispersed in differing amounts of ^2H_2O using NMR methods. Mixtures of galactolipids isolated from chloroplasts were prepared in molar ratios of 1 : 2, 1.2 : 1 and 2 : 1 and codispersed in different amounts of ^2H_2O. Fig. 5 shows the phase properties of the mixture containing a mole ratio of 2 : 1 monogalactosyl- : digalactosyldiacylglycerol, a ratio that is found in the chloroplast thylakoid membrane, over the temperature range 10–40 °C and for water contents up to 14 mol ^2H_2O/mol lipid. At low water content, the hexagonal-II phase is seen to dominate the system. As the concentration of water is increased, a reversed cubic liquid-crystalline phase gradually replaces the hexagonal-II and, in turn, is replaced by a liquid-crystalline lamellar phase in excess water. With increasing proportions of monogalactosyldiacylglycerol, the non-lamellar regions of the phase diagram tend to increase. The reversed cubic liquid-crystalline phase was characterised as a bicontinuous cubic phase structure.

The relevance of water activity in freezing processes can be viewed in the context of the extraction of intracellular water across the cell membranes, where it is crystallised as extracellular ice. The consequence of freeze-induced dehydration is that the water activity is

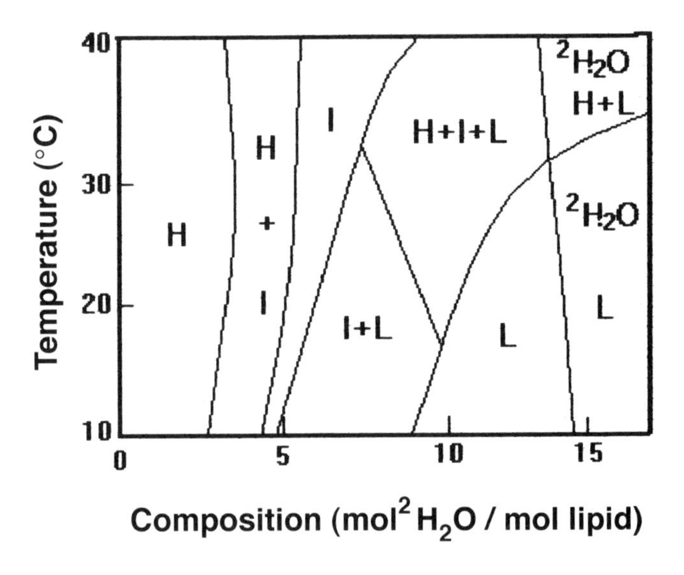

Fig. 5. A composition–temperature phase diagram of a 2 : 1 mol ratio mixture of monogalactosyl- and digalactosyldiacylglycerol dispersed in ^2H_2O. H, hexagonal-II phase; I, reversed cubic phase; L, lamellar phase. All chain configurations are liquid-crystalline. Data from Brentel *et al.* (1985).

reduced to low levels inside the cell and dissolved solutes are concentrated by a process of zone refining at membrane interfaces. Apart from the direct effect of water activity on the stability of lipid phase structure, the concentration of electrolytes is likely to disturb the electrostatic potential at the membrane surface and hence alter the influence of acidic lipids on constraining the non-bilayer lipids into a bilayer configuration.

Genetic scope for engineering frost resistance

Classic genetic methods have established the basic genetic foundation of frost tolerance in many plant species, especially in cereals (Roberts, 1990). Sutka (1994) has concluded on the basis of an analysis of genetic determinants of frost tolerance in wheat that it is controlled by an additive-dominance system in which both additive and non-additive gene action are important in inheritance of the trait. Dominant genes act to reduce frost tolerance, and recessive genes exhibit a reverse trend. At least half the 21 pairs of chromosomes are involved, with chromosomes 5A and 5D being the most frequently implicated.

Membranes as a target for frost tolerance

The physical properties of the lipid matrix reflected in the particular lipid composition of cell membranes are governed by the concerted actions of lipid synthesising and degrading enzymes. In higher plant cells there are numerous morphologically distinct membranes, which possess distinct identities with respect to both the amount and type of proteins and the mixture of polar lipids comprising the membrane matrix. Attempting to establish the factors which govern the regulation of membrane composition via the enzymes of pathways responsible for synthesis of individual components is complicated by the process itself of expansion and differentiation of membranes as well as by the possibility of exchange of lipids between different membranes.

Control of membrane lipid composition

To obtain an understanding of the principles underlying regulation of membrane composition, the response of simple organisms to changes in environmental factors has been examined in some detail. One of the favoured organisms for such studies is the unicellular organism *Acholeplasma laidlawii* which has only one cell membrane. Studies of this organism have shown that the response to changes in environmental conditions, including low temperature, is to tailor the lipid composition so as to preserve a constant surface charge density, a similar phase equilibrium and maintain a spontaneous curvature parameter determined by the polar lipid mixture within relatively narrow limits (Lindblom *et al.*, 1986; Karlsson *et al.*, 1996, 1997). The suggestion that spontaneous curvature of lipid structures is a functionally important membrane parameter, subject to regulation by cells and is one of the constraints controlling lipid composition of the membrane lipid matrix, was first proposed by Gruner (1989). As already noted, interaction of non-bilayer-forming lipids with charged lipids and intrinsic membrane proteins tends to impose a bilayer configuration on lipids that would, when dispersed alone in aqueous systems, form a non-bilayer structure. The proportion of non-bilayer lipid is likely therefore to be influenced by the ratio of lipid : protein as well as to other bilayer-forming lipids.

As noted above, acclimation of rye and oat result in changes in plasma membrane lipid composition, which correlates with ultrastructural changes induced by freezing. The increased proportion of phospholipids relative to cerebrosides in the plasma membranes of acclimated rye cells was rationalised to increase tolerance to freezing by retaining more interfacial water at the membrane surface thereby preventing freeze-induced dehydration and induction of non-lamellar phase transitions (Steponkus, Uemura & Webb, 1993). Uemura & Steponkus (1997) have recently extended their studies to determine the effect of acclimation of rye leaves on the lipid composition of the inner and outer membrane of the chloroplast envelope. Contrary to earlier observations on *Acholeplasma laidlawii*, temperature acclimation of rye plants results in an increase in the proportion of bilayer at the expense of non-bilayer forming lipid in the envelope membranes of the chloroplasts. It was suggested that the increased stability of the lamellar configuration at low temperature was a major factor contributing to improved freezing tolerance. Nevertheless, some caution may be warranted in reaching such a conclusion because other factors such as the extent of unsaturation of the lipids, lipid–protein ratios, etc. also contribute to the overall phase stability of the membrane. This is of particular relevance in the light of a recent report of cold-regulated genes (COR) in *Arabidopsis*, which code for polypeptides targeted to the chloroplasts and enhance freezing tolerance of protoplasts by lowering the temperature of lamellar to non-lamellar phase transition of membrane lipids (Artus *et al.*, 1996).

Examination of the factors that influence activity of enzymes responsible for synthesis of bilayer and non-bilayer-forming lipids has indicated that the level of activity necessary to preserve the phase equilibrium within optimal limits is controlled at the level of the membrane. The enzymes thus appear to be responsive to the physical state of the membrane matrix in which they operate. Thus the maintenance of membrane composition within relatively narrow limits and its adaption to varying environmental conditions exploits biochemical processes that are able to sense the phase state of the lipid bilayer matrix and perform the necessary homeostatic adjustments. This implies that the genetic control of the relative proportions of the different lipid classes is rather indirect and therefore there is probably little scope for direct genetic manipulation of the relative proportions of lipid classes to enhance tolerance of plants to low temperature.

Control of membrane lipid unsaturation

Plants grown at low temperatures show an increased level of unsaturation of their membrane lipids compared to plants grown at higher temperatures (Harwood *et al.*, 1994). The effect of this adjustment on lipid phase equilibrium, however, is often complicated by an associated change in the ratio of lipid : protein in the membranes (Palta *et al.*, 1993). The connection between growth temperature and alteration of lipid unsaturation has been examined in unicellular photosynthetic organisms by observing the response of the organism to saturation of membrane lipids at optimal growth temperatures using membrane-impermeable homogeneous catalysts (Quinn, Joo & Vigh, 1989). In studies of *Dunaliella salina*, Vigh, Horvath and Thompson (1988) hydrogenated cells briefly so that only lipids of the plasma membrane were saturated and observed the effects on growth rates. They found that the cells stopped growing immediatly but growth resumed after about 12 hours, by which time the level of unsaturation of the membrane lipids was restored to their original levels. More recently, Vigh *et al.* (1993) applied the method to *Synechococcus* PCC6803 and showed that hydrogenation of the plasma membrane lipids of the intact cyanobacterium stimulated expression of the desA gene coding for proteins involved in lipid unsaturation.

Nine mutants of the overwintering annual *Arabidopsis thaliana*, defective in membrane lipid unsaturation, have been isolated (Somerville & Browse, 1996). The effect of mutations of two of these genes affecting lipid unsaturation in the chloroplast on chloroplast ultrastructure and chilling sensitivity has been reported. The fad5 mutation involves a single nuclear mutation causing a defect in the activity of chloroplast Δ^9-desaturase responsible for insertion of a double bond into 16 : 0 fatty acyl residues of monogalactosyldiacylglycerol (Kunst, Browse & Somerville, 1989*a*, *b*). The chloroplast membranes of this mutant are characterised by an increased proportion of palmitic acid and a corresponding decrease in unsaturated 16 : 0 fatty acids in the lipids. Another mutation, fad6, results in a defect in the activity of Δ^{12} -desaturase responsible for introducing double bonds into 16-carbon and 18-carbon fatty acids of monogalactosyl- and digalactosyldiacylglycerols (Browse *et al.*, 1989; Hugly *et al.*, 1989). The chloroplast membranes of these mutants contain relatively high proportions of monoenoic fatty acids, as further desaturation of these molecular species is prevented by the defective allele. Both these mutants become sensitive to chilling and are no longer able to acclimate for frost tolerance (Hugly & Somerville, 1992). Another mutation, fad2, causing a defect in Δ^{12} -desaturase affecting the insertion of a double bond into oleoyl residues of extrachloroplastic membranes, also produces plants with a phenotype that does not allow them to survive long exposure to low, non-freezing, temperatures (Miguel *et al.*, 1993). Examination of the ultrastructure of chloroplast membranes of the fad5 and fad6 mutants showed that the protein complexes of the photosynthetic membrane tended to dissociate and become aligned into characteristic arrays within the membrane (Tsvetkova *et al.*, 1994, 1995). The conditions of array formation are illustrated in Fig. 6. Wild-type *Arabidopsis* shows a random orientation of membrane-associated particles in exoplasmic and protoplasmic fracture planes in both appressed and non-appressed regions of the thylakoid membranes when thermally quenched from around the growth temperature (20 °C). Particles form into characteristic arrays in the appressed region of the membrane when the chloroplasts are cooled to 4 °C. This effect is seen at 20 °C in the mutant strains, indicating that the unsaturated lipid is responsible for preventing alignment of particles at the growth temperature. The alignment of particles is likely to be associated with low-temperature photoinhibition typical of photosynthetic membranes containing more saturated lipids (Tasaka *et al.*, 1996), and constraint of photosystem-II complexes in these arrays may inhibit processing of the D1 protein (Oquist *et al.*, 1995).

Several studies from Murata's laboratory (Murata *et al.*, 1992; Murata & Wada, 1995; Nishida & Murata, 1996) reported the prodution of transgenic plants with altered levels of polyunsaturated membrane lipids and sensitivity to exposure to chilling temperatures. The results of these experiments are illustrated in Fig. 7. Tobacco plants, which are considered to have a chilling sensitivity intermediate between Squash and *Arabidopsis* (chilling sensitive and resistant, respectively), when transformed with acyl-ACP : glycerol-3-phosphate acyltransferase from either squash or *Arabidopsis* became more chilling sensitive or resistant, respectively. This was correlated in each case with decreased, or

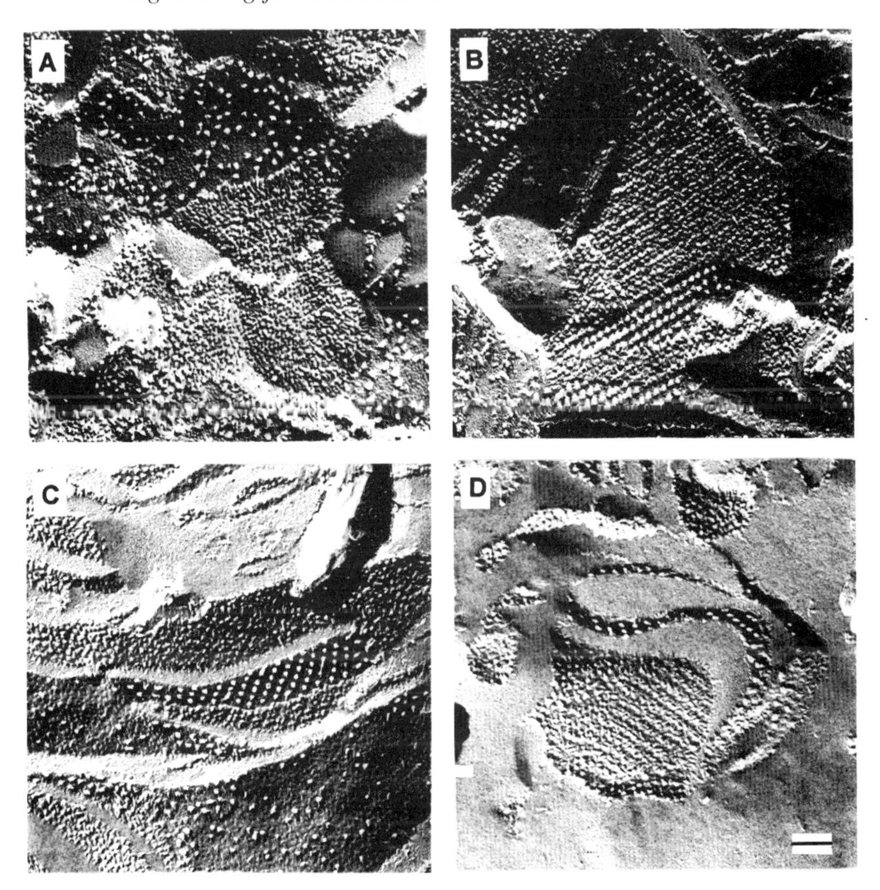

Fig. 6. Electron micrographs of freeze–fracture replicas of chloroplasts showing the effect of temperature on particle distribution in thylakoid membranes. A, Wild-type *Arabidopsis* and B, chloroplasts from plants with fad6 mutation thermally quenched from 20 °C. C, D, Wild type *Arabidopsis* equilibrated and thermally quenched from 4 °C. The magnification bar corresponds to 100 nm.

increased, levels of polyunsaturated membrane lipids. Cold tolerance in tobacco is also enhanced by expression of a gene for chloroplast Δ^3-desaturase from *Arabidopsis* (Kodama *et al.*, 1994, 1995) and Δ^9-desaturase from *Anacystis nidulans* (a chilling sensitive organism) (Ishizakinishizawa *et al.*, 1996). Transformation of *Arabidopsis* with *Escherichia coli* acyl-ACP : glycerol-3-phosphate acyltransferase which

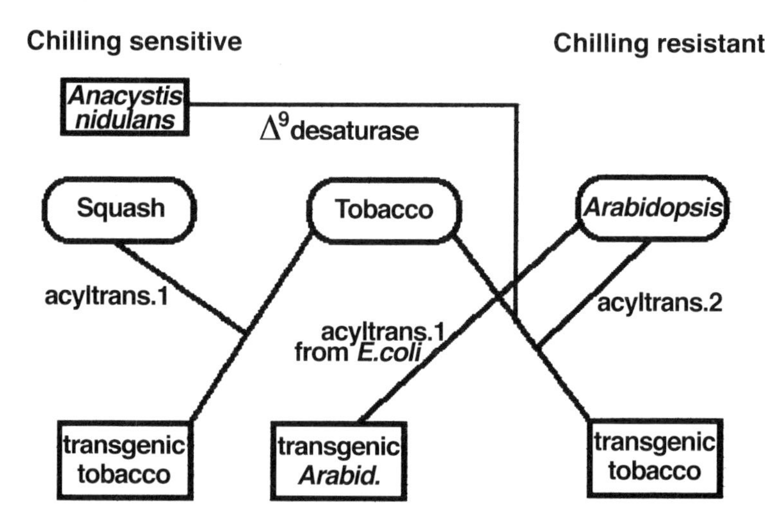

Fig. 7. Genetic manipulation of membrane lipid unsaturation of *Nicotinia tabacum* and *Arabidopsis* and the consequence of exposure of the transgenic plants to chilling conditions.

is unable to discriminate between 16 : 0-ACP and 18 : 0-ACP has been reported by Wolter, Schmidt and Heinz (1992) to increase the proportion of saturated molecular species of phosphatidylglycerol and confer chilling sensitivity on the transgenic plants.

The molecular species of the two acidic lipids of chloroplast thylakoid membranes, phosphatidylglycerol and sulphoquinovosyldiacylglycerol, tend to be more saturated than other lipid classes (Murata & Hoshi, 1984; Tulloch, Heinz & Fischer, 1973). In the case of sulphoquinovosyldiacylglycerol, no correlation was found to exist between the extent of unsaturation of this lipid and the chilling sensitivity of the plant from which it originated. By contrast, several studies reported a correlation between the contents of the 16 : 016 : 0 plus 160 : 16 : 1 (Δ^3-*trans*) molecular species of phosphatidylglycerols and chilling sensitivity (Murata *et al.*, 1982; Murata, 1983; Roughan, 1985). The liquid-crystalline to gel phase transition temperature of these molecular species is above chilling temperature and the rationale explaining the lesion was that the presence of gel phase lipid resulted in an increased solute permeability across the membrane. More recently, Wu and Browse (1995) examined the chilling sensitivity of the fab1 mutant of *Arabidopsis* which is defective in acyl-ACP : glycerol-3-phosphate acyltransferase activity and found that, although the level of saturated molecular species of phosphatidylglycerols represented 43% of the total phospholipid and exceeded that found in many chilling sensitive plants, the mutants were not chilling sensitive. The long-term survival of the mutant strain at 2 °C, however, was affected suggesting that the level of unsaturation of the phosphatidylglycerols may only be one of a number of factors controlling chilling sensitivity in this plant.

It must be emphasised that chilling resistance and frost tolerance may involve separate genetic factors, although it is clear that frost tolerance must involve tolerance also of chilling. Conversely, chilling resistance may not confer tolerance to freezing conditions.

Regulation of compatible solutes

Many organisms accumulate compatible solutes under stress conditions, but the genetic regulation of this trait has not yet been exploited to enhance freezing tolerance. A recent report, however, has demonstrated that chilling tolerance of the cyanobacterium *Synechococcus* PCC 7942 can be enhanced by transformation with the codA gene for choline oxidase from *Arthrobacter* under control of a constitutive promoter (Deshnium *et al.*, 1997). The transformed cells accumulated glycine betaine, reaching 60–80 mM concentration in the cytoplasm, and grew at 20 °C, a temperature lower than for efficient growth by parental cells. Examination of thermotropic phase transition behaviour of the membrane lipids showed that the temperature at which the lipids of the plasma membrane were transformed from the liquid-crystalline to the gel state shifted to lower temperatures. There was no change in the level of unsaturation of the lipids so the reduction in phase transition temperature could be ascribed solely to the high concentration of glycine betaine. Studies of photosynthetic functions also showed that photosynthesis of the transformed cells at 20 °C was not inhibited by light as compared to untransformed cells because they were apparently able to recover from low-temperature photoinhibition.

This demonstration of enhancement of tolerance to low temperature by modulation of the concentration of compatible solutes in cyanobacteria is likely to pave the way for similar experiments in higher plants.

Amelioration of freeze-induced dehydration

Under freezing conditions, frost-tolerant plants exhibit ice nucleation first on the surface of the cells which then subsequently spread from these foci throughout the extracellular areas of the plant tissues (Guy, 1990). Slow cooling of frost-tolerant plants does not usually result in formation of intracellular ice crystals, unlike rapid cooling in which case formation of intracellular ice often proves lethal. The prevention of intracellular ice formation appears to be associated with undercooling and progressive dehydration as water is drawn to the growing ice crystals on the outside (Pearce, 1988).

The formation of small crystals of extracellular ice at relatively high temperatures in frost-tolerant plants is known to be mediated by a special group of complex molecules referred to as ice nucleators. This heterogeneous group of compounds initiate ice formation at specific sites in tissues at temperatures above −20 °C by orienting water molecules into the crystal lattice of ice (Saki & Larcher, 1987). The most efficient ice nucleators are those which induce ice nucleation at the highest temperature (Mueller, Wolber & Warren, 1990; Burke & Lindow, 1990).

There are two sources of ice nucleators: those produced by the plant itself and others produced by epiphytic bacteria colonising the surface of the plant. Some of these bacteria produce highly efficient ice nucleators that are believed to play a role in frost tolerance of annual plant species (Lindow *et al.*, 1978). Endogenous ice nucleators have also been identified independently of epiphytic sources in tissue of perennials such as the stems of *Prunus* (Gross, Proebsting & MacCrindale-Zimmerman, 1988) and *Citrus* fruit (Constantinidou & Menkissoglu, 1992). In overwintering plants the ice nucleation activity observed under frost conditions is likely to be a combination of effects of ice nucleators produced by epiphytic bacteria and those synthesised endogenously by the plant.

A comparison of the ice nucleation efficiency of epiphytic bacteria and endogenous ice nucleators from the leaves of cold-hardened winter rye leaves (*Secale cereale*) is shown in Fig. 8. Ice nucleators from the two sources can be clearly distinguished from the temperature at which ice begins to form in aqueous systems, with the most efficient ice nucleation taking place in the endogenous preparations.

Characterisation of intrinsic ice nucleation factors produced by winter rye has been reported by Brush, Griffith and Mlynarz (1994). Factors with ice nucleation activity extracted from pectin-treated single mesophyll cell suspensions were found to contain proteins, lipids and carbohydrates of which the protein component appeared to be the dominant one. Treatment of the endogenous ice nucleating factors with protease enzymes, sulphydryl reducing agents, urea and exposure to denaturing temperatures reduced the efficiency of ice nucleation. This demonstrated that the protein component of these factors was an important functional ingredient. A particular requirement for free thiol groups was identified, and it was suggested that such groups may interact with ice

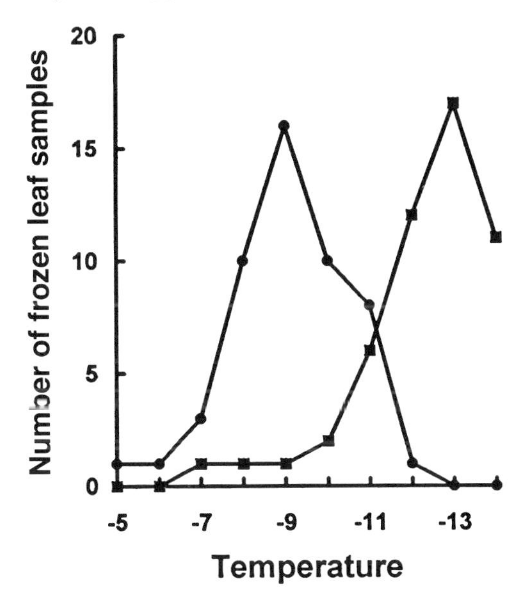

Fig. 8. Ice nucleation assay of acclimated winter rye leaves. Epiphytic ice nucleators were removed from the surface of leaves obtained from rye plants cold-hardened at 5 °C and the activity in nucleating ice formation (■) was compared with the remaining endogenous ice nucleating activity (●). Data obtained from Brush *et al.* (1994).

crystals in the same manner as had been suggested for hydroxyl groups (Kozloff, Schofield & Lute, 1983).

The importance of carbohydrates in ice nucleation activity was inferred from a reduction in ice nucleation activity after treatment with boric or periodic acids. Likewise, the effect of phospholipase-C treatment in reducing ice nucleation activity was suggested to provide evidence for the involvement of phospholipids.

Two species of ice-nucleating bacteria have been studied in some detail: *Pseudomonas syringae* and *Erwinia herbicola*. Studies of the agent responsible for ice nucleating activity in these bacteria have been reported (Kozloff *et al.*, 1991). The factor appears to be correlated with the presence of the phospholipid, phosphatidylinositol. When *Escherichia coli* were transfected with plasmids containing the ice gene DNA, they showed increased levels of activity of phosphatidylinositol synthase and displayed increased ice nucleating activity, which was proportional to the phosphatidylinositol content of their membranes.

Several avenues of opportunity for genetic manipulation of frost tolerance are evident, based on the strategy of ice nucleation activity. A more detailed knowledge of the molecular composition of particular ice nucleating factors is needed, but once this is established, attempts can be made to identify and clone the genes responsible for their production.

Conclusions

There is clear evidence that plants are able to acclimate when exposed to low temperatures to improve tolerance to freezing conditions. The process is seen to involve the up-regulation of specific genes or gene families. The function of many of the gene products remains obscure, but there are obvious effects of some. In respect to the acclimation of membranes, changes in membrane lipid compositon, extent of unsaturation of the fatty acyl chains and lipid : protein ratios have been described. Other strategies that appear to be exploited by plants in acclimation to low temperatures is the up-regulation and synthesis of cold-responsive proteins, some of which appear to affect the phase behaviour of the membrane lipids. A less direct approach is the biosynthesis of compatible solutes with kosmotropic actions on membrane lipid phase behaviour.

Other aspects of genetic regulation of freezing tolerance that are receiving attention include the role of abscisic acid. Recent studies of the treatment of unhardened wheat varieties with abscisic acid revealed that it caused the development of frost tolerance in a way that suggested abscisic acid was involved in an early stage of acclimation (Veisz, Galiba & Sutka, 1996). Abscisic acid has also been implicated in the development of both drought and cold tolerance in *Arabidopsis* and may participate in signal transduction pathways utilising protein kinases (Shinozaki & Yamaguchishinozaki, 1996). Another fruitful line of investigation is the role of heat shock proteins, which are believed to stabilise soluble proteins when they are subjected to unphysiological conditions (Hartl, 1996; Waters, Lee & Vierling, 1996).

Clearly, there are many avenues for investigation which may ultimately lead to regulation of genes responsible for cold tolerance in plants. The achievement of these goals will undoubtedly reap rich economic rewards.

Acknowledgements

The author thanks the numerous collaborators identified in the references cited in this review for their contribution to this work.

References

Artus, N.N., Uemura, M., Steponkus, P.L., Gilmore, S.J., Lin, C.T. & Thomashow, M.F. (1996). Constitutive expression of the cold-regulated *Arabidopsis thaliana* cor15A gene effects both chloroplast and protoplast freezing tolerance. *Proceedings of the National Academy of Sciences, USA*, **93**, 13 404–9.

Blaurock, A.E. (1982). Evidence of bilayer structure and of membrane interactions from X-ray diffraction analysis. *Biochimica et Biophysica Acta*, **650**, 167–207.

Brentel, I., Selstam, E. & Lindblom, G. (1985). Phase equilibria of mixtures of plant galactolipids – the formation of a bicontinuous cubic phase. *Biochimica et Biophysica Acta*, **812**, 816–26.

Browse, J., Kunst, L., Anderson, S., Hugly, S. & Somerville, C. (1989). A mutant of *Arabidopsis* deficient in chloroplast 16 : 1/18 : 1 desaturase. *Plant Physiology*, **90**, 522–9.

Brush, R.A., Griffith, M. & Mlynarz, A. (1994). Characterisation and quantification of intrinsic ice nucleators in winter rye (*Secale cereale*) leaves. *Plant Physiology*, **104**, 725–35.

Burke, M.J. & Lindow, S.E. (1990). Surface properties and size of the ice nucleation site in ice nucleation active bacteria: theoretical considerations. *Cryobiology*, **27**, 80–4.

Carter, D.P. & Staehelin, L.A. (1980). Proteolysis of chloroplast thylakoid membranes. II Evidence for the involvement of the light-harvesting chlorophyll *a*/*b*-protein complex in thylakoid stacking and for the effects of magnesium ions on photosystem II-light-harvesting complex aggregates in the absence of membrane stacking. *Archives of Biochemistry and Biophysics*, **200**, 374–86.

Collins, K.D. & Washabaugh, M.W. (1985). The Hofmeister effect and the behaviour of water at interfaces. *Quarterly Reviews of Biophysics*, **18**, 323–423.

Constantinidou, H.A. & Menkissoglu, O. (1992). Characteristics and importance of heterogeneous ice nuclei associated with citrus fruits. *Journal of Experimental Botany*, **43**, 585–91.

Deshnium, P., Gombos, Z., Nishiyama, Y. & Murata, N. (1997). The action *in vivo* of glycine betaine in enhancement of tolerance of *Synechococcus* sp. strain PCC 7942 to low temperature. *Journal of Bacteriology*, **179**, 339–44.

Ellens, H., Siegel, D.P., Alford, D., Yeagle, P.L., Boni, L., Lis, L.J., Quinn, P.J. & Bentz, J. (1989). Membrane fusion and inverted phases. *Biochemistry*, **28**, 3682–703.

Feige, G.B. (1975). Investigations on the ecology and physiology of the marine blue-green algae lichen *Lichina pygmaea* Ag. III Some aspects of photosynthetic C-fixation under osmoregulatoric conditions. *Zeitschrift für Pflanzenphysiologie*, **77**, 1–15.

Fujikawa, S. (1990). Cryoscanning electron microscope and freeze-replica study on the occurrence of slow freezing injury. *Journal of Electron Microscopy*, **39**, 80–5.

Fujikawa, S. & Steponkus, P.L. (1990). Freeze-induced alterations in the ultrastructure of the plasma membrane of rye protoplasts isolated from cold-acclimated leaves. *Cryobiology*, **27**, 665–6.

Furtado, D., Williams, W.P., Brain, A.P.R. & Quinn, P.J. (1979). Phase separations in membranes of *Anacystis nidulans* grown at different temperatures. *Biochimica et Biophysica Acta*, **555**, 352–7.

George, M.F., Becwar, M.R. & Burke, M.J. (1982). Freezing avoidance by deep undercooling of tissue water in winter-hardy plants. *Cryobiology*, **19**, 628–39.

Gordon-Kam, W.J. & Steponkus, P.L. (1984). Lamellar-to-hexagonal-II phase transition in the plasma membrane of isolated protoplasts after freeze-induced dehydration. *Proceedings of the National Academy of Sciences, USA*, **81**, 6373–7.

Gounaris, K., Sen, A., Brain, A.P.R., Quinn, P.J. & Williams, W.P. (1983). The formation of non-bilayer structures in total polar lipid extracts of chloroplast membranes. *Biochimica et Biophysica Acta*, **728**, 129–39.

Gross, D.C., Proebsting, E.L. & MacCrindale-Zimmerman, H. (1988). Development, distibution and characteristics of intrinsic, nonbacterial ice nuclei in *Prunus* wood. *Plant Physiology*, **88**, 915–22.

Gruner, S.M. (1989). Stability of lyotropic phases with curved interfaces. *Journal of Physical Chemistry*, **93**, 7562–70.

Guy, C.L. (1990). Cold-acclimation and freezing stress tolerance – role of protein metabolism. *Annual Review of Plant Physiology and Plant Molecular Biology*, **41**, 187–223.

Hartl, F.U. (1996). Molecular chaperones in cellular protein folding. *Nature*, **381**, 571–80.

Harwood, J., Jones, A.L., Perry, H.J., Rutter, A.J., Smith, K.L. & Williams, M. (1994). Changes in membrane lipids during temperature adaptation. In *Temperature Adaptation of Biological Membranes*, ed. A.R. Cossins, pp. 107–18. London: Portland Press.

Hirsh, A.G. (1987). Vitrification in plants as a natural form of cryoprotection. *Cryobiology*, **24**, 214–28.

Hughes, M.A. & Dunn, M.A. (1996). The molecular biology of plant acclimation to low temperature. *Journal of Experimental Botany*, **47**, 291–305.

Hugly, S. & Somerville, C. (1992). A role for membrane lipid polyunsaturation in chloroplast biogenesis at low temperature. *Plant Physiology*, **99**, 197–202.

Hugly, S., Kunst, L., Browse, J. & Somerville, C. (1989). Enhanced thermal tolerance of photosynthesis and altered chloroplast ultrastructure in a mutant of *Arabidopsis* deficient in lipid desaturation. *Plant Physiology*, **90**, 1134–42.

Ishizakinishizawa, O., Fujii, T., Azuma, M., Sekiguchi, K., Murata, N., Ohtani, T. & Toguri, T. (1996). Low-temperature resistance of higher plants is significantly enhanced by a nonspecific cyanobacterial desaturase. *Nature Biotechnology*, **14**, 1003–6.

Karlsson, O.P., Rytomaa, M., Dalqvist, A., Kinnunen, P.K.J. & Wieslander, A. (1996). Correlation between bilayer-lipid dynamics and activity of the diglucosyldiacylglycerol synthase from *Acholeplasma laidlawii* membranes. *Biochemistry*, **35**, 10 094–102.

Karlsson, O.P., Dahlqvist, A., Vikstrom, S. & Wieslander, A. (1997). Lipid dependence and basic kinetics of the purified 1,2-diacylglycerol 3-glucosyltransferase from membranes of *Acholeplasma laidlawii*. *Journal of Biological Chemistry*, **272**, 929–36.

Kodama, H., Hamada, T., Horiguchi, G., Nishimura, M. & Iba, K. (1994). Genetic enhancement of cold tolerance by expression of a gene for chloroplast Δ-3 fatty acid desaturase in transgenic tobacco. *Plant Physiology*, **105**, 601–5.

Kodama, H., Horiguchi, G., Nishiuchi, T., Nishimura, M. & Iba, K. (1995). Fatty acid desaturation during chilling acclimation is one of the factors involved in conferring low temperature tolerance to young tobacco leaves. *Plant Physiology*, **107**, 1177–85.

Koynova, R. D., Tenchov, B.G. & Quinn, P.J. (1989). Sugars favour formation of hexagonal (H-II) phase at the expense of the lamellar liquid-crystalline phase in hydrated phosphatidylethanolamines. *Biochimica et Biophysica Acta*, **980**, 377–80.

Kozloff, L.M., Schofield, M.A. & Lute, M. (1983). Ice nucleating activity of *Pseudomonas syringae* and *Erwinia herbicola*. *Journal of Bacteriology*, **153**, 222–31.

Kozloff, L.M., Turner, M.A., Arellano, F. & Lute, M. (1991). Phosphatidylinositol, a phospholipid of ice-nucleating bacteria. *Journal of Bacteriology*, **173**, 2053–60.

Kunst, L., Browse, J. & Somerville, C. (1989*a*) Enhanced thermal tolerance of photosynthesis and altered chloroplast ultrastructure in a mutant of *Arabidopsis* deficient in lipid desaturation. *Plant Physiology*, **90**, 943–7.

Kunst, L., Browse, J. & Somerville, C. (1989*b*) Enhanced thermal tolerance in a mutant of *Arabidopsis* deficient in palmitic acid unsaturation. *Plant Physiology*, **91**, 401–8.

Lindblom, G. & Rilfors, L. (1989). Cubic phases and isotropic structures formed by membrane lipids – possible biological relevance. *Biochimica et Biophysica Acta*, **988**, 221–56.

Lindblom, G., Brentel, I., Sjolund, M., Wikander, G. & Wieslander, A. (1986). Phase equilibria of membrane lipids from *Acholeplasma laidlawii* – importance of a single lipid forming nonlamellar phases. *Biochemistry*, **25**, 7502–10.

Lindow, S.E., Arny, D.C., Upper, C.D. & Barchet, W.R. (1978). The role of bacterial ice nuclei in frost injury to sensitive plants. In

Plant Cold Hardiness and Frost Stress, ed. P.H. Li & A. Sakai, pp. 249–63. New York: Academic Press.

Luzzati, V. (1968). X-ray diffraction studies of lipid-water systems. In *Biological Membranes, Physical Facts and Functions*, ed. D. Chapman, pp. 71–123. London: Academic Press.

McCue, K.F. & Hanson, A.D. (1992). Salt-induced betaine aldehyde dehydrogenase from sugar beet – cDNA cloning and expression. *Plant Molecular Biology*, **18**, 1–11.

Miguel, M., Jas, D., Dooner, H. & Browse, J. (1993). *Arabidopsis* requires polyunsaturated lipids for low-temperature survival. *Proceedings of the National Academy of Sciences, USA*, **901,** 6208–12.

Mueller, G.M., Wolber, P.K. & Warren, G.J. (1990). Clustering of ice nucleating protein correlates with ice nucleating activity. *Cryobiology*, **27**, 416–22.

Murata, N. (1983). Molecular species composition of phosphatidylglycerols from chilling-sensitive and chilling-resistant plants. *Plant and Cell Physiology*, **24**, 81–6.

Murata, N. & Hoshi, H. (1984). Sulfoquinovosyl diacylglycerols in chilling-sensitive and chilling-resistant plants. *Plant and Cell Physiology*, **25**, 1241–5.

Murata, N. & Wada, H. (1995). Acyl lipid desaturases and their importance in the tolerance and acclimatization to cold of cyanobacteria. *The Biochemical Journal*, **308**, 1–8.

Murata, N. & Yamaya, J. (1984). Temperature-dependent phase behaviour of phosphatidylglycerols from chilling-sensitive and chilling-resistant plants. *Plant Physiology*, **17**, 1016–24.

Murata, N., Sato, N., Takahashi, N. & Hamazaki, Y. (1982). Compositions and positional distributions of fatty acids in phospholipids from leaves from chilling-sensitive and chilling-resistant plants. *Plant and Cell Physiology*, **23**, 1071–9.

Murata, N., Ishizaki-Nishizawa, O., Higashi, S., Hayashi, H., Tasaka, Y. & Nishida, I. (1992). Genetically engineered alterations in the chilling sensitivity of plants. *Nature*, **356**, 710–13.

Newsted, W.J., Chibbar, R.N. & Georges, F. (1991). Effect of low-temperature stress on the expression of sucrose synthetase in spring and winter wheat plants – development of a monoclonal antibody against wheat-germ sucrose synthetase. *Biochemistry and Cell Biology*, **69**, 36–41.

Nishida, I. & Murata, N. (1996). Chilling sensitivity in plants and cyanobacteria – the crucial contribution of membrane lipids. *Annual Review of Plant Physiology and Plant Molecular Biology*. **47**, 541–68.

Nolte, K.D., Hanson, A.D. & Gage, D.A. (1997). Proline accumulation and methylation to proline betaine in citrus: implications for genetic engineering of stress resistance. *Journal of the American Society for Horticultural Science*, **122**, 8–13.

Oquist, G., Campbell, D., Clarke, A.K. & Gustafsson, P. (1995). The cyanobacterium *Synechococcus* modulates photosystem-II function in response to excitation stress through D1 exchange. *Photosynthesis Research*, **46**, 151–8.

Palta, J.P., Wittaker, B.D. & Weiss, L.S. (1993). Plasma membrane lipids associated with genetic variability in freezing tolerance and cold acclimation of *Solanum* species. *Plant Physiology*, **103**, 793–803.

Pearce, R.S. (1988). Extracellular ice and cell shape in frost-stressed cereal leaves: a low-temperature scanning electron microscopy study. *Planta*, **175**, 313–24.

Quinn, P.J. (1985). A lipid-phase separation model of low-temperature damage to biological membranes. *Cryobiology*, **22**, 28–46.

Quinn, P.J. (1987). Phase behaviour of binary mixtures of membrane polar lipids in aqueous systems. *Natural Product Reports*, **4**, 129–37.

Quinn, P.J. & Williams, W.P. (1983). The structural role of lipids in photosynthetic membranes. *Biochimica et Biophysica Acta*, **737**, 223–66.

Quinn, P.J., Joo, F. & Vigh, L. (1989). The role of unsaturated lipids in membrane structure and stability. *Progress in Biophysics and Molecular Biology*, **53**, 71–103.

Rhy, S.B., Costa, A., Xin, Z.G & Li, P.H. (1995). Induction of cold-hardiness by salt stress involves synthesis of cold-responsive and abscisic acid-responsive proteins in potato (*Solanum-commersonii* Dun). *Plant and Cell Physiology*, **36**, 145–51.

Roberts, D.W.A. (1990). Identification of loci on chromosome 5A of wheat involved in cold hardiness, vernalization, leaf length, rosette growth habit, and height of hardened plants. *Genome*, **33**, 247–59.

Roughan, P.G. (1985). Phosphatidylglycerol and chilling sensitivity in plants. *Plant Physiology*, **77**, 740–6.

Saki, A. & Larcher, W. (1987). *Frost Survival in Plants*. Berlin: Springer-Verlag.

Seddon, J.M. & Templer, R.H. (1993). Cubic phases of self-assembled amphiphilic aggregates. *Philosophical Transactions of the Royal Society of London, Series B*, **344**, 377–401.

Shinozaki, K. & Yamaguchishinozaki, K. (1996). Molecular responses to drought and cold stress. *Current Opinion in Biotechnology*, **7**, 161–7.

Shipley, G.G., Green, J.P. & Nichols, B.W. (1973). The phase behavour of monogalactosyl-, digalactosyl- and sulphoquinovosyl-diglycerides. *Biochimica et Biophysica Acta*, **311**, 531–44.

Siegel, D.P., Banschbach, J., Alford, D., Ellens, H., Lis, L.J., Quinn, P.J., Yeagle, P.L. & Bentz, J. (1989). Physiological levels of diacylglycerols in phospholipid membranes induce membrane fusion and stabilize inverted phases. *Biochemistry*, **28**, 3703–9.

Smith, H. (1990). Signal perception, differential expression within multigene families and the molecular basis of phenotypic plasticity. *Plant, Cell and Environment*, **13**, 585–94.

Somerville, C. & Browse, J. (1996). Dissecting desaturation – plants prove advantageous. *Trends in Cell Biology*, **6**, 148–53.

Steponkus, P.L., Lynch, D.V. & Uemura, M. (1990). The influence of cold acclimation on the lipid composition and cryobehaviour of the plasma membrane of isolated rye protoplasts. *Philosophical Transactions of the Royal Society of London, Series B*, **326**, 571–83.

Steponkus, P.L., Uemura, M. & Webb, M.S. (1993). A contrast of the cryostability of the plasma membrane of winter rye and spring oat – two species that widely differ in their freezing tolerance and plasma membrane composition. In *Advances in Low-temperature Biology*, vol. 2, ed. P.L. Steponkus, pp. 211–312. London: JAI Press.

Stewart, G.R & Lee, J.A. (1974). The role of proline accumulation in halophytes. *Planta*, **120**, 279–89.

Sutka, J. (1994). Genetic control of frost tolerance in wheat (*Triticum aestivum* L). *Euphytica*, **77**, 277–82.

Tasaka, Y., Gombos, Z., Nishyama, Y., Mohanty, P., Ohba, T., Ohki, K. & Murata, N. (1996). Targeted mutagenesis of acyl lipid desaturases in *Synechocystis*: evidence for the important roles of polyunsaturated membrane lipids in growth, respiration and photosynthesis. *EMBO Journal*, **15**, 6416–25.

Tenchov, B.G. (1985). Nonuniform lipid distribution in membranes. *Progress in Surface Science*, **20**, 273–340.

Thomas, P.G., Brain, A.P.R., Quinn, P.J. & Williams, W.P. (1985). Low pH and phospholipase A2 treatment induce the phase-separation of non-bilayer lipids within pea chloroplast membranes. *Federation of European Biochemical Societies, Letters*, **183**, 161–6.

Treichel, S., Brinckmann, E., Scheitler, B. & von Willert, D.J. (1984). Occurrence of changes in proline content in plants in the Southern Namib desert in relation to increasing and decreasing drought. *Planta*, **162**, 236–42.

Tsvetkova, N.M. & Quinn, P.J. (1994). Compatible solutes modulate membrane lipid phase behaviour. In *Temperature Adaptation of Biological Membranes*, ed. A.R. Cossins, pp. 49–62. London: Portland Press.

Tsvetkova, N.M., Brain, A.P.R. & Quinn, P.J. (1994). Structural characteristics of thylakoid membranes of *Arabidopsis* mutants deficient in lipid fatty acid desaturation. *Biochimica et Biophysica Acta*, **1192**, 263–71.

Tsvetkova, N.M., Apostolova, E.L., Brain, A.P.R., Williams, W.P. & Quinn, P.J. (1995). Factors influencing PSII particle array formation in *Arabidopsis thaliana* chloroplasts and the relationship of such arrays to the thermostability of PSII. *Biochimica et Biophysica Acta*, **1228**, 201–10.

Tulloch, A.P., Heinz, E. & Fischer, W. (1973). Combination and positional distribution of fatty acids in plant sulfolipids. *Hoppe-Seyler's Zeitschrift für Chemie*, **354**, 879–89.

Uemura, M. & Steponkus, P.L. (1989). Effect of cold acclimation on the incidence of 2 forms of freezing-injury in protoplasts isolated from rye leaves. *Plant Physiology*, **91**, 1131–7.

Uemura, M. & Steponkus, P.L. (1997). Effect of cold acclimation on the lipid composition of the inner and outer membrane of the chloroplast envelope isolated from rye leaves. *Plant Physiology*, **114**, 1493–500.

Van Venetie, R. & Verkleij, A.J. (1982). Possible role of non-bilayer lipids in the structure of mitochondria. A freeze-fracture electron microscope study. *Biochimica et Biophysica Acta*, **692**, 397–405.

Veisz, O., Galiba, G. & Sutka, J. (1996). Effect of abscisic acid on the cold hardiness of wheat seedlings. *Journal of Plant Physiology*, **149**, 439–43.

Vigh, L., Horvath, I. & Thompson, G.A. (1988). Recovery of *Dunaliella salina* cells following hydrogenation of lipids in specific membranes by a homogeneous palladium catalyst. *Biochimica et Biophysica Acta*, **937**, 47–50.

Vigh, L., Los, D.A., Horvath, I. & Murata, N. (1993). The primary signal in the biological perception of temperature – Pd-catalysed hydrogenation of membrane lipids stimulated the expression of the *des*A gene in *Synechocystis* PCC6803. *Proceedings of the National Academy of Sciences, USA*, **90**, 9090–4.

Waters, E.R., Lee, G.J. & Vierling, E. (1996). Evolution, structure and function of the small heat-shock proteins in plants. *Journal of Experimental Botany*, **47**, 325–38.

Webb, M.S. & Steponkus, P.L. (1993). Freeze-induced membrane ultrastructural alterations in rye (*Secale cereale*) leaves. *Plant Physiology*, **101**, 955–63.

Wolter, F.P., Schmidt, R. & Heinz, E. (1992). Chilling sensitivity of *Arabidopsis thaliana* with genetically engineered membrane lipids. *EMBO Journal*, **11**, 4685–92.

Wu, J. & Browse, J. (1995). Elevated levels of high-melting-point phosphatidylglycerols do not induce chilling sensitivity in an *Arabidopsis* mutant. *The Plant Cell*, **7**, 17–27.

F. DOMERGUE, J-J. BESSOULE, P. MOREAU,
R. LESSIRE and C. CASSAGNE

8 Recent advances in plant fatty acid elongation

Introduction

The elongation of fatty acids in plant leaves and seeds is an important process, resulting in the formation of very long chain fatty acids (VLCFA), i.e. the fatty acids with 20 or more carbon atoms. The overall process is catalysed by membrane-bound enzymes: the elongases.

In leaves, elongases chiefly catalyse the synthesis of saturated VLCFAs, ubiquitous (but minor) wax components and precursors of most of the other wax components (for reviews see Kolattukudy, Croteau & Buckner, 1976; Tulloch, 1976; Lessire, Abdul-Karim & Cassagne, 1982; von Wettstein-Knowles, 1995; Post-Beittenmiller, 1996). The most common chain lengths of the wax fatty acids lie in the C_{20}–C_{36} range. Leaf VLCFAs are also membrane constituents and have been shown to account for about 5–15% of the acyl moieties of the plasma membrane in higher plants (Lessire, Hartmann-Bouillon & Cassagne, 1982; Moreau *et al.*, 1988*a*). VLCFAs are synthesised by elongases chiefly located in the underlying layer of epidermal cells (Kolattukudy, 1968; Cassagne, 1970; Cassagne & Lessire, 1974).

VLCFAs are not restricted to the external waxes of leaves, and their occurrence has long been known in developing seeds (Downey & Craig, 1964), where they may account for as much as two-thirds of the total fatty acids (esterified either to glycerol as in *Brassica napus*, or to fatty alcohols in the case of jojoba) and even 90% of total fatty acids in *Limnanthes alba* (Hitchcock & Nichols, 1971; Yermanos, 1975; Harwood, 1980; Lardans & Trémolières, 1991). In developing-seed storage lipids, VLCFAs are mostly monounsaturated and belong chiefly to the *cis* (*n*-9) series, i.e. higher homologues of oleic acid (for reviews see Hitchcock & Nichols, 1971; Harwood, 1980; Padley, Gunstone & Harwood, 1994) (*n*-7) series have also been found in the developing seeds from *Sinapis alba* and from various cultivars of *Brassica napus*; their contribution never exceeds 3–20% of the corresponding (*n*-9) isomer and is restricted to *cis* (*n*-7) 18 : 1 (vaccenic acid) and *cis* (*n*-7) 20 : 1 (Applequist, 1969; Harwood, 1980; Mukherjee & Kiewitt, 1980;

Mukherjee, 1986). Besides the major unsaturated fatty acids, unusual ethylenic VLCFAs are also encountered in the seeds of a few species; this is the case of the VLCFA bearing a *cis* double bond between carbons 5 and 6 of the aliphatic chain, as observed in *Limnanthes douglasii* and *Limnanthes alba* (Hitchcock & Nichols, 1971; Harwood, 1980; Lardans & Trémolières, 1991).

Very long monounsaturated fatty acids are of particular importance for the oleochemical industries and the need for renewable raw materials has led to an extensive research of the lipid content of seeds, with special emphasis on their very long chain content and therefore on the enzymes catalysing their biosynthesis: the elongases. Besides the elongation process occurring in seeds, the production of VLCFAs in leaves is also extensively studied because waxes of leaves are a critical barrier to the penetration of xenobiotics into plants, and this is of interest for agrochemical companies.

The question of VLCFA biosynthesis has recently been reviewed (Cassagne *et al.*, 1994, Harwood, 1996). The aim of the present review is to focus on some new developments of this field and to further characterise the different elongases so far identified at different levels: biochemical properties (substrates, enzymatic reactions, products, *in vitro* and *in vivo* regulation of elongating activity), subcellular distribution, purification, molecular biology, and biotechnology.

Two types of acyl elongases

The synthesis of VLCFAs in plant was identified and characterised from various tissues and maturing seeds (for review, see Cassagne *et al.*, 1994; Harwood, 1996). As in animal cells, this synthesis occurs by the condensation of a two-carbon unit donor with a primer having (at least) 18 carbon atoms.

It was shown early that malonyl-CoA (and not acetyl-CoA or acetyl-ACP) is the two-carbon unit donor involved in VLCFA synthesis (Macey & Stumpf, 1968; Kolattukudy & Buckner, 1972). In many cases (Cassagne & Lessire, 1978; Agrawal, Lessire & Stumpf, 1984; Agrawal & Stumpf, 1985*a*, *b*), an addition of ACP has no effect on, or only partially inhibits, the reaction; a slight stimulation was observed only in a particular preparation from germinating pea (Macey & Stumpf, 1968).

Both genetic studies (von Wettstein-Knowles, 1979; Kunst, Taylor & Underhill, 1992) and numerous biochemical investigations (for review, see Cassagne *et al.*, 1994; Harwood, 1996) suggested the existence of several elongases. In most higher plant leaves and seeds, most evidence points to the existence of at least two types of elongases: Type I elongases called 'ATP-dependent elongases' and Type II elongases defined as 'acyl-CoA elongases'.

Both in leaves and seeds, these elongating systems differ in (at least):

- the nature of the elongated primer
- the number of malonyl-CoA molecules involved in the elongation process
- the intracellular location of enzymes
- the effects of various effectors (ATP, CoASH . . .).

Nature of primer

Type I elongases ('ATP-dependent elongases')

An ATP-dependent elongating activity was probably shown for the first time by Kolattukudy and Buckner (1972), who worked on fatty acid elongation by cell-free extracts of epidermis from pea leaves. The existence of this activity was demonstrated in leek by Agrawal *et al.* (1984). VLCFA synthesis occurred when epidermal cell microsomes were incubated with labelled malonyl-CoA, NADPH and ATP in the absence of exogenous acyl-CoAs. Similar results were obtained by using microsomes from aetiolated seedlings (Lessire *et al.*, 1985*c*).This activity exhibited an absolute requirement for ATP, which could not be replaced by any other nucleotide phosphate (Agrawal *et al.*, 1984). Interestingly, by working with the same plant, Evenson and Post-Beittenmiller (1995) have recently shown that ATP-dependent elongase activity is ten-fold higher in microsomes from expanding leaves than in store-bought leek, thus suggesting an induction of this type of elongase in a way reminiscent of that reported years ago by Willemot and Stumpf (1967), Bolton and Harwood (1976*a*) and Walker and Harwood (1986). Evenson and Post-Beittenmiller (1995) did not observe this phenomenon when acyl-CoA elongase activity was measured. This suggests that the activities of the two types of elongases are probably not regulated by the same process. In addition, the results of this study and others suggest that acyl-CoAs are not direct substrates of the ATP-dependent elongase. In good agreement, it was recently shown in developing seeds from *Brassica napus* that the specific radioactivity of oleoyl-CoA decreased in the incubation mixture as a function of time whereas the specific radioactivity of the VLCFA synthesised in the presence of ATP remained constant (Hlousek-Radojcic, Imai & Jaworski, 1995). However, the authors mentioned that their results could be explained by a partitioning of oleoyl-CoA between two pools, as hypothesised years ago by Pollard and Stumpf (1980*a*) who studied *in vivo* VLCFA synthesis by developing seeds of *Tropaeolum majus*.

All these results indicate that type I (ATP-dependent) elongases do not elongate exogenous acyl-CoAs, but they do not shed light on the nature of the substrates.

Type II elongases ('acyl-CoA elongases')

In leaf, the best documented system is that of the microsomal elongases from leek microsomes (for review, see Cassagne *et al.*, 1994). In this system, the saturated acyl-CoAs could be elongated to a series of higher homologues (Lessire *et al.*, 1985*c*), whereas oleoyl-CoA, linoleoyl-CoA and the linolenoyl-CoA are not primers of the elongase (Agrawal & Stumpf, 1985*a*). It was hypothesised that leek epidermal cells had a mechanism for the exclusion of unsaturated acyl-CoAs from its elongation system(s). Whether these unsaturated C_{18}-CoAs are metabolised so rapidly that they are unable to reach the elongases, or whether the acyl-CoA elongases simply do not accept them, remained obscure until it was shown that the purified enzyme devoid of transacylase activity elongated almost any saturated acyl-CoA in the C_{12-24} range as well as oleoyl-CoA (Lessire, Bessoule & Cassagne, 1989*a*, *b*).

In seeds, the case of jojoba (*Simmondsia chinensis*) appears unique in that it is able to elongate almost equally well both long chain acyl-CoAs and C_{18}-ACP. $C_{18:1}$ (*n*-9) *cis* CoA and higher homologues ($C_{20:1}$-CoA and $C_{22:1}$-CoA) were elongated as well as C_{18}-CoA, whereas short chain acyl-CoAs were not, or only poorly, elongated (Pollard *et al.*, 1979). *In vivo*, however, the jojoba seed does not contain saturated VLCFAs or alcohols, so that either the specificity of the elongases is modified *in vitro*, or the C_{18}-CoA and/or C_{18}-ACP are not available *in vivo*, because of the high activity of the C_{18}-ACP desaturase.

The elongase of developing honesty seeds (Fehling & Mukherjee, 1991) accepts numerous acyl-CoAs as substrates in the C_{14}-C_{24} range, with a marked preference for the saturated or monounsaturated C_{18} and C_{20}-CoAs (no difference in elongation rate was observed when using saturated or unsaturated acyl-CoAs). *In vivo* incubations of immature seeds of meadowfoam with palmitic acid, stearic, oleic and (*n*-9) eicosenoic acid led to the synthesis of labelled C_{20} and C_{22} fatty acids containing the unusual $\Delta5$ double bond (Pollard & Stumpf, 1980*b*). Palmitic acid seemed to play a prominent role as a precursor of $C_{20:1}\Delta5$ biosynthesis. A cell-free homogenate exhibited an equal ability to form C_{20} and $C_{20:1}\Delta5$ fatty acyl chains from palmitoyl-CoA and stearoyl-CoA, whereas C_{20}-CoA was poorly elongated (Pollard & Stumpf, 1980*b*).

$C_{18:1}$-CoA and $C_{20:1}$-CoA were the best substrates of elongases from homogenates of developing embryos of rapeseed, but all saturated acyl-CoAs (C_{14}-C_{22}) could be accepted, *in vitro*, by the elongase from rapeseed (Créach, Lessire & Cassagne, 1992).

Number of molecules of malonyl-CoA involved in VLCFA synthesis

In leek leaves, C_{18}-CoA and malonyl-CoA incorporations are identical for C_{20} synthesis, whereas malonyl-CoA incorporation into C_{22}, C_{24} and C_{26} are two, three and four times higher, respectively, than that, of C_{18}-CoA. This shows that, when exogenous acyl-CoAs are used, no elongation of endogenous primer occurs and that there is a sequential elongation of the primer (C_{18}-CoA) by malonyl-CoA (Lessire *et al.*, 1985*c*).

In seeds, although comparison of elongation rates of $C_{18:1}$-CoA and malonyl-CoA led to different results (Murphy & Mukherjee, 1988; Taylor *et al.*, 1992*a*), it was assumed that the biosynthesis of $C_{20:1}$ and then $C_{22:1}$ involved successive condensation of malonyl-CoA with oleoyl-CoA. Clear-cut results have been reported from rapeseed: the ratio of the moles of [2-^{14}C] malonyl-CoA incorporated into each fatty acid to the moles of [1-^{14}C] oleoyl-CoA incorporated into the corresponding fatty acid strongly supported the view that the elongation mechanism (leading to the synthesis of unsaturated VLCFA in *Brassica napus*) involves successive additions of malonyl-CoA to oleoyl-CoA (Créach, Lessire & Cassagne, 1993*a*).

In marked contrast with acyl-CoA elongases, VLCFAs are synthesised by ATP-dependent elongases by a unique malonyl-CoA condensation with an unknown primer (Kolattukudy & Buckner, 1972; Agrawal, Lessire & Stumpf, 1984). Both in leek and pea, the degradation analyses of ^{14}C labelled C_{18}, C_{20}, C_{22} and C_{24} fatty acids synthesised from [2-^{14}C] malonyl-CoA in the presence of ATP by microsomes, suggested that they were formed mostly by elongation of their immediate precursor acids, i.e. C_{16}, C_{18}, C_{20} and C_{22}, respectively, which have been shown to be present in microsomes from leek epidermis (Agrawal *et al.*, 1984). This is due to the fact that, whatever the nature of the substrate, the products are released from the elongases as acyl-CoAs (see below). In the case of acyl-CoA elongase, the product may be further elongated, whereas in the case of ATP-dependent elongase, the product (acyl-CoA) is not accepted as a substrate by the condensing enzyme, and the elongation stops.

Intracellular location

Acyl-CoA elongases were found very early in a crude microsomal pellet (Macey & Stumpf, 1968). Both in leek epidermis and aetiolated maize coleoptiles, the stearoyl-CoA elongase was associated with an ER enriched fraction (Cassagne & Lessire, 1978; Lessire *et al.*, 1982). Later, a second elongase, the eicosanoyl-CoA elongase, was found in the Golgi apparatus (Moreau *et al.*, 1988*b*). In this study, the stearoyl-CoA elongase peaked together with the CDP-choline diacylglycerol phosphotransferase, as did the eicosanoyl-CoA elongase with the latent IDP-ase. Interestingly, the ATP-dependent formation of VLCFA was not found in the ER- and Golgi-enriched fractions, but was associated with distinct heavy membranes. Moreover, no elongating activity was detected in pure right-side out plasma membrane fractions prepared by phase partition (Moreau *et al.*, 1988*a*, *b*). The ATP-dependent elongase was concentrated in the lower (dextran-enriched) phase which contained endomembranes and inside-out plasma membrane vesicles unable to elongate acyl-CoAs. Hence, the ATP-dependent elongase could either be located within endomembranes (other than ER- and Golgi-enriched fractions) or at the cytosolic face of the plasma membrane. A location in the plasma membrane of 8-week-old bonus barley was suggested early in genetic studies by Von Wettstein-Knowles (1979). In plant, unlike in yeast (Bessoule *et al.*, 1987, 1988), elongating activities were not found associated with the organelles (Bessoule, 1989).

In seeds, the location of the elongases remains unclear. The most active subcellular fractions were found to be the 2000***g*** and/or the 15 000***g*** pellets obtained from homogenates of developing seeds of honesty, nasturtium and mustard (Murphy & Mukherjee, 1988). In meadowfoam, saturated acyl-CoAs were elongated best in a 100 000***g*** pellet, whereas oleoyl-CoA was more efficiently elongated in a 15 000***g*** pellet (Lardans & Trémolières, 1992). In developing rape seeds, the elongating activities were also mostly associated with a 15 000***g*** pellet (Créach *et al.*, 1992). Whitfield, Murphy and Hills (1993) found that the location of fatty acid elongase in developing seeds of honesty and rape was likely to be the endoplasmic reticulum. Contrasting results were obtained in jojoba (Pollard *et al.*, 1979), where elongase activities were found in the so-called 'floating pad' fractions. Interestingly, Hlousek-Radojcic *et al.* (1995) reinvestigated the question of the ATP-dependent elongase in rapeseeds. The authors showed that the highest specific activity was associated with the oil-body fraction, and that this fraction was devoid of acyl-CoA (ATP-independent) elongating activity. If confirmed, this result would indicate a clear-cut separation of the sites of both types of elongating activities.

As the structure of oil-bodies begins to be more and more understood (Huang, 1996; Napier, Stobart & Shewry, 1996), it is clear that associated ER domains might be tightly associated with the oil-bodies (see, for example, Settlage, Wilson & Kwanyuen, 1995) and could be responsible, at least in part, for the elongating activities observed in oil-bodies from *Brassica napus* (F. Domergue, unpublished results).

Differential effects of ATP and CoASH

It was recently shown (Domergue, 1997) that ATP at a concentration allowing maximal ATP-dependent elongation in rapeseed, had a dramatic inhibitory effect on acyl-CoA elongation. On the other hand, a CoASH addition had no effect on oleoyl-CoA elongation but led to a drastic inhibition of ATP-dependent elongation. Hence, depending upon acyl-CoA, ATP, CoASH concentrations, a modulation of the relative activities of the two types of elongases could be achieved *in vivo*.

Since the occurrence of several reactions could produce confusing results, special care must be paid to the conditions used to determine the level of elongating activity *in vitro*: Type I (ATP dependent) elongase should be measured in the presence of labelled malonyl-CoA and ATP but in the absence of exogenous acyl-CoAs, and the activity of Type II (acyl-CoA) elongase (such as the icosanoyl-CoA synthase (EC. 2.3.1.1) has to be determined by using acyl-CoA and malonyl-CoA but not ATP.

Acyl-CoA elongation: intermediate reactions and reductant requirement

Since the pioneering work of Nugteren (1965), it is generally admitted that the acyl-CoA elongation process in animals involves the following reactions:

- condensation of acyl-CoA to malonyl-CoA to form a 3-oxoacyl-CoA
- reduction of the 3-oxoacyl-CoA
- dehydration of the resulting 3-hydroxyacyl-CoA
- reduction of the (E) 2,3 enoyl-CoA to form the saturated higher homologue acyl-CoA.

In higher plants, this reaction sequence has been investigated and the occurrence of the postulated intermediates has been demonstrated in the case of leaf and seed elongases.

3-ketoacyl-CoA synthase (KCS)

The involvement of a 3-ketoacyl-CoA intermediate in the acyl-CoA elongation process was demonstrated by using an acyl-CoA elongase purified from leek epidermal cells (Lessire *et al.*, 1989*a*, *b*): products of elongation were released from the acyl-CoA elongase as CoA esters. The absence of reductant led to an increase in two long-chain methyl-ketones deriving from the non-enzymatic decarboxylation of 3-ketoacyl-CoAs during saponification. Similarly, a TLC and radio GLC analysis of the elongation products synthesised by a particulate fraction from *Lunaria annua* clearly showed the presence of 3-hydroxy $C_{20:1}$-CoA; 3-hydroxy $C_{22:1}$-CoA; (E) 2,3 $C_{20:2}$-CoA; (E) 2,3 $C_{22:2}$-CoA, and methyl ketones (Fehling & Mukherjee, 1991). As in the case of leek, the methyl ketone yield was increased when NADPH and NADH were omitted from the reaction mixture.

The enzyme catalysing the condensation reaction (3-ketoacyl-CoA synthase) was studied using membranes from leek leaves (Schneider *et al.*, 1993) and developing rape seeds (Domergue, 1997). The condensing enzyme from leek exclusively uses $C_{18:0}$-CoA and $C_{20:0}$-CoA and, interestingly, the unsaturated acyl-CoAs were not elongated. The activity of condensing enzyme from rapeseeds was maximum with 50 μM malonyl-CoA and 20 μM oleoyl-CoA. When saturated/or unsaturated acyl-CoAs, ranging from C_{12} to C_{22}, were used as precursors, the highest activities were obtained with $C_{18:0}$-CoA and $C_{18:1}$-CoA and led to the highest activities, whereas short-chain acyl-CoAs were poor substrates (Domergue, 1997). The substrate preference of partially purified condensing enzyme from jojoba seeds has also been determined (Lassner, Lardizabal & Metz, 1996). This enzyme showed a maximal activity with monounsaturated and saturated C_{18} and C_{20}-CoA and also exhibited some activity with $C_{22:1}$-CoA.

Since the condensing enzymes have a different preference for the acyl-CoAs and catalyse the limiting step of the process (Lessire *et al.*, 1992), they were considered as key enzymes in determining the quantity, chain length and unsaturation of VLCFA synthesised by the acyl-CoA elongase (see also Knutzon and Knauf, this volume). This hypothesis is now supported by several experiments. The 3-ketoacyl-CoA synthase cDNA from jojoba complements the canola fatty acid elongation mutation in transgenic rapeseeds (Lassner *et al.*, 1996). Similarly, when a condensing enzyme gene from *Lunaria annua* (containing $C_{24:1}$ fatty acid as a major component of its seed oil) was introduced into high erucic acid rapeseed (HEAR), the resulting oil contained approximately 20% $C_{24:1}$ (Lassner, 1997). This important role of the condensing enzyme was underlined by Millar and Kunst (1997). By expressing FAE1 in yeast and in tissues of *Arabidopsis* and tobacco where no significant quantities of VLCFA were found, they demonstrated that the introduction of FAE1 alone is sufficient for the production of VLCFAs.

3-ketoacyl-CoA reductase (KCR)

Owing to the lack of substrate (to our knowledge, 3-ketoacyl-CoAs have never been synthesised), this reaction has never been studied directly. However, using an antibody raised against a hydroxyacyl-CoA dehydratase from rat liver (Cook *et al.*, 1992), Lessire *et al.* (1993) were able to block the elongation process at the level of the 3-hydroxyacyl-CoA which accumulated as the end product. It was shown that the rate of the reaction was at least equal to that of the condensation.

Hydroxyacyl-CoA dehydratase (HCD)

The presence of 3-hydroxyacyl-CoA dehydratase in leek microsomes has been investigated by measuring the reverse reaction using (E) 2,3 $C_{16:1}$-CoA as substrate (Lessire *et al.*, 1992), but the characterisation of the dehydratase involved in the overall elongation process was not possible until chemically synthesised [1-^{14}C] 3-hydroxy $C_{20:0}$-CoA was available (Lucet-Levannier *et al.*, 1995). The dehydratase activity has been measured using either microsomes from aetiolated leek seedlings or 15 000***g*** particulate fraction from rapeseeds. The 3-hydroxyacyl-CoA was not the substrate of the condensing enzyme and entered the elongase at the level of the dehydratase only when Triton X100 was added. This result suggested a limited accessibility of the active site of the dehydratase in the native enzyme, and that the hydroxyl group of the molecule plays a central role in the specificity of the binding to the dehydratase. Kinetic studies showed a very high dehydration rate of (R) 3-hydroxyacyl-CoA (S. Chevalier & R. Lessire, unpublished data).

The 3-hydroxyacyl-CoA dehydratase activity measured using synthetic substrates was inhibited by adding antibodies raised against a 3-hydroxyacyl-CoA dehydratase purified from rat liver (Cook *et al.*, 1992). These antibodies were used for immunoblotting experiments, and evidenced a protein with an apparent molecular mass of 65 kDa. This led to the suggestion that the dehydratase component of the elongation complex had been identified (Lessire *et al.*, 1993).

2,3 enoyl-CoA reductase (ECR)

The *trans*-2,3 enoyl-CoA reductase activity was measured in leek microsomes. (E) 2,3 $C_{18:1}$-CoA was a better substrate than (E) $C_{12:1}$-CoA and (E) $C_{16:1}$-CoA (Lessire *et al.*, 1992). Fowler *et al.* (1995) have reported the presence of two membrane-associated enoyl-CoA reductases in *Brassica napus*, one using NADH and the second NADPH.

Very recently, an intermediate of the elongation process ((E) 2,3 eicosenoyl-CoA) was chemically synthesised (Lucet-Levannier *et al.*, 1995) and used as substrate for the leek acyl-CoA elongase. (E) 2,3 icosanoyl-CoA was not accepted by the condensing enzyme but by the enoyl-CoA reductase. A VLCFA synthesis from this intermediate required that Triton X-100 and the resulting products should be saturated straight-chain acyl-CoAs. (E) 2,3 $C_{20:1}$-CoA reduction was also studied as a function of NADH and NADPH concentrations. The results indicated that the enoyl-CoA reductase of the elongation complex has a marked preference for the NADPH (Spinner *et al.*, 1995).

Reductant requirement in overall elongation process

That the biosynthesis of VLCFAs requires the presence of reductants is no longer a matter of debate: it has been shown repeatedly that NADH and NADPH allow the formation of VLCFA, whereas their omission from the reaction mixture leads invariably to the formation of 3-ketoacyl-CoA (see above). However, neither the mechanism of the reduction, nor the nature of the reductant required for the two proposed reduction steps, have been definitely established. A variety of different responses has been observed, but, in numerous cases, NADPH is the 'preferred' reductant. Agrawal and Stumpf (1985*b*) have established that, in a cell-free extract from *Brassica juncea* developing seeds, the oleoyl-CoA elongation to eicosenoyl-CoA accepts NADPH or NADH equally well, whereas the elongation of $C_{20:1}$ to $C_{22:1}$ exhibits a pronounced preference for NADPH. This result was taken as an indication of the presence of several elongases in *B. juncea*. Similar observations and conclusions were made in the case of *Sinapis alba*, *Lunaria annua* and *Tropaeolum majus* (Murphy & Mukherjee, 1989).

Using *Brassica napus* embryos, Taylor *et al.* (1992*a*) indicated a higher synthesis of VLCFA in the presence of NADPH than in the presence of NADH. Identical conclusions were drawn from a study carried out using a particulate fraction from *Brassica napus* developing seeds: an additional slight preference for NADPH in terms of overall VLCFA formation was evidenced (R. Lessire, unpublished observations). NADPH also seems to be the reductant used for the stearoyl-CoA elongation by the microsomes from developing seeds of meadowfoam, and NADH has no effect (Lardans & Trémolières, 1992).

In microsomes from leek epidermal cells, the acyl-CoA elongase relies on the provision of NADPH and, to a lesser extent, NADH. The omission of NADPH had a marked effect on the overall fatty acid synthesis and on the accumulation of intermediates, and seemed particularly important in the first reduction step (Lessire *et al.*, 1993).

The fact that all the intermediates and products are CoA derivatives (whatever the form of type I and type II elongases) shows that no covalent link between the acyl substrates, intermediates and products of the elongases is likely to occur. This was recently shown using antibodies raised against acyl-CoAs (Maneta-Peyret *et al.*, 1997).

Products of elongation and their transport

Nature of elongation products

The first indication that in seeds, the products of the type 2 acyl-CoA elongase were released as acyl-CoAs was obtained in jojoba seeds (Pollard *et al.*, 1979). This was confirmed, in developing rapeseeds: very long chain acyl-CoAs were rapidly formed from labelled oleoyl-CoA by a 15 000*g* particulate fraction of developing rapeseeds before appearing in triacylglycerols (Taylor *et al.*, 1992*a*). When glycerol 3-phosphate was supplied in time-course experiments, [^{14}C] $C_{20:1}$ and [^{14}C] $C_{22:1}$ were detected in TAG and to a lesser extent, in Lyso PA, PA, DG and PC as well as acyl-CoA and FFA pools. On the other hand, Créach *et al.* (1993*a*), using a particulate fraction from developing rapeseeds, showed that the unsaturated very long chain acyl moieties were released from the elongase as acyl-CoAs, and then esterified to triacylglycerols. More recently, Hlousek-Radojcic *et al.* (1995) analysed the VLCFA-CoAs derived from the elongation of [^{14}C] oleoyl-CoA in oil bodies and showed the presence of $C_{20:1}$-CoA and $C_{22:1}$-CoA.

In leaves and using microsomes from leek epidermis, Abdul-Karim, Lessire and Cassagne (1982), obtained the first evidence for the release of very long chain acyl-CoAs followed by the insertion of their acyl moieties into phospholipids and neutral lipids. A kinetic study of the elongation of stearoyl-CoA by leek epidermis microsomes was followed by the resolution of the malonyl-CoA, long chain acyl-CoAs, PC, PE, glycolipids and neutral lipids by HPTLC of the whole reaction mixture (Cassagne & Lessire, 1982). After a 10 min incubation, all the labelled VLCFAs were found in the acyl-CoA fraction; then, they progressively appeared in phospholipids (chiefly PC) and in neutral lipids. Another analysis was carried out with aetiolated leek seedling microsomes incubated with [1-^{14}C] stearoyl-CoA (Lessire *et al.*, 1985*a*). That very long chain acyl-CoAs were the products of the acyl-CoA elongase was demonstrated by the fact that, at the shortest time of incubation (5 min), all the label of the VLCFAs was restricted to the acyl-CoA fraction. No label was associated with PC or neutral lipids. Using longer incubation times, VLCFAs were found in PC and NL. In addition, the very high level of elongation (34% in the acyl-CoAs after 60 min) made it possible to show the progressive appearance of the very long chain acyl-CoAs (C_{20}-CoA after a 5 min lag, C_{22}-CoA after 10 min, C_{24}-CoA after 40 min), as well as the (until then unnoticed) fact that only C_{20} could be inserted into PC. In neutral lipids, the C_{20} appeared after a lag of 20 min and the C_{22} after 40 min.

The nature of the products released from Type I ATP-dependent elongase was established using aetiolated leek seedlings. Schneider *et al.* (1996) demonstrated that [^{14}C] Acyl-CoAs (and neutral lipids +FFA) were the first components detectable during the first 10 minutes of incubation.

Finally, using solubilised and/or partially purified enzymes from different sources (leaves and seeds), it was definitely established that the products of the acyl-CoA elongase are VLCFA-CoAs (Lessire *et al.*, 1989*a*, *b*; Fehling *et al.*, 1992; Créach *et al.*, 1993*b*).

Intracellular and extracellular transfer of VLCFA

The model of 7-day-old aetiolated leek seedlings (Moreau *et al.*, 1984) made it possible, through pulse-chase experiments, to demonstrate the transfer of VLCFAs from ER to the plasma membrane *in vivo* (Moreau *et al.*, 1986, 1988*a*) via the Golgi apparatus and intermediate vesicles (Bertho *et al.*, 1991).

The kinetic parameters of the transfer were compatible with a vesicular process (Moreau *et al.*, 1988*b*) which was stopped by monensin at the level of the Golgi apparatus (Bertho *et al.*, 1989, 1991).

The lipid transport through the vesicular pathway was regulated according to the chain length of the fatty acids (Bertho *et al.*, 1991). Briefly, the transfer of VLCFA-containing lipids to the plasma membrane was blocked by monensin, whereas that of the C_{16}- and C_{18}- containing lipids was not. Lowering the temperature to 12 °C also had a drastic effect on the lipid traffic. At 12 °C, the transfer of VLCFA-containing lipids was blocked at the Golgi apparatus level, so that VLCFAs were accumulated in the ER and Golgi. This temperature block did not affect the C16- and C18-containing lipids (Moreau *et al.*, 1994; Sturbois-Balcerzak *et al.*, 1995). This sorting was probably not due to the polar head of the polar lipids since VLCFAs were chiefly found in PC and PE, but whereas VLCFA-PC and VLCFA-PE were blocked at the Golgi level at low temperature, C16- and C18-PC or PE reached the plasma membrane. Two conclusions should be drawn from these studies: (i) VLCFA biosynthesis requires the cooperation of at least two compartments: ER and Golgi which contain C18-CoA and C20-CoA elongase respectively; (ii) the products (acyl-CoAs) are probably not transferred directly to the next membranes (via partition of acyl-CoAs or acyl-CoA binding proteins), but the VLCFA moieties are probably acylated to PC and PE, which are sorted through the ER-Golgi-plasma membrane pathway.

Solubilisation, purification and 'molecular biology' of plant elongases

Solubilisation and purification

The solubilisation of plant acyl-CoA elongase was first studied using leek microsomes and more recently using developing seeds. In most cases, Triton X-100 is the best detergent to solubilise the elongases without any inhibition of their activities.

Using a Triton X-100-to-protein ratio (w/w) of about 2, Agrawal and Stumpf (1985*a*) solubilised elongases from microsomes of leek epidermal cells. Using the same material and a detergent-to-protein ratio of 1, Lessire *et al.* (1985*b*) solubilised and separated two elongases:

- a $C_{18:0}$-CoA elongase peaking at 0.51 M sucrose and showing, in the presence of Triton X-100 micelles, an apparent molecular mass of 350 kDa by filtration on Sephacryl-S300
- a $C_{20:0}$-CoA elongase peaking at 0.62 M sucrose with an apparent molecular mass of 600 kDa.

The solubilised $C_{18:0}$-CoA elongase was then partially purified using two chromatographic steps on DEAE 23 and Ultrogel AcA34 columns (Bessoule, Lessire & Cassagne, 1989*a*). The elongase had an apparent MW of at least 300 kDa after filtration on Ultrogel and was resolved in four major proteins after analysis by SDS–PAGE. The purified proteins were highly delipidated. Some PS and PE remained associated with the activity. The products synthesised by the purified fraction were saturated acyl-CoAs, indicating that the four activities constituting the acyl-CoA elongase (i.e. 3-ketoacyl-CoA synthase, 3-ketoacyl-CoA reductase, 3-hydroxyacyl-CoA dehydratase and trans-2,3-enoyl-CoA reductase) were present.

Polyclonal antibodies raised against the purified elongase (Bessoule *et al.*, 1992) were able to immunoprecipitate 1% of the solubilised proteins, which exhibited high elongase activity. During the purification procedure, only the fractions containing the elongating activity were immunogenic; among the proteins of the partially purified fraction, only the 65 kDa protein was recognised by an antibody raised against the purified 3-hydroxyacyl-CoA dehydratase from rat liver (see above).

Recently, Evenson and Post-Beittenmiller (1995) showed that elongating activity was inhibited by 0.1% of all the detergents used. Nevertheless, they solubilised this complex using Z 3–10 at a 4% detergent-to-protein ratio and 25% of the elongase activity was recovered in the soluble fraction. After gel filtration the elongating activity was eluted in two peaks corresponding to molecular masses of 29.5 and 220 kDa.

Acyl-CoA elongases from developing seeds have also been solubilised, usually from a particulate 15 000***g*** fraction, using the procedures developed for leek elongases. Murphy and Mukherjee (1989) successfully solubilised the elongase from a particulate fraction from developing seeds of *Sinapis alba* and *Lunaria annua*. The elongase of *Lunaria annua* was partially purified by a three-step procedure (Fehling *et al.*, 1992). The four enzymes constituting the acyl-CoA elongase were present in the active fraction after chromatography on DEAE-cellulose, but only the 3-ketoacyl-CoA synthase and 3-ketoacyl-CoA reductase were present after Ultrogel AcA34. The elongation activities of saturated acyl-CoAs located in the microsomal pellet of *Limnanthes alba* seeds (100 000***g***) was solubilised using Triton X-100 and was partially purified on DEAE cellulose (Lardans & Trémolières, 1992).

The rapeseed acyl-CoA elongases have been solubilised from a 15 000***g*** particulate fraction (Créach & Lessire, 1993*b*). The elongation complex was still functional, even though accumulation of intermediates suggested that the elongases were affected during the process. The $C_{18:1}$-CoA elongase partially purified using three successive chromatographic columns: C_{16}-CoA agarose, Ultrogel AcA34 and DEAE 52 (Créach, 1994; Créach, Domergue & Lessire, 1995; Domergue, 1997) was highly similar to leek leaf C_{18}-CoA elongase (Bessoule *et al.*, 1989*a*): four enzymatic activities were present, and the apparent MW on Ultrogel filtration was around 350 kDa (but very interestingly around 600 kDa after Sephacryl S300). The partially purified complex, was highly dilipidated and was composed of several proteins of 55–65 kDa according to the SDS–PAGE analysis.

Imai *et al.* (1995) solubilised the ketoacyl-CoA synthase (KAS or condensing enzyme) of the elongation system involved in the synthesis of VLCFA in *Brassica napus* seeds. They used the oil body fraction and 25 mM zwittergent 3–10 to solubilise 85% of the elongation activity. Nevertheless, a chromatography on Superose 6 and an analysis of the products revealed that only the condensing enzyme had been solubilised. The dissociation of the elongation complex could be due either to the use of the oil body fraction instead of the 15 000***g*** pellet and/or of the detergent which was not Triton X-100. Lassner *et al.* (1996) described similar results after solubilisation of the microsomal pellet from jojoba embryos using CHAPS. After purification on a Superose 12 column, the elongation complex was shown to be dissociated, and the activity of the condensing enzyme was eluted in fractions corresponding to a molecular mass of 138 kDa. As the purified protein showed a molecular mass of 57 kDa on SDS gel, this higher value could reflect a 'dimerisation' of the condensing enzyme.

Even if none of the assays summarised above produced the obtention of highly purified acyl-CoA elongases, these results strongly support that the acyl-CoA elongase is an oligomeric protein composed of four different subunits in the molecular range of 50–70 kDa. This is in agreement with the elongation mechanism involving four different activities and with the molecular masses of the purified subunits (dehydratase and KAS).

'Molecular biology' of elongases

The sequence of the 3-ketaoacyl-CoA synthase involved in the elongation of fatty acid in *Arabidopsis* seeds was the first ever characterised (James *et al.*, 1995). cDNA was cloned by direct transposon tagging with the maize element activator. These authors isolated cDNA encoding for a protein with a predicted mass of 56 kDa; i.e. in the size range of the proteins obtained after purification of the leek elongase complex (Bessoule *et al.*, 1989*a*). The predicted sequence of the protein (FAE1 protein) shared homology with other condensing enzymes using malonyl-CoA (chalcone and stilbene synthases, and KAS III, a condensing enzyme of plant fatty acid synthase). This supports the notion that FAE1 is a gene encoding the condensing enzyme of plant elongases.

This assumption was strengthened by a further study carried out by Lassner *et al.* (1996). These authors partially purified the condensing enzyme of the elongating complex from jojoba embryos. Peptides were generated from purified proteins and subjected to microsequencing. From the resulting sequences, primers for reverse transcription and PCR experiments were designed. A library was screened and a cDNA of interest was isolated. It encoded a 57–59 kDa protein sharing a 52% amino acid identity with the *Arabidopsis* FAE1 gene product. Hydrophobicity analysis showed that the protein encoded by the isolated cDNA is an integral membrane protein containing between four and seven *trans*-membrane domains. The jojoba cDNA was further ligated adjacent to a regulatory sequence derived from a *Brassica rapa* napin gene, and used to transform low erucic acid rapeseed. The resulting transgenic seed oil contained a high level of VLCFAs. Moreover, it was shown that transformation of canola with the isolated jojoba cDNA restored the condensing activity to developing embryos *in vitro*.

In addition, and as mentioned above (see *II.1 KCS*), Millar and Kunst (1997) demonstrated that the introduction of FAE1 alone is sufficient for the production of VLCFAs both in yeast and plants.

Recently, two sequences homologous to the FAE1 gene were isolated from a *Brassica napus* immature embryo cDNA library. The sequences of the two cDNAs were highly homologous yet distinct (Barrett *et al.*, 1998).

Regulation of elongating activity

Regulation *in vitro* in absence of inhibitor

Type II (acyl-CoA) elongases are membrane-bound enzymes using amphiphilic substrates (long chain acyl-CoAs), and their kinetic parameters are strictly dependent upon the local (membrane) availability of the substrates, which has to be determined. The first important parameter to consider is the partition coefficient of the substrates and the effect of metabolism (other than the metabolism under study) on the distribution of these substrates between the membrane and aqueous phases. It was shown that the partition of acyl-CoAs, between microsomes and the aqueous phase is dependent upon the amount of proteins: the binding of a constant amount of acyl-CoAs as a function of the membrane protein amount was hyperbolic and the density of the substrate in the membrane firstly increased for a lower amount of membranes, reached a maximum and then decreased (Juguelin, Bessoule & Cassagne, 1991). This must be taken into account because it is advisable to consider the density of the substrate within the membrane rather than its 'global concentration' in the incubation mixture in order to analyse the enzyme activity.

A kinetic model has been developed and applied to the case of elongases from *Allium porrum* microsome's (Bessoule, Lessire & Cassagne, 1989*b*; Schneider *et al.*, 1993). It has been demonstrated that:

(i) The concentration of long chain acyl-CoAs (S_m) within the membranes is several hundred times higher than that present in the aqueous phase (S_{aq}). This may help to understand the decreased activity for a given total acyl-CoA concentration when the enzyme is solubilised: in the absence of a bilayer structure, the apparent K_m will be strongly increased. In good agreement, the value of the K_m for the purified enzyme is higher than 100 μM. The insertion into a phospholipid bilayer results in the increase of the actual acyl-CoA concentration in the vicinity of the enzyme and in the significant decrease of the apparent K_m (1.7 μM, quite similar to that observed for the microsomal native enzyme). An interesting development of this mathematical analysis is the indication that the poor activity of the purified, delipidated enzyme may reflect the low concentration of the substrate (S_{aq}) compared to S_m rather than a denaturation of the elongase. Conversely, when the purified elongase is incorporated into lipid vesicles, the observed increase of the specific activity could be due to the decrease of the apparent K_m induced by a change of the enzyme environment and not necessarily to an increase of the catalytic constant ('renaturation').

(ii) As observed for many enzymes using acyl-CoAs, the curve of the activity of the elongase as a function of the acyl-CoA concentration is sigmoidal. Mathematical analysis and the resulting model (Bessoule *et al.*, 1989*b*) made it possible to interpret the transformation of the sigmoid curve of the elongating activity observed when the elongase was measured in native microsomes (i.e. when the membrane-bound acyl-CoAs were also the substrates of acyl-CoA thioesterases and acyl-CoA transacylase), to a hyperbolic curve when purified elongase was included in the phospholipid vesicles, i.e. when no other enzymes using acyl-CoAs were present in the membrane.

(iii) These considerations do not apply to the case of hydrosoluble malonyl-CoA. Accordingly, the apparent K_m for malonyl-CoA seems to be constant whatever the environmental conditions of the acyl-CoA elongase (over 50 μM for both microsomal and purified elongases from *Allium porrum* epidermal cells or rapeseeds. In fairly close agreement, the elongating activity increases as a function of malonyl-CoA concentration, at least up to 100 μM (Lessire *et al.*, 1989*a*; Créach *et al.*, 1993*a*).

(iv) The global concentration of acyl-CoAs interferes with the level of activity not only at low (as discussed in (ii)) but also at high concentrations: in microsomes from leek seedlings and in developing seeds of *Brassica napus* (Schneider *et al.*, 1993; Domergue, 1997), the variations of the acyl-CoA elongases as a function of acyl-CoA concentrations are typical of an inhibition due to an excess of substrate. The results so far obtained have been interpreted by the occurrence on the condensing enzyme of a second inhibitory site of binding of acyl-CoAs.

Inhibitors of VLCFA biosynthesis

Some years ago, it was shown that several molecules, such as fluoride and thiocarbamates, are potent inhibitors of fatty acid elongation in plants (for example Harwood & Stumpf, 1971; Bolton & Harwood, 1976*b*; Abulnaja & Harwood, 1991; Barrett & Harwood, 1993). More recently, various other potential inhibitors of elongating systems have been studied. Briefly, two types of molecules have been investigated:

(i) the potential inhibitors having a common functional (polar) head and variable aliphatic chain lengths;
(ii) the potential inhibitors having variable functional groups but a constant aliphatic chain length (20 carbon atoms).

The cerulenin family

There is a large body of evidence that cerulenin is a potent inhibitor of fatty acid synthases (FAS) in animal cells, plant cells and bacteria. Cerulenin acts chiefly on the condensing enzyme of the FAS after covalent binding to the active site of the enzyme.

In seeds, no cerulenin inhibition was detected when a particulate fraction of rapeseeds was used as enzyme source (F. Domergue, unpublished results), while in leek leaves Agrawal *et al.* (1984) reported a selective inhibition of the ATP-dependent formation of C_{22}–C_{28} acids by cerulenin. A similar effect was detected in rapidly expanding leaves by Evenson and Post-Beittenmiller (1995) who reported the inhibition of elongase by 2 μM, cerulenin and showed that two polypeptides (65 and 88 kDa) could be labelled with [^{3}H] cerulenin.

A comprehensive study of the effect of cerulenin on the elongation of acyl-CoAs by leek seedling microsomes showed that the 3-ketoacyl-CoA synthase of the elongase(s) is the target of the drug. The kinetic data were best explained by a non-competitive inhibition of the elongase upon cerulenin addition (Schneider *et al.*, 1993). Two analogs of cerulenin, Cer 16 (with 16 carbon atoms) and Cer 18 (with 18 carbon atoms) have also been used. It was shown that Cer 16 and 18 were able to inhibit both acyl-CoA elongases and the 'ATP-dependent' elongase. The analogues acted on the first reaction of the elongation process, i.e on the 3-oxoacyl-CoA synthase. Interestingly, Cer 18 had no effect on *de novo* fatty acid synthesis (Schneider *et al.*, 1996).

The potential inhibitors of elongases having a constant side chain and variable functional groups

Since the above results indicate the critical role played by side chain length, a series of molecules having 20 carbon atoms and various functional groups was chemically synthesised to analyse their effect on the various reactions catalysed by the acyl-CoA elongases. Although the following results have to be considered very carefully simply because of their preliminary nature, they already provide important information.

(i) Analogues of the 3-oxoacyl-CoA derivatives and particularly the 'dioxolane' family, ethylene-acetals of the icosanoyl-CoA and of the 3-hydroxyacyl-CoA derivatives inhibit the elongation process.
(ii) When the 'dioxolane' family is checked, no effect is observed unless the molecule is thioesterified to CoASH.
(iii) The 'dioxolane-CoA' seems to inhibit specifically the condensation step.
(iv) The 'dioxolane-CoA' inhibits equally well both the acyl-CoA elongases and 'ATP-dependent elongase' from leek seedlings.
(v) The 'dioxolane-CoA' seems to be a potent inhibitor of the elongases from rabbit sciatic nerves and its effect is highly similar to that observed with the elongases from leek seedlings.
(vi) Very preliminary results have been obtained concerning the effect of 'dioxolane-CoAs' on the FAS from leek seedlings. An almost linear stimulation of the activity seems to be induced by the addition to the reaction mixture of up to 100 μM 'dioxolane-CoA' (Schneider *et al.*, 1996).

Regulation of elongating activity *in vivo*

Besides the regulation of the VLCFA synthesis by elongases (see above 'Acyl-CoA elongation mechanism, condensation reaction'), the nature and the amount of VLCFA in plant cells could be controlled by several enzymes and other factors.

ACCases

In plant, acetyl-CoA carboxylases are located both in plastids and in cytosol. Whereas in cytosol this enzyme occurs in a homodimeric form ('eukaryotic structure'), in most plant species (i.e. other than *graminae*) plastid ACCases are found in a heteromeric form ('prokaryotic structure'). In *graminae* the plastid ACCases also have an eukaryotic structure (e.g. Alban, Baldet & Douce, 1994; Konishi *et al.*, 1996; Konishi & Sasaki, 1994; Sasaki, Konishi & Nagano, 1995)

ACCases are involved in the regulation of fatty acid synthesis in plant cells *in vivo* (Roesler *et al.*, 1997). Consequently, plastid ACCases must regulate the amount of VLCFA synthesised at least by controlling the amount of stearic or oleic acid produced by plastids, and further used as substrates (as CoA-esters) by elongases. To our knowledge the synthesis of CoA-esters by acyl-CoA synthetase has never been described as a regulatory step in fatty acid metabolism (elongation, transacylation, . . .).

Cytosolic ACCase synthesises the second substrate of the elongases: malonyl-CoA. Because elongases have a low affinity for malonyl-CoA (see above), the intracellular concentration of this compound is an important parameter of the VLCFA synthesis. The level of activity of cytosolic ACCase is then also a critical parameter for the VLCFA synthesis.

Acyl-ACP thioesterases

These enzymes play an essential role in chain termination during *de novo* fatty acid synthesis, e.g. in plants that accumulate petroselinic acid, a petroselinoyl-ACP thioesterase (distinct from the oleoyl-ACP thioesterase) was isolated (Dörmann, Frentzen & Ohlrogge, 1994), and transformation of *Arabidopsis* with a lauroyl-ACP thioesterase cDNA (from California Bay) resulted in the accumulation of laurate into seeds of this plant (Voelker *et al.*, 1992; Yuan, Voelker & Hawkins, 1995; Eccleston *et al.*, 1996). Moreover, expression of a palmitoyl-ACP thioesterase in *Brassica napus* led to the production of a 16 : 0-rich oil (Jones, Davies & Voelker, 1995).

More than 20 years ago, it was shown that both elongation and decarboxylation of saturated fatty acids (precursors wax molecules) occur in leek epidermal cells rather than in other tissues (Kolattukudy, 1968; Cassagne, 1970; Cassagne & Lessire, 1974). Very interestingly, an acyl-ACP thioesterase specific for stearoyl-ACP was recently purified from this tissue, and was shown to be expressed primarily, if not exclusively, in epidermal tissue (Liu & Post-Beittenmiller, 1995). This enzyme could be important for the production of high amounts of saturated VLCFA in plant epidermis.

Other enzymes

Other genes and enzymes play a role in controlling the amount and the nature of VLCFAs in plant cells (see, for example, Lemieux *et al.*,

1990). For instance, a decrease in stearoyl-ACP desaturase activity may also increase the level of saturated VLCFAs (Knutzon *et al.*, 1992). Katavic *et al.* (1995) reported the characterisation of a seed lipid mutant with a reduced level of 20 : 1 (and 18 : 1). They suggested that the mutation resulted in a reduced DAG acyltransferase activity, so that triacylglycerol biosynthesis was also reduced. Lysophosphatidic acyltransferase is also involved in the limitation of the erucic acid content in oil: as this enzyme in *Brassica napus* does not accept VLCFA, triacylglycerols do not contain erucic acid esterified to the sn-2 position of the glycerol backbone. Therefore, a maximal theoretical value of 66% of VLCFA could be expected in these oils. To obtain higher values, transformation of *Brassica napus* with a lysophosphatidic acyltransferase from Limnanthes (which accept VLCFAs as substrate) was recently carried out (for references: see below: VLCFA biotechnology).

Other parameters

Only a few studies have been devoted to the regulation of VLCFA synthesis in leaves (von Wettstein-Knowles, 1995; Harwood, 1996). The modulation of VLCFA synthesis by light has been shown in maize (von Wettstein-Knowles, Avato & Mikkelsen, 1980). The label distribution differed according to the nature of the segment suggesting that light promotes C_{32} acid synthesis. Using aetiolated leek seedlings Moreau *et al.* (1984) demonstrated that VLCFA synthesis was maximal using 7-day-old seedlings.

The regulation of oil seed content and fatty acid composition has been extensively studied particularly in *Brassica napus*. Downey and Craig (1964) demonstrated that the erucic acid content was controlled in an additive manner by two distinct genes. Moreover, studies concerning lipid synthesis and fatty acid composition during seed development have been reported describing three different steps in the oil accumulation, with a maximum of erucic acid content between 7 and 9 weeks after anthesis (Fowler & Downey, 1970; Norton & Harris, 1983; Murkherjee, 1983; Murphy & Cummins, 1989; Lardans & Trémolières, 1991; Perry & Harwood, 1993). Accordingly, Créach *et al.* (1992) showed that cell-free extracts prepared from developing rapeseeds presented the highest $C_{18:1}$-CoA elongase activity 6–8 weeks after anthesis.

Abscisic acid (ABA) has been suggested as a potential control signal during the reserve accumulation phase of *Brassica napus* seed development (Finkelstein *et al.*, 1985). Seed storage-protein regulation and lipid accumulation have been studied and the response to exogenous ABA addition demonstrated (Delisle & Crouch, 1989; Taylor *et al.*, 1990). The effects of ABA on lipid synthesis by early cotyledon stage embryos of *B. napus* were analysed by Finkelstein and Somerville (1989). Under all conditions tested, exogenous ABA, present in the culture media (0.1 to 10 μM), stimulated the accumulation of $C_{20:1}$ and $C_{22:1}$ acids. The modulation of very long chain fatty acid content has also been reported by Holbrook, Majus and Taylor (1992). Microspore-derived embryos treated with 10 μM ABA had a total fatty acid content 40% higher than the control. The levels of $C_{20:1}$ and $C_{22:1}$ acids increased 3.4 fold after ABA treatment, corresponding to the stimulation of the $C_{18:1}$-CoA elongation. The major effect was the 3.5-fold increase of [1-^{14}C] $C_{22:1}$ synthesis. Finally, Zou *et al.* (1995) demonstrated that VLCFA synthesis was highly stimulated when microspore-derived embryos of *Brassica napus* were grown in the presence of (+)-ABA or 8′-OH ABA.

ABA was not the only modulating factor of the oil content and fatty composition of the seed. Osmotic pressure also influenced the distribution of fatty acids between the pathways leading to desaturation or elongation (Finkelstein & Somerville, 1989). The level of erucic acid in developing seeds and microspore derived embryos of oil seed rape was also dependent upon temperature (Canvin, 1965; Albrecht, Möllers & Rëbbelen, 1994; Wilmer, Helsper & Van der Plas, 1996). When comparing the oil composition of seeds grown on plants kept at 15 or 25 °C, more oil and $C_{22:1}$ acid accumulated in seeds grown at 15 °C than those grown at 25 °C. Growth temperature and ABA effects upon erucic acid level in rapeseed oil have been studied simultaneously (Wilmer *et al.*, 1997, 1998). It has been shown that temperature and ABA act independently. In addition, these authors have found that differences in total elongase activity in microspore-derived embryos grown under different culture conditions correlate closely with the amount of erucic acid.

VLCFA biotechnology

Biotechnological alterations of plant lipids developed greatly in recent years (for reviews: Knauf, 1987; Battey, Schmid & Ohlrogge, 1989; Hills & Murphy, 1991; Taylor, 1991; Murphy, 1992, 1994; Taylor *et al.*, 1992*b*; Ohlrogge, 1994; Töpfer, Martini & Schell, 1995; Slabas, Simon & Elborough, 1995; Miquel & Browse, 1995; Budziszewski, Croft & Hildebrand, 1996).

After focusing on the medium and long chain fatty acids, new approaches are being developed to increase VLCFA production (Murphy, 1992; Leonard, 1994), particularly the erucic acid ($C_{22:1}$,Δ13) which is an important renewable raw material used in plastic film manufacture (Leonard, 1994), in the synthesis of nylon 13,13, and in the lubricants and emollient industries.

In the case of rapeseed, lysophosphatidic acid acyl transferase (LPAAT) discriminates against very long chain *acyl*-CoAs (Bernerth & Frentzen, 1990; Frentzen, this volume; Taylor *et al.*, 1991). To increase the erucic acid content of rapeseed oil, strategies have been developed for introducing into *Brassica napus* a gene encoding a LPAAT able to esterify the acyl moiety of erucoyl-CoA to the OH in position 2 of the glycerol backbone, as in *Limnanthes alba* (Cao, Oo & Huang, 1990; Löhden, Berneth & Frentzen, 1990; Hills & Murphy, 1991; Taylor, 1991; Taylor *et al.*, 1992*b*; Murphy, 1994; Ohlrogge, 1994).

cDNAs encoding LPAAT were isolated from developing seeds of *Limnanthes alba* and *Limnanthes douglasii* and expressed in developing seeds of high erucic acid rapeseed (HEAR) (Lassner *et al.*, 1995; Brough *et al.*, 1997). The introduction of a cDNA encoding the LPAAT from yeast into *Arabidopsis* has also recently been reported (Zou *et al.*, 1997). Since the introduction of the cDNA encoding the 3-ketoacyl-CoA synthase of the acyl-CoA elongase complex in a canola variety has been successful in restoring the erucic acid mutation (Lassner *et al.*, 1996), the LPAATs and condensing enzyme genes could be combined together in the same plant to explore the possibility of developing oils with extremely high VLCFA content.

Two recent studies tending to increase the oil yield have been reported. The homomeric form of acetyl CoA carboxylase (ACCase) of *Arabidopsis* has been targeted successfully to plastids of *Brassica napus*. When ACCase was over-expressed in plastids of developing rapeseed embryos, the fatty acid content of the seeds increased by approximately 5% (Roesler *et al.*, 1997). Oleic acid increased from 15.4 to 20%, whereas the erucic acid yield was unchanged. The second approach (Zou *et al.*, 1997) is to introduce a yeast LPAAT into rapeseed which leads to transgenic plants exhibiting up to 25% increases in oil content.

Concluding remarks

Although much has been done in the past years to improve our knowledge of the fatty acyl elongation in higher plants, many problems are still unresolved. This is largely due to difficulties inherent to the complex – and as yet unclear – structural organisation of the membrane-bound elongases, to their multiple intracellular location in leaf and probably in seeds, to the amphiphilic nature of at least one of their substrates, and finally to the necessary modulation of their activity by surrounding (and often competing) enzymes.

(i) From a structural point of view, elongases seem to be composed of an association of 4 distinct subunits. This is most logical since acyl-CoA substrate is not covalently bound to any subunit (i.e. no transacylation steps are required), and only four reactions have to be catalysed. However, this seemingly simple situation has to be further analysed in greater detail. Whereas C18-CoA and C20-CoA elongases have distinct molecular weight (350 kDa or 600 kDa), intracellular location (ER or Golgi apparatus) and substrate specificity, it is not established whether the limiting step of the whole process (namely the condensation step) is catalysed by structurally distinct enzymes. Even if it is likely that at least two condensing enzymes exist, the fact that the structural differences (conferring different specificities) are lost upon purification suggests different organisations of the same subunits. Virtually nothing is known about this 'dimerisation' of acyl-CoA elongases.

(ii) Besides the respective structures of the C18-CoA and C20-CoA elongases, the differences between acyl-CoA elongases and ATP-dependent elongases, as well as their respective 'functions' in various plant cells, remain totally unknown. In addition, the role of ATP in the type I elongases remains mysterious: whereas Agrawal *et al.* (1984) proposed that ATP was required to form acyl-CoA, recent contributions (Hlousek-Radojcic *et al.*, 1995) have convincingly shown that acyl-CoAs are probably not the direct substrates. The nature of this substrate remains to be elucidated.

(iii) Leaf elongases which play a key role in wax production are also of interest from two points of view: (a) agricultural and pharmaceutical industries require increasing amounts of natural waxes which, for their vast majority, are imported into Europe; (b) leaf wax acts as a barrier limiting the penetration of xenobiotics, and strategies are being devised to lower that barrier by acting at the level of wax formation. In both cases a modulation of elongating activities in leaves could prove important.

(iv) The main point to be developed in the next few years is to increase the production of very long chain fatty acids in seeds and particularly of erucic acid. The data so far indicate that the (over)expression of FAE1 (i.e. KCS) induces a synthesis of VLCFA in plants, irrespective of the expression of additional enzymes such as KCR, HCD and ECR (Millar & Kunst, 1997). Besides pointing to the complex questions of structural organisation, expression, functioning and regulation of membrane-bound elongases, these results (see also Knutzon & Knauf, this volume) strongly suggest that VLCFA production in plants should be controlled in a very near future.

Acknowledgements

The work of the authors presented in this paper was supported by Conseil Régional d'Aquitaine, Cetiom and Onidol. Part of the study was conducted under the Bioavenir programme/groupe de recherches 'Barrières Cuticulaires' financed by Rhône-Poulenc.

The authors are greatly indebted to Drs: Dooner H., Hills M.J., Jaworski J., Murphy D.J., Ohlrogge J.B., Post-Beittenmiller D., Somerville C.R and Taylor D.C. for providing information and to Drs Kunst L., Lassner M.W., Roscoe T. and Wilmer J.A., for sending us manuscripts prior to publication.

References

Abdul-Karim, T., Lessire, R. & Cassagne, C. (1982). Involvement of an acyl-coenzyme A transacylase in the insertion of C_{20}–C_{30} fatty acids in the microsomal lipids from leek epidermal cells. *Physiologie Végétale*, **20**, 679–89.

Abulnaja, K.O. & Harwood, J.L. (1991). Thiocarbamate herbicides inhibit fatty acid elongation in a variety of monocotyledons. *Phytochemistry*, **30**(5), 1445–7.

Agrawal, V.P., Lessire, R. & Stumpf, P.K. (1984). Biosynthesis of very long chain fatty acids in microsomes from epidermal cells of *Allium porrum*, L. *Archives of Biochemistry and Biophysics*, **230**, 580–9.

Agrawal, V.P. & Stumpf, P.K. (1985*a*). Characterization and solubilization of an acyl chain elongation system in microsomes of leek epidermal cells. *Archives of Biochemistry and Biophysics*, **240**, 154–65.

Agrawal, V.P. & Stumpf, P.K. (1985*b*). Elongation systems involved in the biosynthesis of erucic acid from oleic acid in developing *Brassica juncea* seeds. *Lipids*, **20**, 361–6.

Alban, C., Baldet, P. & Douce, R. (1994). Localization and characterization of two structurally different forms of acetyl-CoA carboxylase in young pea leaves, of which one is sensitive to aryloxyphenoxypropionate herbicides. *Biochemical Journal*, **300**, 557–65.

Albrecht, S., Möllers, C. & Rëbbelen, G. (1994). Selection for fatty acid composition in microspore-derived embryoids (MDE) of rapeseed, *Brassica napus* (L.). *Journal of Plant Physiology*, **143**, 526–9.

Applequist, L.A. (1969). Lipids in *Cruciferae*. IV. Fatty acid patterns in single seeds populations of various *Cruciferae* and in different tissues of *Brassica napus*, L. *Hereditas*, **61**, 9–44.

Barrett, P.B. & Harwood, J.L. (1993). The basis of thiocarbamate action on surface lipid synthesis in plants. *Weeds*, **3C-2**, 183–8.

Barrett, P.B., Delourme, R., Renard, M., Domergue, F., Lessire, R., Delseny, M. & Roscoe, T.J. (1998). The rapeseed FAE1 gene is linked to the E1 locus associated with variation in the content of erucic acid. *Theoretical and Applied Genetics*, **96**, 177–86.

Battey, J.F., Schmid, K.M. & Ohlrogge, J.B. (1989). Genetic engineering for plant oils: potential and limitations. *TIBTECH*, **7**, 122–5.

Bernerth, R. & Frentzen, M. (1990). Utilization of erucoyl-CoA by acyltransferases from developing seeds of *Brassica napus* (L.) involved in triacylglycerol biosynthesis. *Plant Science*, **67**, 21–8.

Bertho, P., Moreau, P., Juguelin, H., Gautier, M. & Cassagne, C. (1989). Monensin-induced accumulation of neosynthesized lipids and fatty acids in a Golgi fraction prepared from etiolated leek seedlings. *Biochimica et Biophysica Acta*, **978**, 91–6.

Bertho, P., Moreau, P., Morré, D.J. & Cassagne, C. (1991). Monensin blocks the transfer of very long chain fatty acid containing lipids to the plasma membrane of leek seedlings. Evidence for lipid sorting based on fatty acyl chain length. *Biochimica et Biophysica Acta*, **1070**, 127–34.

Bessoule, J.J. (1989). Les acyl-CoA élongases d'*Allium porrum*: solubilisation, purification partielle, essais de reconstitution et analyses théorique de leur fonctionnement. PhD thesis, Université Bordeaux, 2.

Bessoule, J.J., Lessire, R., Rigoulet, M., Guérin, B. & Cassagne, C. (1987). Fatty acid synthesis in mitochondria from *Saccharomyces cerevisiae FEBS Letters*, **214**(1), 158–62.

Bessoule, J.J., Lessire, R., Rigoulet, M., Guérin, B. & Cassagne, C. (1988). Localization of the synthesis of very-long-chain fatty acid in mitochondria from *Saccharomyces cerevisiae*. *European Journal of Biochemistry*, **177**, 207–11.

Bessoule, J.J., Lessire, R. & Cassagne, C. (1989*a*). Partial purification of the acyl-CoA elongase of *Allium porrum* leaves. *Archives of Biochemistry and Biophysics*, **268**, 475–84.

Bessoule, J.J., Lessire, R. & Cassagne, C. (1989*b*). Theoretical analysis of the activity of membrane-bound enzymes using amphiphilic or hydrophobic substrates. Application to the acyl-CoA elongases from *Allium porrum* cells and to their purification. *Biochimica et Biophysica Acta*, **983**, 35–41.

Bessoule, J.J., Créach, A., Lessire, R. & Cassagne, C. (1992). Evaluation of the amount of acyl-CoA elongases in leek (*Allium porrum* L.) leaves. *Biochimica et Biophysica Acta*, **1117**, 78–82.

Bolton, P. & Harwood, J.L. (1976*a*). Fatty acid synthesis in aged potato slices. *Phytochemistry*, **15**, 1501–6.

Bolton, P. & Harwood, J.L. (1976*b*) Effect of thiocarbamate herbicides on fatty acid synthesis by potato. *Phytochemistry*, **15**, 1507–9.

Brough, C.L., Coventry, J., Christie, W., Kroon, J., Barsby, T. &

Slabas, A. (1997). Engineering trierucin into oilseed rape by the introduction of a 1-acyl-sn-glycerol-3-phosphate acyltransferase from *Limnanthes douglassii*. In *Physiology, Biochemistry and Molecular Biology of Plant Lipids*, ed. J.P. Williams, M.U. Khan & N.W. Lem, pp. 392–4. Dordrecht, Germany: Kluwer Academic Publishers.

Budziszewski, G.J., Croft, K.P.C. & Hildebrand, D.F. (1996). Uses of biotechnology in modifying plant lipids. *Lipids*, **31**, 557–69.

Canvin, D.T. (1965). The effect of temperature on the oil content and fatty acid composition of the oils from several oil seed crops. *Canadian Journal of Botany*, **43**, 63–9.

Cao, Y.Z., Oo, K.C. & Huang, A.H.C. (1990). Lysophosphatidate acyltransferase in the microsomes from maturing seeds of meadowfoam (*Limnanthes alba*). *Plant Physiology*, **94**, 1199–206.

Cassagne, C. (1970). Physiologie végétale. Incorporation d'acétate radioactif dans les hydrocarbures des feuilles d'*Allium porrum*, L. *Comptes-Rendus à l'Académie des Sciences, Paris*, **270**, 3055–8.

Cassagne, C. & Lessire, R. (1974). Studies on alkane biosynthesis in epidermis of *Allium porrum*, L. leaves. *Archives of Biochemistry and Biophysics*, **165**, 274–80.

Cassagne, C. & Lessire, R. (1978). Biosynthesis of saturated very long chain fatty acids by purified membrane fractions from leek epidermal cells. *Archives of Biochemistry and Biophysics*, **191**, 146–52.

Cassagne, C. & Lessire, R. (1982). On the nature and further metabolism of the products of the microsomal elongase(s) from leek epidermal cells. In *Biochemistry and Metabolism of Plant Lipids*, ed. J.F.G.M. Wintermans & P.J.C. Kuiper, pp. 79–82. Amsterdam: Elsevier Biomedical Press.

Cassagne, C., Lessire, R., Bessoule, J.J., Moreau, P., Créach, A., Schneider, F. & Sturbois-Balcerzak, B. (1994). Biosynthesis of very long chain fatty acids in higher plants. *Progress in Lipid Research*, **33**, 55–69.

Cook, L., Nagi, M.N., Suneja, S.K., Hand, A.R. & Cinti, D.L. (1992). Evidence that β hydroxyacyl-CoA dehydrase purified from rat liver microsomes is of peroxisomal origin. *Biochemical Journal*, **287**, 91–100.

Créach, A. (1994). Biosynthèse de l'acide érucique dans la graine de colza (*Brassica napus, L.*) en cours de maturation. Solubilisation et purification partielle de la 18 : 1-CoA élongase. PhD thesis, Université Bordeaux, 2.

Créach, A., Lessire, R. & Cassagne, C. (1992). Biosynthesis of very long-chain fatty acids by subcellular fractions of developing rapeseeds. In *Metabolism, Structure and Utilization of Plant Lipids*, ed. A. Cherif, D.B. Miled-Daoud, B. Marzouk, A. Smaoui & M. Zarrouk, pp. 428–31. Tunis: CNP.

Créach, A., Lessire, R. & Cassagne, C. (1993*a*). Kinetics of $C_{18:1}$-CoA elongases and transacylation in rapeseeds. *Plant Physiology and Biochemistry*, **31**(6), 923–30.
Créach, A. & Lessire, R. (1993*b*). Solubilization of acyl-CoA elongases from developing rapeseed (*Brassica napus*, L.). *Journal of the American Oil Chemists' Society*, **70**(11), 1129–33.
Créach, A., Domergue, F. & Lessire, R. (1995). Study of the partially purified C18:1-CoA elongase from developing rapeseeds (*Brassica napus*, L.). In *Plant Lipid Metabolism*, ed. J.C. Kader & P. Mazliak, pp. 121–3. The Netherlands: Kluwer Academic Publishers.
Delisle, A.J. & Crouch, M.L. (1989). Seed storage protein transcription and RNA levels in *Brassica napus* during development and in response to exogenous abscissic acid. *Plant Physiology*, **91**, 617–23.
Domergue, F. (1997). Caractérisation et purification de la 3-hydroxy-acyl-CoA déhydratase et de la 3-cétoacyl-CoA synthase, composantes de l'oléoyl-CoA élongase de *Brassica napus*. PhD thesis, Université Victor Ségalen, Bordeaux, 2.
Dörmann, P., Frentzen, M. & Ohlrogge, J.B. (1994). Specificities of the acyl–acyl carrier protein (ACP) thioesterase and glycerol-3-phosphate acyltransferase for octadecenoyl-ACP isomers. *Plant Physiology*, **104**, 839–44.
Downey, R.K. & Craig, B.M. (1964). Genetic control of fatty acid biosynthesis in rapeseed (*Brassica napus*, L.). *Journal of the American Oil Chemists' Society*, **41**, 475–8.
Eccleston, V.S., Crammer, A.M., Voelker, T.A. & Ohlrogge, J.B. (1996). Medium-chain fatty acid biosynthesis and utilization in *Brassica napus* plants expressing lauroyl-acyl carrier protein thioesterase. *Planta*, **198**, 46–53.
Evenson, K.J. & Post-Beittenmiller, D. (1995). Fatty acid elongating activity in rapidly expanding leek epidermis. *Plant Physiology*, **109**, 707–16.
Fehling, G. & Mukherjee, K.D. (1991). Acyl-CoA elongase from a higher plant (*Lunaria annua*): metabolic intermediates of very-long-chain acyl-CoA products and substrate specificity. *Biochimica et Biophysica Acta*, **1082**, 239–47.
Fehling, E., Lessire, R., Cassagne, C. & Mukherjee, K.D. (1992). Solubilization and partial purification of constituents of acyl-CoA elongase from *Lunaria annua*. *Biochimica et Biophysica Acta*, **1126**, 88–94.
Finkelstein, R. & Somerville, C. (1989). Abscisic acid or high osmoticum promote accumulation of long-chain fatty acids in developing embryos of *Brassica napus*. *Plant Science*, **61**, 213–17.
Finkelstein, R.R., Tenbarge, K.M., Shumway, J.E. & Crouch, M.L. (1985). Role of ABA in maturation of rapeseed embryos. *Plant Physiology*, **78**, 630–6.

Fowler, A., Simon, J.W., Fawcett, T. & Slabas, A.R. (1995). Soluble and membrane associated enoyl reductases in *Brassica napus*. In *Plant Lipid Metabolism*, ed. J.C. Kader & P. Mazliak, pp. 93–5. Netherlands: Kluwer Academic Publishers.

Fowler, D.B. & Downey, R.K. (1970). Lipid and morphological changes in developing rapeseed, *Brassica napus*, L. *Canadian Journal of Plant Sciences*, **50**, 233–47.

Harwood, J. (1980). Plant acyl lipids: structure, distribution, and analysis. In *The Biochemistry of Plants*, ed. P.K. Stumpf & E.E. Conn, pp. 1–55. New York: Academic Press.

Harwood, J.L. (1996). Recent advances in the biosynthesis of plant fatty acids. *Biochimica et Biophysica Acta*, **1301**, 7–56.

Harwood, J.L. & Stumpf, P.K. (1971). Fat metabolism in higher plants. XLIII. Control of fatty acid synthesis in germinating seeds. *Archives of Biochemistry and Biophysics*, **142**, 281–91.

Hills, M.J. & Murphy, D.J. (1991). Biotechnology of oilseeds. *Biotechnology and Genetic Engineering Reviews*, **9**, 1–45.

Hitchcock, C. & Nichols, B.W. (1971). The lipid and fatty acid composition of specific tissues. In *Plant Lipid Biochemistry*, pp. 1–44. London: Academic Press.

Hlousek-Radojcic, A., Imai, H. & Jaworski, J.G. (1995). Oleoyl-CoA is not an immediate substrate for fatty acid elongation in developing seeds of *Brassica napus*. *Plant Journal*, **8**(6), 803–9.

Holbrook, L.A., Majus, J.R. & Taylor, D.C. (1992). Abscisic acid induction of elongase activity, biosynthesis and accumulation of very long chain monounsaturated fatty acids and oil body proteins in microspore-derived embryos of *Brassica napus*, L. cv Reston. *Plant Science*, **84**, 99–115.

Huang, A.H.C. (1996). Oleosins and oil bodies in seeds and other organs. *Plant Physiology*, **110**, 1055–61.

Imai, H., Hlousek-Radojcic, A., Mathis, A. & Jaworski, J. (1995). Elongation system involved in the biosynthesis of very long chain fatty acids in *Brassica napus* seeds: characterization and solubilization. In *Plant Lipid Metabolism*, ed. J.C. Kader & P. Mazliak, pp. 118–20. The Netherlands: Kluwer Academic Publishers.

James, D.W. Jr., Lim, E., Keller, J., Plooy, I., Ralston, E. & Dooner, H.K. (1995). Directed tagging of the *Arabidopsis* fatty acid elongation1 (FAE1) gene with the maize transposon activator. *Plant Cell*, **7**, 309–19.

Jones, A., Davies, H.M. & Voelker, T.A. (1995). Palmitoyl-acyl carrier protein (ACP) thioesterase and the evolutionary origin of plant acyl-ACP thioesterases. *Plant Cell*, **7**, 359–71.

Juguelin, H., Bessoule, J.J. & Cassagne, C. (1991). Interaction of amphiphilic substrates (acyl-CoAs) and their metabolites (free fatty acids) with microsomes from mouse sciatic nerves. *Biochimica et Biophysica Acta*, **1068**, 41–51.

Katavic, V., Reed, D.W., Taylor, D.C., Giblin, E.M., Barton, D.L., Zou, J., MacKenzie, S.L., Covello, P.S. & Kunst, L. (1995). Alteration of seed fatty acid composition by an ethyl methanesulfonate-induced mutation in *Arabidopsis thaliana* affecting diacylglycerol acyltransferase activity. *Plant Physiology*, **108**, 399–409.

Knauf, V.C. (1987). The application of genetic engineering to oilseed crops. *TIBTECH*, **5**, 40–7.

Knutzon, D.S., Thompson, G.A., Radke, S.E., Johnson, W.B., Knauf, V.C. & Kridl, J.C. (1992). Modification of *Brassica* seed oil by antisense expression of a stearoyl-acyl carrier protein desaturase gene. *Proceedings the National Academy of Sciences, USA*, **89**, 2624–8.

Kolattukudy, P.E. (1968). Further evidence for an elongation–decarboxylation mechanism in the biosynthesis of paraffins in leaves. *Plant Physiology*, **43**, 375.

Kolattukudy, P.E. & Buckner, J.S. (1972). Chain elongation of fatty acid by cell-free extracts of epidermis from pea leaves (*Pisum sativum*). *Biochemical and Biophysical Research Communications*, **46**(2), 801–7.

Kollatukudy, P.E., Croteau, R. & Buckner, J.S. (1976). Biochemistry of plant waxes. In *Chemistry and Biochemistry of Natural Waxes*, ed. P.E. Kolattukudy, pp. 289–347. Amsterdam: Elsevier.

Konishi, T. & Sasaki, Y. (1994). Compartmentalization of two forms of acetyl-CoA carboxylase in plants and the origin of their tolerance toward herbicides. *Proceedings of the National Academy of Sciences, USA*, **91**, 3598–601.

Konishi, T., Shinohara, K., Yamada, K. & Sasaki, Y. (1996). Acetyl-CoA carboxylase in higher plants: most plants other than gramineae have both the prokaryotic and the eukaryotic forms of this enzyme. *Plant Cell Physiology*, **37**(2), 117–22.

Kunst, L., Taylor, D.C. & Underhill, E.W. (1992). Fatty acid elongation in developing seeds of *Arabidopsis thaliana*. *Plant Physiology and Biochemistry,* 30, 425–34.

Lardans, A. & Trémolières, A. (1991). Accumulation of C_{20} and C_{22} unsaturated fatty acids in triacylglycerols from developing seeds of *Limnanthes alba*. *Phytochemistry*, **30**(12), 3955–61.

Lardans, A. & Trémolières, A. (1992). Fatty acid elongation activities in subcellular fractions of developing seeds of *Limnanthes alba*. *Phytochemistry*, **31**, 121–7.

Lassner, M.W., Levering, C.K., Davies, M. & Knutzon, D.S. (1995). Lysophosphatidic acid acyltransferase from meadowfoam mediates insertion of erucic acid at the sn-2 position of triacylglycerol in transgenic rapeseed oil. *Plant Physiology*, **109**, 1389–94.

Lassner, M.W., Lardizabal, K. & Metz, J.G. (1996). A jojoba β-ketoacyl-CoA synthase cDNA complements the canola fatty acid elongation mutation in transgenic plants. *Plant Cell*, **8**, 281–92.

Lassner, M.W. (1997). Transgenic oilseed crops: a transition from basic research to product development. *Lipid Technology*, **9**(1), 5–9.
Lemieux, B., Miquel, M., Somerville, C. & Browse, J. (1990). Mutants of *Arabidopsis* with alterations in seed lipid fatty acid composition. *Theoretical and Applied Genetics*, **80**, 234–40.
Leonard, C. (1994). Sources and commercial applications of high-erucic vegetable oils. *Lipid Technology*, **4**, 79–83.
Lessire, R., Hartmann-Bouillon, M.A. & Cassagne, C. (1982). Very long chain fatty acids: occurrence and biosynthesis in membrane fractions from etiolated maize coleoptiles. *Phytochemistry*, **21**(1), 55–9.
Lessire, R., Abdul-Karim, T. & Cassagne, C. (1982). Origin of the wax very long chain fatty acids in leek, *Allium porrum* L., leaves: a plausible model. In *The Plant Cuticle*, ed. D. F. Cutler, K. L. Alvin & C. E. Price, pp. 167–80, London: Academic Press.
Lessire, R., Juguelin, H., Moreau, P. & Cassagne, C. (1985*a*). Nature of the reaction product of [1-^{14}C] stearoyl-CoA elongation by etiolated leek seedling microsomes. *Archives of Biochemistry and Biophysics*, **239**, 260–9.
Lessire, R., Bessoule, J.J. & Cassagne, C. (1985*b*). Solubilization of C_{18}-CoA and C_{20}-CoA elongases from *Allium porrum*, L. epidermal cell microsomes. *FEBS Letters*, **187**, 314–20.
Lessire, R., Juguelin, H., Moreau, P. & Cassagne, C. (1985*c*). Elongation of acyl-CoAs by microsomes from etiolated leek seedlings. *Phytochemistry*, **24**(6), 1187–92.
Lessire, R., Bessoule, J.J. & Cassagne, C. (1989*a*). Properties of partially purified acyl-CoA elongase from *Allium porrum* leaves. In *Biological Role of Plant Lipids*, ed. P.A. Biacs, K. Gruiz & T. Kremmer, pp. 131–4. New York and London: Akadémiai Kiado Budapest and Plennum Publishing Corporation.
Lessire, R., Bessoule, J.J. & Cassagne, C. (1989*b*). Involvement of a β-ketoacyl-CoA intermediate in acyl-CoA elongation by an acyl-CoA elongase purified from leek epidermal cells. *Biochimica et Biophysica Acta*, **1006**, 35–40.
Lessire, R., Schneider, F., Bessoule, J.J., Cook, L., Cinti, D. & Cassagne, C. (1992). Characterization of the intermediate reactions involved in the leek icosanoyl-CoA synthase. In *Metabolism, Structure and Utilization of Plant Lipid*, ed. A. Cherif, D. Ben Miled, B. Marzouk, A. Smaoui & Zarrouk, pp. 144–7. CNP Tunis, Tunisia.
Lessire, R., Bessoule, J.J., Cook, L., Cinti, D.L. & Cassagne, C. (1993). Occurrence and characterization of a dehydratase enzyme in the leek icosanoyl-CoA synthase complex. *Biochimica et Biophysica Acta*, **1169**, 243–9.
Liu, D. & Post-Beittenmiller, D. (1995). Discovery of an epidermal stearoyl-acyl carrier protein thioesterase. *Journal of Biological Chemistry*, **270**(28), 16 962–9.

Löhden, I., Bernerth, R. & Frentzen, M. (1990). Acyl-CoA: 1-acylglycerol-3-phosphate acyltransferase from developing seeds of *Limnanthes douglasii* (R. Br.) and *Brassica napus* (L.). In *Plant Lipid Biochemistry, Structure and Utilization*, ed. J.L. Harwood & P.J. Quinn, pp. 175–7. London: Portland Press Ltd.

Lucet-Levannier, K., Lellouche, J.P. & Mioskowski, C. (1995). Polysilylated coenzyme A for a high-yielding preparation of very lipophilic acyl coenzyme A in anhydrous organic solvents. *Journal of American Chemical Society*, **117**, 7546–7.

Macey, M.J.K. & Stumpf, P.K. (1968). Fat metabolism in higher plants XXXVI: long-chain fatty acid synthesis in germinating peas. *Plant Physiology*, **43**, 1637–47.

Maneta-Peyret, L., Sturbois-Balcerzak, B., Cassagne, C. & Moreau, P. (1997). Antibodies to long-chain acyl-CoAs. A new tool for lipid biochemistry. *Biochimica et Biophysica Acta*, **1389**(1), 50–6.

Millar, A.A. & Kunst, L. (1997). Very-long chain fatty acid biosynthesis is controlled through the expression and specificity of the condensing enzyme. *Plant Journal*, **12**(1), 121–31.

Miquel, M. & Browse, J. (1995). Molecular biology of oilseed modification. *Inform*, **6**, 108–11.

Moreau, P., Juguelin, H., Lessire, R. & Cassagne, C. (1984). *In vivo* incorporation of acetate into the acyl moieties of polar lipids from etiolated leek seedlings. *Phytochemistry*, **23**, 67–71.

Moreau, P., Juguelin, H., Lessire, R. & Cassagne, C. (1986). Intermembrane transfer of long chain fatty acid synthesized by etiolated leek seedlings. *Phytochemistry*, **25**, 387–91.

Moreau, P., Juguelin, H., Lessire, R. & Cassagne, C. (1988*a*). Plasma membrane biogenesis in higher plants: *in vivo* transfer of lipids to the plasma membrane. *Phytochemistry*, **27**, 1631–8.

Moreau, P., Bertho, P., Juguelin, H. & Lessire, R. (1988*b*). Intracellular transport of very long chain fatty acids in etiolated leek seedlings. *Plant Physiology and Biochemistry*, **26**, 173–8.

Moreau, P., Sturbois-B., Morré, D.J. & Cassagne, C. (1994). Effect of low temperatures on the transfer of phospholipids with various acyl-chain lengths to the plasma membrane of leek cells. *Biochimica et Biophysica Acta*, **1194**, 239–46.

Mukherjee, K.D. & Kiewitt, I. (1980). Formation of (n-9) and (n-7) *cis*-monounsaturated fatty acids in seeds of higher plants. *Planta*, **149**, 461–3.

Mukherjee, K.D. (1983). Lipid biosynthesis in developing mustard seed. Formation of triacylglycerol from endogenous and exogenous fatty acids. *Plant Physiology*, **73**, 929–34.

Mukherjee, K.D. (1986). Elongation of (n-9) and (n-7) *cis*-monounsaturated and saturated fatty acids in seeds of *Sinapis alba*. *Lipids*, **21**, 347–52.

Murphy, D.J. (1992). Modifying oilseed crops for non-edible products. *TIBTECH*, **10**, 84–7.
Murphy, D.J. (1994). Transgenic plants – a future source of novel edible and industrial oils. *Lipid Technology*, **4**, 84–91.
Murphy, D.J. & Cummins, I. (1989). Biosynthesis of seed strorage products during embryogenesis in rapeseed, *Brassica napus* L. *Journal of Plant Physiology*, **135**, 63–9.
Murphy, D.J. & Mukherjee, K.D. (1988). Biosynthesis of very long monounsaturated fatty acids by subcellular fractions of developing seeds. *FEBS Letters*, **230**, 101–4.
Murphy, D.J. & Mukherjee, K.D. (1989). Elongases synthesizing very long chain monounsaturated fatty acids in developing oilseeds and their solubilization. *Zeitschrift für Naturforschung*, **44**, 629–34.
Napier, J.A., Stobart, A.K. & Shewry, P.R. (1996). The structure and biogenesis of plant oil bodies: the role of the ER membrane and the oleosin class of proteins. *Plant Molecular Biology*, **31**, 945–56.
Norton, G. & Harris, J.F. (1983). Triacylglycerols in oilseed rape during seed development. *Phytochemistry*, **22**, 2703–7.
Nugteren, D.H. (1965). The enzymic chain elongation of fatty acids by rat-liver microsomes. *Biochimica et Biophysica Acta*, **106**, 280–90.
Ohlrogge, J.B. (1994). Design of new plant products: engineering of fatty acid metabolism. *Plant Physiology*, **104**, 821–6.
Padley, F.B., Gunstone, F.D. & Harwood, J.L. (1994). Occurrence and characteristics of oils and fats. In *The Lipid Handbook*, ed. F.D. Gunstone, J.L. Harwood & F.B. Padley, pp. 47–223. London: Chapman & Hall.
Perry, H.J. & Harwood, J.L. (1993). Changes in the lipid content of developing seeds of *Brassica napus*. *Phytochemistry*, **32**, 1411–15.
Pollard, M.R. & Stumpf, P.K. (1980*a*) Biosynthesis of C_{20} and C_{22} fatty acids by developing seeds of *Limnanthes alba*. *Plant Physiology*, **66**, 641–8.
Pollard, M.R. & Stumpf, P.K. (1980*b*) Long chain (C_{20} and C_{22}) fatty acids biosynthesis in developing seed of *Tropealum majus*. An *in vivo* study. *Plant Physiology*, **66**, 649–55.
Pollard, M.R., McKeon, T., Gupta, L.M. & Stumpf, P.K. (1979). Studies on biosynthesis of waxes by developing jojoba seed. II. Demonstration of wax biosynthesis by cell-free homogenates. *Lipids*, **14**, 651–62.
Post-Beittenmiller, D. (1996). Biochemistry and molecular biology of wax production in plants. *Annual Review of Plant Physiology and Plant Molecular Biology*, **47**, 405–30.
Roesler, K., Shintani, D., Savage, L., Boddupalli, S. & Ohlrogge, J.B. (1997). Targeting of the *Arabidopsis* homomeric acetyl-coenzyme A carboxylase to plastids of rapeseeds. *Plant Physiology*, **113**, 75–81.

Sasaki, Y., Konishi, T. & Nagano, Y. (1995). The compartmentation of acetyl-coenzyme A carboxylase in plants. *Plant Physiology*, **108**, 445–9.

Schneider, F. (1995). Vers une modulation des cires épicuticulaires chez *Allium porrum*, L. Inhibition du transfert et de la biosynthèse des acides gras à très longue chaîne. PhD thesis, Université de Bordeaux, 2.

Schneider, F., Lessire, R., Bessoule, J.J., Juguelin, H. & Cassagne, C. (1993). Effect of cerulenin on the synthesis of very-long-chain fatty acids in microsomes from leek seedlings. *Biochimica et Biophysica Acta*, **1152**, 243–52.

Schneider, F., Boiron-Sargueil, F., Bessoule, J.J., Lessire, R., Moreau, P., Levannier-Lucet, K., Lellouche, J.P., Mioskowsky, C. & Cassagne, C. (1996). Towards a specific inhibition of plant leaf elongases. In *Oils, Fats, Lipids*, vol. 1, pp. 95–100. P.J. Barnes & Associates, Publishers.

Settlage, S.B., Wilson, R.F. & Kwanyuen, P. (1995). Localization of diacylglycerol acyltransferase of oil body associated endoplasmic reticulum. *Plant Physiology and Biochemistry*, **33**(4), 399–407.

Slabas, A.R., Simon, J.W. & Elborough, K.M. (1995). Information needed to create new oil crops. *Inform*, **6**, 159–66.

Sturbois-Balcerzak, B., Morré, D.J., Loreau, O., Noel, J.P., Moreau, P. & Cassagne, C. (1995). Effects of low temperatures on the transfer of phospholipids to the plasma membrane and on the morphology of the ER-Golgi apparatus-plasma membrane pathway of leek cells. *Plant Physiology and Biochemistry*, **33**(6), 625–37.

Spinner, C., Levannier, K., Lessire, R., Lellouche, J.P., Mioskowski, C. & Cassagne, C. (1995). Characterization of the trans 2–3 enoyl-CoA reductase of the acyl-CoA elongase from leek (*Allium porrum*, L.). In *Plant Lipid Metabolism*, ed. J.C. Kader & P. Mazliak, pp. 124–6. The Netherlands: Kluwer Academic Publishers.

Taylor, D.C., Weber, N., Underhill, E.W., Pomeroy, M.K., Keller, W.A., Scowcroft, W.R., Wilen, R.W., Moloney, M.M. & Holbrook, L.A. (1990). Storage-protein regulation and lipid accumulation in microspore embryos of *Brassica napus* L. *Planta*, **181**, 18–26.

Taylor, D.C., Weber, N., Barton, D.L., Underhill, E.W., Hogge, L.R., Weselake, R.J. & Pomeroy, M.K. (1991). Triacylglycerol bioassembly in microspore-derived embryos of *Brassica napus*, L. cv Reston. *Plant Physiology*, **97**, 65–8.

Taylor, D.C., Barton, D.L., Rioux, K.P., MacKenzie, S.L., Reed, D.W., Underhill, E.W. Pomeroy, M.K. & Weber, N. (1992*a*). Biosynthesis of acyl lipids containing very-long chain fatty acids in microspore-derived and zygotic embryos of *Brassica napus*, L. cv Reston. *Plant Physiology*, **99**, 1609–18.

Taylor, D.C., Magus, J.R., Bhella, R., Zou, J., MacKenzie, S.L., Giblin, E.M., Pass, E.W. & Crosly, W.L. (1992*b*). Biosynthesis of

triacylglycerols in *Brassica napus*, L. cv. Reston; Target: Trierucin. In *Seeds Oils for the Future*, ed. S.L. MacKenzie & D.C. Taylor, pp. 77–102. Champaign, USA: AOCS Press.
Taylor, P. (1991). Plant lipid biotechnology. *Lipid Technology*, **1**, 9–15.
Töpfer, R., Martini, N. & Schell, J. (1995). Modification of plant lipid synthesis. *Science*, **268**, 681–6.
Tulloch, A.P. (1976). Chemistry of waxes of higher plants. In *Chemistry and Biochemistry of Natural Waxes*, ed. P.E. Kolattukudy, pp. 235–87. Amsterdam: Elsevier.
Voelker, T.A., Worrell, A.C., Anderson, L., Bleibaum, J., Fan, C., Hawkins, D.J., Radke, S.E. & Davies, H.M. (1992). Fatty acid biosynthesis redirected to medium chains in transgenic oilseed plants. *Science*, **257**, 72–4.
Von Wettstein-Knowles, P.M. (1979). Genetics and biosynthesis of plant epicuticular waxes. In *Advances in the Biochemistry and Physiology of Plant Lipids*, ed. L.A. Appelqvist & C. Lijenberg, pp. 1–26. Elsevier: North-Holland Biomedical Press.
Von Wettstein-Knowles, P.M. (1995). Biosynthesis and genetics of waxes. In *Waxes: Chemistry, Molecular Biology and Functions*, ed. R.J. Hamilton, pp. 91–130. Alloury, Ayr, Scotland: Oily Press.
Von Wettstein-Knowles, P.M. Avato, P. & Mikkelsen, J.D. (1980). Light promotes synthesis of the very long fatty acyl chains in maize wax. In *Biogenesis and Function of Plant Lipids*, ed. P. Mazliak, P. Benveniste, C. Costes &, R. Douce, pp. 271–4. Elsevier: North Holland Biomedical Press.
Walker, K.A. & Harwood, J.L. (1986). Evidence for separate elongation enzymes for very-long-chain-fatty acid synthesis in potato (*Solanum tuberosum*). *Biochemical Journal*, **237**, 41–6.
Whitfield, H.V., Murphy, D.J. & Hills, M.J. (1993). Sub-cellular localization of fatty acid elongase in developing seeds of *Lunaria annua* and *Brassica napus*. *Phytochemistry*, **32**, 255–8.
Willemot, C. & Stumpf, P.K. (1967). Fat metabolism in higher plants. XXXIII. Development of fatty acid synthetase during the 'aging' of storage tissue slices. *Canadian Journal of Botany*, **45**, 579–84.
Wilmer, J.A., Helsper, J.P.F.G. & Van der Plas, L.H.W. (1996). Effect of growth temperature on erucic acid levels in seeds and microspore-derived embryos of oilseed rape, *Brassica napus* L. *Journal of Plant Physiology*, **147**, 486–92.
Wilmer, J.A., Helsper, J.P.F.G. & Van der Plas, L.H.W. (1997). Effects of abscisic acid and temperature on erucic acid accumulation in oilseed rape (*Brassica napus*, L.). *Journal of Plant Physiology*, **150**, 414–20.
Willmer, J.A., Lessire, R., Helsper, J.P.F.G. & van der Plas, L.H.W. (1998). Regulation of elongase activity by abscissic acid and

temperature in microscope-derived embryos of oilseed rape (*Brassica napus*). *Physiologia Plantarum*, **102**, 185–91.

Yermanos, D.M. (1975). Composition of jojoba seed during development. *Journal of the American Oil Chemists' Society*, **52**, 115–20.

Yuan, L., Voelker, T.A. & Hawkins, D.J. (1995). Modification of the substrate specificity of an acyl–acyl carrier protein thioesterase by protein engineering. *Proceedings of the National Academy of Sciences, USA*, **92**, 10 639–43.

Zou, J., Abrams, G.D., Barton, D.L., Taylor, D.C., Pomeroy, M.K. & Abrams, S.R. (1995). Induction of lipid and oleosin biosynthesis by (+)-abscisic acid and its metabolites in microspore-derived embryos of *Brassica napus*, L. cv Reston. *Plant Physiology*, **108**, 563–71.

Zou, J.T., Katavic, V., Giblin, E.M., Barton, D.L., MacKenzie, S.L., Keller, W.A., Hu, X. & Taylor, D.C. (1997). Modification of seed oil content and acyl composition in the brassicaceae by expression of a yeast sn-2 acyltransferase gene. *The Plant Cell*, **9**, 909–23.

PART III: Complex lipids: assembly, genetic manipulation and environmental aspects

A.K. STOBART, S. STYMNE, P.R. SHEWRY
and J. NAPIER

9 Triacylglycerol biosynthesis

Introduction

Vegetable oils and fats have many applications in the food, chemical and pharmaceutical industries. Triacylglycerol, the major constituent of the oil, consists of a glycerol backbone to which a fatty acid acyl group is esterified to each of the three hydroxy groups. The quality of the oil, and hence its end use, is determined by this fatty acid composition and, to some extent, the positional acyl distribution in the molecule. Most commercial annual oil-crops have seed-oils largely composed of the five major fatty acids, palmitic (16 : 0), stearic (18 : 0), oleic (18 : 1Δ9), linoleic (18 : 2Δ9,12) and alpha-linolenic (18 : 3Δ9,12,15) acid. These are the 'house-keeping' fatty acids, so-called because they are also found in the membrane lipids of all plant cells. The fatty acid composition of membrane lipid in leaves and roots, however, is under strict control and differs little between plant species whereas the relative abundance of a fatty acid in the storage triacylglycerol can show distinct variations between different oil-crops. Although the natural variation in the fatty acid composition of oil-crop species gives a range of oil quality, there would be many advantages if an oleaceous species could be tailored to produce a 'designer' oil to suit a particular end use (Stobart, Stymne & Shewry, 1992).

The increased understanding of plant gene expression (Verma, 1992) and the successful characterisation of cDNA encoding genes for the synthesis of many fatty acids is leading to the manipulation of oil quality and the production of transgenic varieties to suit commercial needs. The Kingdom Plantae offers an enormous genetic resource and several hundred uncommon fatty acids have been identified in the seed oils of numerous species (Badami & Patil, 1981; Hilditch & Williams, 1964). Many of these would be of commercial value if available in quantity and at moderate cost. Thus the genetic engineering of oil quality would not only help satisfy the demand of the oleochemical industry but also create new and novel oils designed for precise end use. Here, triacylglycerol biosynthesis is described and, where appropriate, its relationship to the formation of fatty acids in the endoplasmic reticulum discussed.

Triacylglycerol assembly

Saturated fatty acids up to the C18 level and oleic acid (18 : 1Δ9) are synthesised in the plastid. All further modifications of the acyl-chain and the assembly of triacylglycerol are accomplished by membrane-bound enzymes localised in the endoplasmic reticulum (ER). The oil is released from the endoplasmic reticulum membranes and stored as discrete oil-bodies. Maximum oil deposition in the seed usually occurs over a narrow time window after pollination. In species such as sunflower and safflower over 60% of the oil is laid down over a period of a few days some 15 days after flowering and when the cells of the cotyledon are undergoing cell expansion. Seed cotyledons in this phase of development, yield microsomal preparations capable of massive oil assembly *in vitro*. Given suitable substrates, microsomal preparations produce triacylglycerol in mass and this accumulates as visible oil-droplets in the incubation medium (Stobart, Stymne & Hoglund, 1986) .

The assembly of the triacylglycerol proceeds through three consecutive acylation steps (Fig. 1). In the first, the acyl moiety from acyl-CoA is esterified to the *sn*-1 position of glycerol 3-phosphate to yield lysophosphatidic acid. The glycerol 3-phosphate acyltransferase (GPAT) catalysing this reaction, generally has a broad acyl specificity and will accept both saturated and unsaturated fatty acids (Ichihara, 1984; Bafor, Stobart & Stymne, 1990*a*). In some species, however, which produce an unusual seed-specific fatty acid, the GPAT may exhibit a greater specificity for the acid than the enzyme from plants which lack it (Dutta, Appleqvist & Stymne, 1992; Bafor & Stymne, 1992). A puzzling exception is erucic acid in high erucic rape and turnip rape. The GPAT from rape *in vitro* has very low activity with erucoyl-CoA and this is in a species particularly rich in erucate at the *sn*-1 position of the triacylglycerol (Bafor *et al.*, 1990a; Taylor *et al.*, 1991).

The enzyme, lysophosphatidic acid acyl transferase (LPAAT), which catalyses the acylation of lysophosphatidic acid at the *sn*-2 position to yield phosphatidic acid has, in most oil seeds, a strict selectivity for unsaturated fatty acids (Griffiths, Stobart & Stymne, 1985; Ichihara, Asahi & Fujii, 1987). As a general rule, therefore, plant triacylglycerols have mainly unsaturated fatty acids at position *sn*-2. Interesting exceptions are plants with triacylglycerols rich in medium chain fatty acids (C8–C14) and especially in the genus *Cuphea* (Graham, Hirsinger &

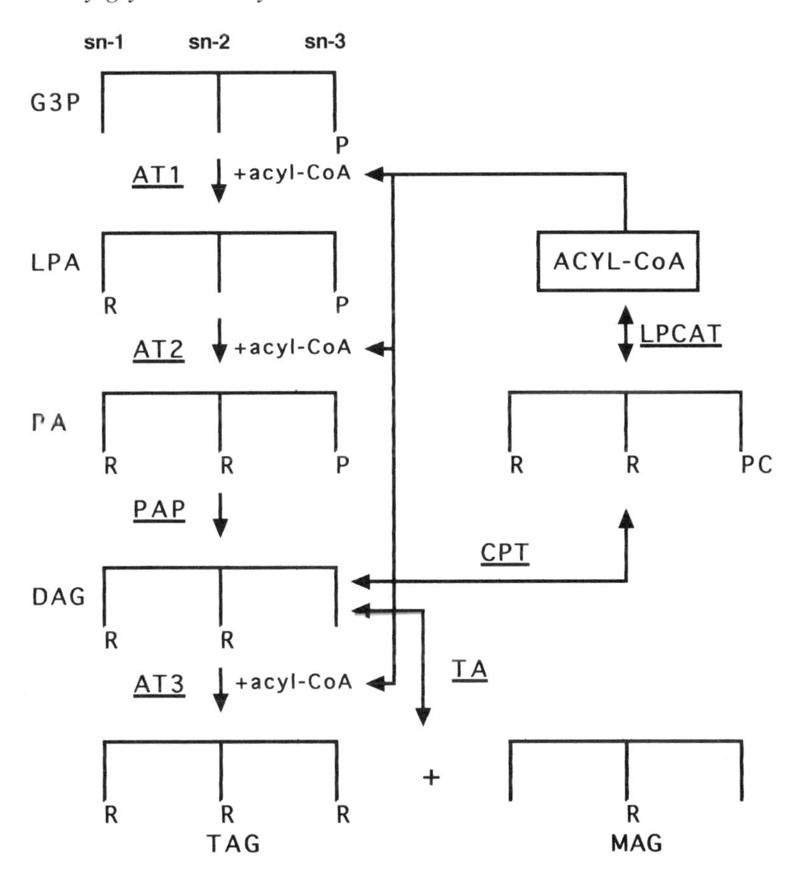

Fig. 1. Glycerol 3-phosphate (G3P) undergoes acylation to yield lysophosphatidic acid (LPA) and phosphatidic acid (PA) through the action of glycerol phosphate acyltransferase (AT1) and lysophosphatidic acid acyltransferase (AT2), respectively. A phosphatidic acid phosphohydrolase (PAP) gives rise to diacylglycerol (DAG) which undergoes acylation, catalysed by diacylglycerol acyltransferase (AT3), to form triacylglycerol (TAG). Oleic acid (18: 1) from the acyl-CoA pool can enter *sn*-2 of *sn*-phosphatidylcholine (PC), through the activity of lysophosphatidylcholine acyltransferase (LPCAT), for desaturation to C-18 polyunsaturated fatty acids. The desaturated fatty acid products in phosphatidylcholine enter the acyl-CoA pool by the reverse reaction of LPCAT. During the movement of glycerol backbone towards triacylglycerol through the so-called Kennedy pathway, diacylglycerol can freely interconvert with phosphatidylcholine through cholinephosphotransferase (CPT) activity. Diacylglycerols also give rise to triacylglycerol and monoacylglycerol(MAG) by a transacylation (TA), which is freely reversible.

Röbbelen, 1981). The acyl specificities of the LPAAT from three different *Cuphea* species have been investigated and have similar properties. Medium chain acyl-CoA substrates are readily utilised. Long chain acyl-CoA esters, however, were only used if the acyl acceptor, lysophosphatidic acid, contained a long chain acyl group (Bafor *et al.*, 1990*b*; Bafor & Stymne, 1992). The LPAATs from oil seed species which accumulate triacylglycerols with only long chain fatty acids lack these properties. These enzymes utilise medium chain lysophosphatidates poorly and acylate them at a considerable higher rate with long chain acyl-CoA than medium chain acyl-CoA (Oo & Huang, 1989). The rather special properties of the *Cuphea* LPAAT enables the cell to synthesise phosphatidic acid in two major molecular species consisting of medium chain and long chain acyl groups, respectively, and with little of the mixed acyl species. This feature is probably a necessity in *Cuphea* in order to maintain correct membrane lipid assembly and function (see below).

The phosphatidic acid is converted to diacylglycerol and inorganic phosphate through the action of the enzyme, phosphatidate phosphohydrolase. In oil seeds, the enzyme appears to be present in both a soluble and membrane-bound form and these may have a regulatory function in lipid synthesis (Ichihara, Murota & Fujii, 1990). Recently good purification of this enzyme has been achieved and is currently under investigation (Pearce & Slabas, 1997).

The final step in triacylglycerol biosynthesis is the acylation of the diacylglycerol at the *sn*-3 position and catalysed by diacylglycerol acyltransferase (DAGAT). The acylation of diacylglycerol is the only dedicated step in triacylglycerol assembly since the diacylglycerols can also be used in membrane lipid biosynthesis. There are many plant species with unusual fatty acids in their seed oil. These are often major constituents in the triacylglycerol and are almost absent in the seed membrane lipid. It is probable that the presence of many of these unusual acids in membrane lipid would severely impair proper membrane function. How particular acyl species of diacylglycerol are channelled almost exclusively to triacylglycerol is not fully understood. Although experimental data suggest that the mechanisms might differ depending on the type of 'triacylglycerol specific' fatty acid, it appears that the DAGAT must play an important role in this regulation. Diacylglycerols with acyl groups that are absent in the seed phospholipid have to be either excluded from the formation of membrane lipid and/or the unusual fatty acids are rapidly removed from synthesised complex polar lipid. It is, of course, possible that phospholipid is synthesised from a diacylglycerol pool spatially separated from that used in triacylglycerol production. Although experimental evidence so far favours the first two proposals, the observation that purified membrane preparations from oil seeds can be obtained that appear only to catalyse the progressive acylation of the glycerol backbone and oil assembly is of interest and requires further investigation (Lacey & Hills, 1996).

In *Cuphea* seeds, rich in medium chain fatty acids, the molecular species of phosphatidic acid, as mentioned above, consist of either long chain or medium chain fatty acids. Diacylglycerol species will also exist therefore, which are identical in composition to the phosphatidic acid pools. *Cuphea lanceolata* and *Cuphea procumbens* have seed triacylglycerols with 90% capric acid (C10 : 0) whereas membrane lipids, such as phosphatidylcholine, contain only trace amounts of this acid. It is noteworthy that, although long chain acyl groups represent only 10% triacylglycerol acid, they predominate in the diacylglycerol pool (Bafor *et al.*, 1990*b*). This suggests a selective utilisation of medium chain diacylglycerol by the DAGAT leaving the diacylglycerol pool enriched in long chain acyl species for phospholipid production. Such DAGAT acyl selectivity was demonstrated in microsomal fractions from developing oil seeds (Wiberg, Tillberg & Stymne, 1992) The *Cuphea* DAGAT was highly selective for dicaproyl-glycerol against dioleoylglycerol when oleoyl-CoA was an acyl donor and was extremely efficient in constructing tricaproyl-glycerol. A comparison with the sunflower enzyme, suggests that the selectivity for medium chain diacylglycerol and medium chain acyl-CoA by the *Cuphea* DAGAT is a specialisation of the enzyme in tissues accumulating high levels of medium chain triacylglycerol. Castor bean endosperm accumulates triacylglycerol rich in the hydroxylated fatty acid, ricinoleic acid, whereas the phospholipids contain only trace amounts of this acid. The DAGAT of castor, when presented with a mixture of diricinoleoyl- and dioleoylglycerols, selected the ricinoleoyl species (Bafor *et al.*, 1991).

The substrate selectivity properties therefore, of the DAGAT enzyme in seeds accumulating triacylglycerol specific fatty acids, explains, to some extent, the diversion of diacylglycerol moieties, which contain the unusual fatty acid, away from phospholipid synthesis. This, however, may not be the complete explanation. Ricinoleic acid and the epoxy acid, vernolic acid, are synthesised from phospholipid acyl substrates, and then removed by specific phospholipase enzymes of the A-type to yield free fatty acids and lysophospholipids (see below). The lyso derivative can be reacylated with an acceptable common fatty acid (Bafor *et al.*, 1991). If such oxygenated acyl groups were channelled into phospholipid from diacylglycerol, they would be removed in the same way. In tissues accumulating medium chain fatty acids, it is likely that the phospholipid synthesising enzymes are selective against diacylglycerols with medium chain acyl groups, although it cannot be excluded that these acids, if entering the phospholipid, are also recognised by specific phospholipases.

Diacylglycerol acyltransferase activity has been reported in oil-body preparations from the developing mesocarp of oil-palm and this enzyme had different acyl specificity to the enzyme in the microsomal fraction (Oo & Chew, 1992). Germinating soy bean tissue also appears to be rich in a protein with DAGAT activity (Kwanyuen & Wilson, 1986). The significance of these enzymes in lipid metabolism requires elucidation.

Biosynthesis of triacylglycerol fatty acids

Palmitic, stearic and oleic acid are synthesised in the plastid and made available to the endoplasmic reticulum for triacylglycerol assembly and the formation of other fatty acids. Anti-sense constructs of the Δ9 stearoyl-ACP desaturase, with seed-specific promoters, inserted into oilseed rape, brought about a dramatic increase in stearate in the rape oil from the normal $<2\%$ to 40% and with a corresponding decrease in oleate (Knutzon *et al.*, 1992). Triacylglycerols with high stearate are of commercial interest as cocoa butter substitutes in confectionery fat. Oleate, from the plastid, is channelled into phospholipid, mainly phosphatidylcholine of the endoplasmic reticulum and there, depending on the plant species, sequentially desaturated by a Δ12 and Δ15 desaturase into linoleate and α-linolenate (Stymne & Stobart, 1987; Stymne, Tonnet & Green, 1992). Similar desaturation steps occur in the plastid galactolipids and are catalysed by desaturase enzymes encoded by genes which are separate from those encoding the endoplasmic reticulum proteins (Somerville & Browse, 1991). It is only the endoplasmic reticulum desaturases, however, that appear to be responsible for the production of the polyunsaturated fatty acids in the storage triacylglycerol. The Δ12 desaturation in oil seeds occurs with oleate at both the *sn*-1 and *sn*-2 positions of *sn*-phosphatidylcholine, whereas Δ15 desaturation in linseed exhibits a strong preference for substrate at position *sn*-2 (Slack, Roughan & Browse, 1979; Stobart & Stymne, 1985*a*; Griffiths *et al.*, 1985; Stymne *et al.*, 1992). The oil-seed endoplasmic reticulum Δ12 desaturase is cyanide sensitive and requires NAD(P)H and molecular oxygen. Spectrophotometric evidence shows that the safflower microsomal Δ12-desaturation is mediated through the cytochrome *b*5-cytochrome *b*5 reductase system (Smith *et al.*, 1990) and this was confirmed with antibodies to cytochrome *b*5, which inhibited desaturase activity (Kearns, Hugly & Somerville, 1991; Smith, 1992). Seed microsomal cytochrome *b*5 has been molecularly characterised for studies on desaturation and polyunsaturated fatty acid synthesis (Smith *et al.*, 1994*a*,*b*; Napier *et al.*, 1995).

Mutagenesis and conventional plant breeding programmes (Griffiths, Stymne & Stobart, 1988*a*) have produced several oil-seed varieties with blocks in both oleate and linoleate desaturase activity resulting in dramatic changes in the unsaturation level of the seed triacylglycerol (Soldatov, 1976; Green,1986; Powell *et al.*, 1990). The seeds develop normally and are viable. There were no fatty acid lesions in the other parts of the plant, indicating tissue-specific Δ12 and Δ15 desaturase genes (Powell *et al.*, 1990; Sperling *et al.*, 1990; Tonnet & Green, 1987). Such observations bode well for manipulating the polyunsaturation in oil seeds by genetic engineering. A notable exception, however, is the report that the same endoplasmic reticulum desaturase genes seem to be expressed throughout the whole plant in *Arabidopsis* (Lemieux *et al.*, 1990).

Desaturation reactions involved in the synthesis of a number of other, more uncommon, polyunsaturated fatty acids also involves endoplasmic reticulum-bound enzymes. Only γ-linolenic acid (18 : 3Δ6,9,12), however, has been studied in detail. Polyunsaturated fatty acids with a Δ6 double bond are prevalent in the *Boraginaceae* . Here, the dominating Δ6 fatty acid in leaf tissue is an octadecatetraenoic acid (stearidonic acid, 18 : 4Δ6,9,12,15) and this is particularly enriched in the galactolipids (Stymne, Griffiths & Stobart, 1987). In seed-oil, however, γ-linolenate predominates and in borage accounts for some 25% of the triacylglycerol fatty acid. The seed γ-linolenic acid is synthesised by a Δ6 NADH-dependent desaturation of linoleate esterified to largely microsomal phosphatidylcholine (Griffiths, Stobart & Stymne, 1988*b*). The Δ6 desaturase has a strong preference for linoleate at position *sn*-2 and thus the enzyme seems to have a similar substrate requirement as the linseed Δ15 desaturase. Alpha-linolenic acid acts as a substrate for the Δ6 desaturase and the formation of 18 : 4(Δ6,9,12,15). On the other hand, γ-linolenic acid does not appear to serve as a substrate for Δ15 desaturation and 18 : 4 synthesis (Griffiths *et al.*, 1996). In this context it is also of interest that borage seed cotyledons when exposed to the light produce chlorophyll and alpha-linolenate and under these conditions 18 : 4 accumulates. These observations support the suggestion that phospholipid desaturases do not recognise the acyl substrate from the methyl end, but align with the first double bond towards the carboxylic end (Griffiths *et al.*, 1989).

Gamma-linolenic acid is considered of therapeutic value in alleviating a wide range of clinical and atopic disorders (Horrobin, 1990) and γ-linolenate-rich oil from a number of species(borage, evening primrose, black currant and some fungi; Gunstone, 1992) is available for this purpose. The inclusion of a Δ6 desaturase gene in a high linoleate oil-crop could yield γ-linolenate oil for medical and dietary use and at a competitive price. In borage the membrane lipids are also enriched in γ-linolenate and the Δ6 desaturase gene appears to be expressed in the whole plant. It can be assumed therefore, that its expression in a transgenic plant would not adversely affect membrane function and this appears to be the case. Recently a cDNA has been isolated encoding the Δ6-desaturase from the developing seeds of borage (Sayanova *et al.*, 1997). Its function has been confirmed by ectopic expression in transgenic tobacco plants. The transformants grow well and contained appreciable quantities of γ-linolenic acid (13%) and stearidonic acid (10%). The polypeptide encoded by the Δ6 desaturase cDNA is unusual and contains diagnostic sequences related to cytochrome *b*5 at the N-terminus. The microsomal form of borage cytochrome *b*5 is 32% identical to the N-terminal extension of the Δ6 desaturase. Microsomal desaturases (Δ12, Δ15) that have been cloned from higher plants do not contain this cytochrome *b*5 domain (Okuley *et al.*, 1994), even though the requirement for cytochrome *b*5 has been demonstrated *in vitro* (Smith *et al.*, 1990). The borage Δ6 desaturase therefore, appears to represent a new class of fatty acid desaturase enzymes in higher plants existing as cytochrome *b*5 fusion proteins (Napier *et al.*, 1997). A related cDNA sequence with some 60% homology to the borage fusion protein has also been characterised from sunflower (Sperling, Schmidt & Heinz, 1995). It would seem, however, unlikely that Δ6 desaturation is the enzymic function encoded by the sunflower cDNA, since this species does not contain Δ6 desaturated fatty acids. It is interesting that none of the other plant desaturases (Δ12, Δ15; Okuley *et al.*, 1994) or the closely related hydroxylase from castor (Loo *et al.*, 1995) have a cytochrome *b*5 domain and use the 'free' microsomal form of the protein. It is possible that desaturation is more efficient with an endogenous haem binding domain in the protein and this awaits investigation. Also, it may be significant that the Δ6 desaturase from borage differs from the Δ12 and Δ15 desaturase enzymes in that it introduces a centre of unsaturation between an existing double bond and the carboxy carbon, i.e. a front end desaturase (Aitzetmuller & Tsevegsuren, 1994) as opposed to double bond formation towards the methyl end of the carbon chain. The borage Δ6 desaturase is the only higher plant front end desaturase characterised. It will be of interest therefore, to isolate and examine other similar desaturases as they become available.

Since phospholipids are the prime substrate for the production of linoleic and linolenic acids that are destined for plant triacylglycerol assembly, there must be efficient mechanisms by which the oleate is channelled into phospholipid for desaturation and the desaturated acyl groups made available for triacylglycerol biosynthesis. Among the phospholipids present in the endoplasmic reticulum, it is only phosphatidylcholine that undergoes rapid turnover and it is this lipid that is important in providing polyunsaturated fatty acids for triacylglycerol formation. An equilibration reaction between acyl groups in phosphatidylcholine and the acyl-CoA pool occurs in oil seeds and by this reaction oleate from oleoyl-CoA enters position *sn*-2 of phosphatidylcholine for desaturation. Concomitantly, the products from that position enter the acyl-CoA pool for the acylation of all three positions of the glycerol backbone. The reaction can be catalysed by the combined forward and reverse reactions of an acyl-CoA : lysophosphatidylcholine acyltransferase (LPCAT) (Stymne & Stobart, 1984).

Diacylglycerol may be utilised for phosphatidylcholine synthesis by a CDP-choline : diacylglycerol cholinephosphotransferase(CDP) catalysed reaction. In oil seeds, this appears to be reversible and thus the phosphatidylcholine can also re-convert with diacylglycerol (Slack *et al.*, 1985). In the combined forward and reverse reactions of the cholinephosphotransferase, the oleate at both positions of diacylglycerol can enter phosphatidylcholine for desaturation and the polyunsaturated products transferred to the diacylglycerol pool for triacylglycerol synthesis (Stobart & Stymne, 1985*b*). The relative importance of the two reactions, the acyl exchange between acyl-CoA and phosphatidylcholine and the phosphatidylcholine–diacylglycerol interconversion, in providing polyunsaturated fatty acids for triacylglycerol assembly is unclear and might differ between plant species (Griffiths, Stymne & Stobart, 1988*c*, 1988*d*). Recently the CDP has been cloned from soybean (Monks *et al.*, 1997) and this should lead to a better understanding of the role of this enzyme in influencing oil quality.

Hydroxy and epoxy fatty acids

Oils which contain hydroxy and epoxy fatty acids have been found in the seeds of a number of plant species (Badami & Patil, 1981). These fatty acids are of considerable industrial interest, but it is only one species, the castor bean, that is commercially exploited. The oil of castor bean endosperm contains some 90% ricinoleic acid (12-hydroxy-18 : 1Δ9; Achaya, Craig & Youngs, 1964). The ricinoleate is formed by hydroxy-substitution at the Δ12 carbon of oleate by a microsomal-bound hydroxylase and requires NAD(P)H and molecular oxygen. The enzyme is sensitive to cyanide but not carbon monoxide (Moreau & Stumpf, 1981) and utilises cytochrome *b*5 as part of the electron transport system (Smith *et al.*, 1992*a*). The hydroxylase uses as substrate the oleate at position *sn*-2 of microsomal phospholipids, and particularly phosphatidylcholine (Bafor *et al.*, 1991). The castor hydroxylase appears to differ from other animal and plant hydroxylases since these require P-450 (White & Coon, 1980; Donaldson & Luster, 1991). The biochemical similarity of the plant Δ12 desaturase and the Δ12 hydroxylase suggested an evolutionary relationship between the two enzymes (Stymne & Stobart, 1993). Since then the hydroxylase has been molecularly characterised and shown to be a fatty acyl desaturase homologue (van de Loo *et al.*, 1995). Expression of hydroxylase cDNA from castor bean in transgenic *Arabidopsis* gives rise in the seeds not only to ricinoleic acid but also to lesquerolic and densipolic acids (Broun & Somerville, 1997).

Although phosphatidylcholine serves as the substrate for the hydroxylase in castor bean, only some 5% ricinoleate is present in this complex lipid *in vivo*. The phosphatidylcholine ricinoleate is rapidly hydrolysed by a microsomal phospholipase to yield free ricinoleic acid and, presumably, lysophosphatidylcholine (Bafor *et al.*, 1991). The phospholipase shows little activity with oleoyl and linoleoyl acyl groups. The free ricinoleate formed could be activated to a acyl-CoA ester by a microsomal acyl-CoA synthetase and then utilised in triacylglycerol assembly via the glycerol 3-phosphate pathway. It is apparent that phospholipases with high activity for phospholipids which contain ricinoleate are not only present in membranes from species capable of synthesising ricinoleate. Microsomal membranes from developing cotyledons of rape and safflower catalyse the efficient hydrolysis of ricinoleate, as well as linoleate hydroperoxides, from phosphatidylcholine (Banas, Johansson & Stymne, 1992). Such phospholipase enzymes appear to occur generally in plants and function in the removal of oxygenated fatty acids from membrane lipid, perhaps to maintain membrane integrity and correct membrane function.

The epoxy fatty acid, vernolic acid (12-epoxy-18 : 1Δ9), accumulates in *Euphorbia lagascae* seed. Some 70% of the fatty acid in the endosperm triacylglycerol is vernolic acid, whereas the phospholipid contains only trace amounts. The vernolic acid is synthesised by epoxidation of the Δ12 double bond of the linoleate esterified to phosphatidylcholine (Smith *et al.*, 1992*b*). The epoxidase, which is microsomal, has a strict requirement for NADPH as the co-reductant. The enzyme is strongly inhibited by carbon monoxide, whilst cyanide is ineffective indicating a P-450-mediated reaction. Although alpha-

Hydroxylase CN^- sensitiv aber CO insensitiv
Epoxydase CO " " CN^- "

linolenic acid is not found in the *Euphorbia* endosperm, this could also serve as an efficient substrate for the epoxidase and give rise to 12-epoxy-18 : 2Δ9,15 (Bafor *et al.*, 1993). The vernoleate synthesised in the phospholipids of the *Euphorbia* microsomes is released as free vernolic acid, indicating that vernolic acid is removed from the membrane lipid by phospholipase action with a high specificity for oxygenated fatty acids and similar to that observed in castor bean endosperm. The epoxidase acts on linoleate at both esterified positions of phosphatidylcholine (Bafor *et al.*, 1993) and this implies the presence of phospholipase(s) in the *Euphorbia* membranes that can recognise both of these sites.

Since phospholipases, which can remove ricinoleate and other uncommon fatty acids, are present in other species (Bafor *et al.*, 1991, 1993; Stahl, Banas & Stymne, 1995), it is unlikely that such acids would accumulate in membrane lipids of transgenic oil crops. The glycerol acylating enzymes of both rape and safflower utilise ricinoleoyl substrates (Stymne *et al.*, 1990). The creation of transgenic plants by the introduction of a hydroxylase/epoxidase gene (van de Loo *et al.*, 1995; Broun & Somerville, 1997) should lead therefore, to the development of new crops producing oils for industrial applications.

Very long chain fatty acids

Many seed oils contain high levels of fatty acid longer than C18. The original varieties of rape and turnip-rape, as in most *Cruciferae* species, contain erucic acid (22 : 1Δ13). In modern varieties, however, the erucate, was eliminated through breeding programmes after the observation that dietary erucoyl oil induced heart muscle lesions in test animals. The erucic acid is synthesised by elongation of oleic acid with malonyl-CoA via the formation of 20 : 1Δ11 by similar reactions to *de novo* fatty acid synthesis. The enzymes which catalyse the elongation, however, appear to be membrane bound and localised in the endoplasmic reticulum (Fehling *et al.*, 1992). In the *Cruciferae*, the erucic acid in confined to positions *sn*-1 and *sn*-3 of the triacylglycerol. The absence of this acid from position *sn*-2 has been attributed to the inability of the rape lysophosphatidate acyltransferase(LPAAT) to utilise erucoyl-CoA (Löhden, Bernerth & Frentzen, 1990; Bernerth & Frentzen, 1990). Meadowfoam (*Limnanthes alba*) is also extremely rich in long chain fatty acids, with the majority being 20 : 1Δ5. The Δ5 double bond is formed by direct desaturation of 20 : 0 but the subcellular localisation of the elongation of 18 : 0 to 20 : 0 and the Δ5 desaturase are unknown (Pollard & Stumpf, 1980; Moreau, Pollard & Stumpf, 1981). The seeds of meadowfoam also contain some erucic acid and this is largely confined to position *sn*-2 of the triacylglycerol. It has been shown that the meadowfoam LPAAT enzyme can, contrary to the corresponding rape enzyme, efficiently acylate the 1-acyl-glycerol 3-phosphate acceptor with erucoyl-CoA (Löhden *et al.*, 1990; Cao, Oo & Huang, 1990). Introduction of the LPAAT cDNA from meadowfoam into rape has now been achieved giving rise to triacylglycerols with erucic acid at the *sn*-2 position (Lassner *et al.*, 1995; Brough *et al.*, 1997). It should be feasible therefore, to obtain an oil with a higher erucate content than the 66% that now seems the theoretical limit.

Transacylation reactions

Some observations suggest that other events to those described may also occur during oil assembly. In previous experiments with safflower microsomal membranes, triacylglycerols appeared to form from diacylglycerol, but in the absence of acyl-CoA, and hence not by the direct acylation through diacylglycerol acyltransferase activity (Stobart & Stymne, 1985*b*). At that time no explanation was offered for this. Also, it has been shown recently that preformed triacylglycerol in the embryos of sunflower at an early stage of development can become further enriched with linoleate (Garces, Sarmiento & Mancha, 1994) suggesting that fatty acids esterified to triacylglycerol, and presumably before stabilisation in the oil-body, were available to the Δ12-desaturase of the endoplasmic reticulum. Further, Mancha and Stymne (1997) have demonstrated that there is considerable remodelling of *in situ* synthesised triacylglycerol involving acyl groups from polar lipids in microsomal preparations from the developing castor bean endosperm. In attempts to reconcile these observations with our present understanding of oil assembly, the utilisation by microsomal membranes of radiolabelled neutral lipid substrates has been investigated. Safflower microsomes were found to catalyse the interconversion of mono-, di-, and triacylglycerol (Stobart, Stymne & Mancha, 1996; Stobart *et al.*, 1997). From these studies it was clear that diacylglycerol gave rise to triacylglycerol and monoacylglycerol as well as to phosphatidylcholine. Radioactivity was transferred from triacylglycerols to diacylglycerol and monoacylglycerol was rapidly converted to di-, and triacylglycerol. These interconversions occur in the absence of acyl-CoA and hence do not involve diacylglycerol acyltransferase activity. The evidence is consistent with the operation of a diacylglycerol–diacylglycerol transacylation yielding monoacylglycerol and triacylglycerol, the reaction being freely reversible. A similar transacylase reaction has been characterised in rat intestinal microsomes, where it is involved in the net synthesis of monoacylglycerol and triacylglycerol (Lehner & Kuksis, 1993). The activity of the transacylase in safflower microsomes is of the same order of magnitude as the diacylglycerol acyltransferase. Hence, in any diacylglycerol acyltransferase assays using radioactive diacylglycerol, it would not be possible to discriminate triacylglycerol formation via transacylation or direct acylation (see Vogel & Browse, 1996).

Experimental evidence also shows that the diacylglycerol, arising directly by the so-called Kennedy pathway, or via transacylation, rapidly interconverts with phosphatidylcholine, probably through the reversible action of choline phosphotransferase (Slack *et al.*, 1985). Any oleoyl substrate entering phosphatidylcholine via choline phosphotransferase is readily desaturated and the polyunsaturated fatty acid products become incorporated into the triacylglycerol oil. Transacylation, therefore, can account for the further enrichment of preformed triacylglycerol with polyunsaturated fatty acid observed *in vivo* (Garces *et al.*, 1992). It cannot be ruled out, of course, that such a desaturation can occur *in vivo* whilst the oil is still associated with the endoplasmic reticulum, and presumably before stabilisation in the oil body. In this respect a solubilised $\Delta 12$-desaturase from the chloroplast was found to utilise, to some degree, other substrates (free fatty acid, acyl-CoA and the complex lipid monogalactosyldiacylglycerol; Schmidt & Heinz, 1993). It is clear, however, that the consequence of any reaction which gives rise to diacylglycerol will also bring about the further enrichment of the glycerol backbone with C18-polyunsaturated fatty acids. Transacylation could also account for the previous observation with microsomal membranes that radioactivity in diacylglycerol would accumulate in triacylglycerol in the absence of acyl-CoA (Stobart & Stymne, 1985*b*), and that *in situ* synthesised triacylglycerols are remodelled with acyl groups from polar lipids (Mancha & Stymne, 1997). Our current understanding of triacylglycerol assembly and turnover *in vitro* in developing oil-seeds and the relationship to microsomal desaturation is outlined in Fig. 1. It is also apparent that microsomal membranes catalyse the production of lysophosphatidylcholine from monoacylglycerol (Stobart *et al.*, 1997). This appears to occur via a phosphocholine transfer between phosphatidylcholine and the monoacylglycerol, the products being diacylglycerol and lysophosphatidylcholine. The lyso derivative can be rapidly reacylated from acyl-CoA (Stymne & Stobart, 1984; Sperling & Heinz, 1993). Such a combination of reactions could yield triacylglycerol without the participation of diacylglycerol acyltransferase and without any accumulation of monoacylglycerol.

If transacylation occurs *in vivo*, what might be its biological significance? Clearly, it can provide even further opportunity for desaturation and the enrichment of the oil with C18-polyunsaturated fatty acids. The reactions which channel oleate to phosphatidylcholine for its subsequent desaturation help to concentrate substrate and give a greater chance(s) for desaturation. It is possible that desaturation on complex lipids, such as phosphatidylcholine, is not an efficient process and that a number of mechanisms have evolved, including transacylation, which enable the rapid formation of polyunsaturated fatty acids for storage and membrane lipid assembly. One might also speculate that the transacylation may play some role in the assembly of triacylglycerols containing over 70% of a particular fatty acid and particularly those of an unusual nature perhaps by giving rise to species of diacylglycerol which are immediately acylated directly with the required fatty acid. Transacylation, however, could have a more profound biochemical importance. It may help to regulate the size of an active diacylglycerol pool, a possible necessity which prevents the detrimental net movement of phosphatidylcholine to diacylglycerol (via cholinephosphotransferase) and the consumption of endoplasmic reticulum structural lipid during the period of extremely rapid oil deposition that is found in the developing seed.

Concluding remarks

In this review we have attempted to indicate our present understanding of triacylglycerol assembly and its relationship to the biosynthesis of fatty acids both common and uncommon. Where appropriate, the feasibility of manipulating oil composition in oil crops has been discussed. In 1987 (Stymne & Stobart, 1987) it was stated that, eventually through the techniques of recombinant DNA technology, it should be possible to tailor oils for research and applied purposes. In the intervening years great progress has been made in this direction and now, and at an ever-increasing rate, experimental transgenic plants with changed fatty acid profiles are being produced. It is therefore, only a matter of time before oleaceous agricultural crops with new and novel designer oils become available for exploitation.

Although fatty acid biosynthesis and oil assembly have commercial applications, studies are yielding knowledge of a more fundamental nature. For instance, the discovery of plant phospholipases which act on oxygenated acyl groups in membrane lipids has wider significance. The general occurrence of such lipase activity might indicate its possible role in membrane repair and the specific removal of oxidised acyl groups from membrane lipids. It was suggested (Sevanian & Kim,

1985; van Kuizk, Sevanian & Handelman, 1987) that phospholipase A2 in animals played an important role in the protection of membranes from peroxidative damage by selectively removing the offending fatty acid thus preventing free radical and aldehyde production. For many years it has been considered that specific phospholipase enzymes were involved in membrane lipid turnover and remodelling. The evidence from seed oxygenated fatty acid biosynthesis helps to support these proposals and provides an insight as to how plants protect their membranes from oxidative damage. In this regard it is also pertinent that acyl-CoA : lysophosphatidylcholine acyltransferase (LPCAT) in plant microsomal membranes is extremely active. This enzyme, together with specific phospholipase A2, would remodel membrane lipid and prevent the deleterious build up of harmful fatty acids and lysophospholipids. Likewise, the newly discovered transacylase in oil seeds whilst playing a role in governing the final fatty acid quality of the triacylglycerol may have fundamental biological significance and help to maintain the phospholipid integrity of the endoplasmic reticulum.

The evolutionary origin(s) of the enzymes which catalyse the biosynthesis of the unusual fatty acids found in plant triacylglycerols is also of interest. Many of these fatty acids only occur in the seed oil of a limited number of species and often in totally unrelated plant families. It is interesting therefore, to consider how the capacity to synthesise such a myriad of oil fatty acids has evolved. It is possible that the genes which code for the specific enzymes required in the synthesis of unusual acids occur generally in plants but are, in most cases, expressed at a very low level and under tight control. The random insertion of an efficient seed specific promoter could then have brought about high rates of expression in particular families and species. In this context it is interesting that coronaric acid (9-epoxy-18 : 1Δ12), a stereoisomer of vernolic acid occurring in seed oil from certain species (Badami & Patil, 1981), is synthesised by extracts from germinating pea seedlings and has a growth stimulating effect when added exogenously to plants (Grechkin *et al.*, 1989). It is, of course, also possible that genes for seed-specific fatty acids have evolved from genes that code for enzymes occurring generally in plants. The evolutionary relationship between the Δ12 hydroxylase in castor bean and the Δ12 desaturase is a case in point. The sequence similarity between the gene responsible for petroselinic acid synthesis and the Δ9 desaturase gene (Cahoon, Shanklin & Ohlrogge, 1992) could be another example.

In the next few years, as more and more genes are cloned and sequenced, many answers to the above should be forthcoming. The vast increase in knowledge therefore, which is rapidly accumulating, will not only help considerably in the production of valuable plant products but also provide a deeper insight into many important aspects of plant biology.

References

Achaya, K.T., Craig, B.M. & Youngs, C.G. (1964). Component fatty acids and glycerides of castor oil. *Journal of the American Oil Chemists Society*, **41**, 783–4.

Aitzetmuller, K. & Tsevegsuren, N. (1994). Occurrence of γ-linolenic acid in *Ranunculaceae* seed oils. *Plant Physiology*, **143**, 578–80.

Badami, R.C. & Patil, K.B. (1981). Structure and occurrence of unusual fatty acids in minor seed oils. *Progress in Lipid Research*, **19**, 119–53.

Bafor, M. & Stymne, S. (1992). Substrate specificities of glycerol acylating enzymes from developing embryos of two *Cuphea* species. *Phytochemistry*, **31**, 2973–6.

Bafor, M., Stobart, A.K. & Stymne, S. (1990*a*). Properties of the glycerol acylating enzymes in microsomal preparations from the developing seeds of safflower and turnip rape and their ability to assemble cocoa-butter type fats. *Journal of the American Oil Chemists Society*, **67**, 217–25.

Bafor, M., Jonsson, L., Stobart, A.K. & Stymne, S. (1990*b*). Regulation of triacylglycerol biosynthesis in embryos and microsomal preparations from the developing seeds of *Cuphea lanceolata*. *The Biochemical Journal*, **272**, 31–8.

Bafor, M., Smith, M.A., Jonsson, L., Stobart, K. & Stymne, S. (1991). Ricinoleic acid biosynthesis and triacylglycerol assembly in microsomal preparations from developing castor-bean endosperm. *The Biochemical Journal*, **280**, 507–14.

Bafor, M., Smith, M.A., Jonsson, L., Stobart, A.K. & Stymne, S. (1993). Biosynthesis of vernoleate (*cis*-12-epoxyoctadeca-*cis*-9-enoate) in microsomal preparations from developing endosperm of *Euphorbia lagascae*. *Archives of Biochemistry and Biophysics*, **303**, 145–51.

Banas, A., Johansson, I. & Stymne, S. (1992). Plant microsomal phospholipases exhibit preference for phosphatidylcholine with oxygenated acyl groups. *Plant Science*, **84**, 137–4.

Bernerth, R. & Frentzen, M. (1990). Utilisation of erucoyl-CoA by acyltransferases from developing seeds of *Brassica napus* involved in triacylglycerol biosynthesis. *Plant Science*, **67**, 21–8.

Brough, C.L., Coventry, J., Christie, W., Kroon, J., Barsby, T. & Slabas, A. (1997). Engineering trierucin into oilseed rape by the introduction of a 1-acyl-sn-glycerol-3-phosphate acyltransferase from *Limnanthes douglasii*. In *Physiology, Biochemistry and*

Molecular Biology of Plant Lipids, ed. J.P. Williams, M.U. Kahn & N.W. Lem, pp. 392–4. Dordrecht, Netherlands: Kluwer Academic Press.

Broun, P. & Somerville, C. (1997). Accumulation of ricinoleic, lesquerolic and densipolic acids in seeds of transgenic *Arabidopsis* plants that express a fatty acyl hydroxylase cDNA from castor bean. *American Journal of Plant Physiology*, **113**, 933–42.

Cahoon, E.B., Shanklin, J. & Ohlrogge, J.B. (1992). Expression of a novel coriander desaturase results in petroselinic acid production in transgenic tobacco. *Proceedings of the National Academy of Sciences, USA*, **89**, 11 184–8.

Cao, Y-Z., Oo, A.C. & Huang, A.H.C. (1990). Lysophosphatidic acid acyltransferase in the microsomes from the maturing seeds of meadowfoam (*Limnanthes alba*). *American Journal of Plant Physiology*, **94**, 1199–206.

Donaldson, R.P. & Luster, D.G. (1991). Multiple forms of cytochrome P-450. *American Journal of Plant Physiology*, **96**, 669–74.

Dutta, P.C., Appelqvist, L-Å. & Stymne, S. (1992). Utilisation of petroselinate (C18 : 1Δ6) by glycerol acylation enzymes in microsomal preparations of developing embryos of carrot, safflower and oil-seed rape. *Plant Science Letters*, **81**, 57–64.

Fehling, E., Lessire, R., Cassagne, C. & Mukherjee, K.D. (1992). Solubilisation and partial purification of constituents of acyl-CoA elongase from *Lunaria annua. Biochimica et Biophysica Acta*, **1126**, 88–94.

Garces, R., Sarmiento, C. & Mancha, M. (1994). Oleate from triacylglycerols is desaturated in cold-induced developing sunflower seeds. *Planta*, **193**, 507–14.

Graham, S.A., Hirsinger, F. & Röbbelen, G. (1981). Fatty acids of *Cuphea* seed lipids and their taxonomic significance. *American Journal of Botany*, **68**, 908–17.

Grechkin, A.N., Gafarova, T.E., Korolev, O.S., Kuramshin, R.A. & Tarchevsky, I.A. (1989). The monoxygenase pathway of linoleic acid oxidation in pea seedlings. In *Biological Role of Plant Lipids*, ed. P.A. Biacs, K. Gruuiz & T. Kemmer, pp. 151–3. Budapest: Académiai Kiadó, Plenum Press.

Green, A. (1986). A mutant genotype of flax containing very low levels of linolenic acid in its seed oil. *Canadian Journal of Plant Science*, **66**, 499–503.

Griffiths, G., Stobart, A.K. & Stymne S. (1985). The acylation of sn-glycerol 3-phosphate and the metabolism of phosphatidate in microsomal preparations from the developing cotyledons of safflower seed. *The Biochemical Journal*, **230**, 379–88.

Griffiths, G., Stymne, S. & Stobart, A.K. (1988*a*).Triglyceride biosynthesis. In *Biotechnology for the Fats and Oils Industry*, ed.

T.H. Applewhite, pp. 23–9. Champaign, IL: American Oil Chemists Society.

Griffiths, G., Stobart, A.K. & Stymne, S. (1988*b*). Δ6 and Δ12 desaturase activities and phosphatidic acid formation in microsomal preparations from the developing cotyledons of common borage. *The Biochemical Journal*, **252**, 641–7.

Griffiths, G., Stymne, S. & Stobart, A.K. (1988*c*).Utilisation of fatty acid substrates in triacylglycerol biosynthesis by tissue slices of developing safflower and sunflower cotyledons. *Planta*, **173**, 309–16.

Griffiths, G., Stymne, S. & Stobart, A.K. (1988*d*). Phosphatidylcholine and its relationship to triacylglycerol biosynthesis in oil tissues. *Phytochemistry*, **27**, 2089–94.

Griffiths, G., Brechany, E.Y., Christie, W.W., Stymne, S. & Stobart, A.K. (1989). Synthesis of octadecatetraenoic acid in borage. In *Biological Role of Plant Lipids*, ed. P.A. Biacs, K. Gruuiz & T. Kemmer, pp. 151–3. Budapest: Académiai Kiadó, Plenum Press.

Griffiths, G., Jackson, F., Brechany, E.Y., Christie, W.W., Stymne S. & Stobart, A.K. (1996). Octadecatetraenoic acid (C18 : 4; stearidonic acid) biosynthesis in borage (*Borago officinalis*). *Phytochemistry*, **43,** 381–6.

Gunstone, F.D. (1992). γ-Linolenic acid-occurrence and physical and chemical properties. *Progress in Lipid Research*, **31**, 145–61.

Hilditch, T.P. & Williams, P.N. (1964). *The Chemical Constitution of Natural Fats*. London: Chapman and Hall.

Horrobin, D.F. (1990). *Omega-6 Essential Fatty Acids: Pathology and Roles in Clinical Medicine*. New York: Wiley-Liss.

Ichihara, K. (1984). Glycerol 3-phosphate acyltransferase in a particulate fraction from maturing safflower seeds. *Archives of Biochemistry and Biophysics*, **232**, 685–98.

Ichihara, K., Asahi, T. & Fujii, S. (1987). 1-Acyl-*sn*-glycerol-3-phosphate acyltransferase in maturing safflower seeds and its contribution to the non-random fatty acid distribution of triacylglycerol. *European Journal of Biochemistry*, **167**, 339–47.

Ichihara, K., Murota, N. & Fujii, S. (1990). Intracellular translocation of phosphatidate phosphatase in maturing safflower seeds: a possible mechanism of feedforward control of triacylglycerol biosynthesis for fatty acids. *Biochimica et Biophysica Acta*, **1043**, 227–34.

Kearns, E.V., Hugly, S. & Somerville, C.R. (1991). Role of cytochrome b5 in Δ12 desaturation of oleic acid by microsomes of safflower. *Archives of Biochemistry and Biophysics*, **284**, 431–5.

Knutzon, D.S., Thompson, G.A., Radke, S.E., Johnson, W.B., Knauf, V.C. & Kridl, J.C. (1992). Modification of Brassica seed oil by antisense expression of a stearoyl-acyl carrier protein desaturase gene. *Proceedings of the National Academy of Sciences, USA*, **89**, 2624–8.

Kwanyuen, P. & Wilson, R.F. (1986). Isolation and purification of diacylglycerol acyltransferase from germinating soybean cotyledons. *Biochimica et Biophysica Acta*, **877**, 238–45.

Lacey, D. & Hills, M.J. (1996). Heterogeneity of the endoplasmic reticulum with respect to lipid synthesis in developing seeds of *Brassica napus*. *Planta*, **199**, 545–51.

Lassner, M.W., Levering, C.K., Davies, H.M. & Knutzon, D.S. (1995). Lysophosphatidic acid acyltransferase from meadowfoam mediates insertion of erucic acid at the *sn*-2 position of triacylglycerol in transgenic rapeseed oil. *American Journal of Plant Physiology*, **109**, 1389–94.

Lehner, R. & Kuksis, A. (1993). Triacylglycerols synthesis by a diacylglycerol transacylase from rat intestinal microsomes. *Journal of Biological Chemistry*, **268**, 8781–6.

Lemieux, B., Miquel, M., Somerville, C. & Browse, J. (1990). Mutants of *Arabidopsis* with alteration in seed lipid fatty acid composition. *Journal of Theoretical and Applied Genetics*, **80**, 234–40.

Löhden, I., Bernerth, R. & Frentzen, M. (1990). Acyl-CoA : 1-acylglycerol -3-phosphate acyltransferase from developing seeds of *Limnanthes douglasii* and *Brassica napus*. In *Plant Lipid Biochemistry, Structure and Utilization*, ed. P.J. Quinn & J.L. Harwood, pp. 175–7. London: Portland Press.

Mancha, M. & Stymne, S. (1997). Remodelling of triacylglycerols in microsomal preparations from developing castor bean endosperm. *Planta*, **203**, 51–7.

Monks, D.D., Goode, J.H., Dinsmore, P.K. & Dewey, R.E. (1997). Phosphatidylcholine biosynthesis in soybean: cloning and characterisation of genes encoding enzymes of the nucleotide pathway. In *Physiology, Biochemistry and Molecular Biology of Plant Lipids*, ed. J.P. Williams, M.U. Kahn & N.W. Lem, pp. 110–12. Dordrecht, Netherlands: Kluwer Academic Press.

Moreau, R.A. & Stumpf, P.K. (1981). Recent studies of the enzymatic synthesis of ricinoleic acid by developing castor beans. *American Journal of Plant Physiology*, **67**, 672–6.

Moreau, R.A., Pollard, M.R. & Stumpf, P.K. (1981). Properties of a $\Delta 5$ fatty acyl-CoA desaturase in the cotyledons of developing *Limnanthes alba*. *Archives of Biochemistry and Biophysics*, **209**, 376–84.

Napier, J.A., Smith, M.A., Stobart, A.K. & Shewry, P.R. (1995). Isolation of a cDNA encoding a cytochrome b5 specifically expressed in developing tobacco seeds. *Planta*, **197**, 200–2.

Napier, J.A., Sayanova, O., Stobart, A.K. & Shewry, P.R. (1997). A new class of cytochrome b5 fusion proteins. *The Biochemical Journal*, **328**, 717–18.

Okuley, J., Lightner, J., Feldmann, K., Yadav, N., Lark, E. & Browse,

J. (1994). *Arabidopsis* FAD2 gene encodes the enzyme that is essential for polyunsaturated lipid synthesis. *Plant Cell*, **6**, 147–8.

Oo, K-C. & Huang, A.H.C. (1989). Lysophosphatidate acyltransferase activities in the microsomes from palm endosperm, maize scutellum, and rapeseed cotyledon of maturing seeds. *American Journal of Plant Physiology*, **91**, 1288–95.

Oo, K-C. & Chew, Y-H. (1992). Diacylglycerol acyltransferase in microsomes and oil bodies of oil palm mesocarp. *Plant Cell Physiology*, **33**, 189–95.

Pearce, M. & Slabas, A.R. (1997). Purification and characterisation of phosphatidate phosphatase from avocado. In *Physiology, Biochemistry and Molecular Biology of Plant Lipids*, ed. J.P. Williams, M.U. Kahn & N.W. Lem, pp. 140–2. Dordrecht, Netherlands: Kluwer Academic Press.

Pollard, M.R. & Stumpf, P.K. (1980). Biosynthesis of C20 and C22 fatty acids by developing seeds of *Limnanthes alba*. *American Journal of Plant Physiology*, **66**, 649–55.

Powell, G., Abott, A., Knauft, D. & Barth, J. (1990). Oil desaturation in developing peanut seeds: studies of lipid desaturation in a peanut mutant that accumulates high levels of oleic acid. In *Plant Lipid Biochemistry, Structure and Utilization*, ed. P. J. Quinn and J. L. Harwood, pp. 131–3. London: Portland Press.

Sayanova, O., Smith, M.A., Lapinskas, P., Stobart, A.K., Dobson, G., Christie, W.W., Shewry, P.R. & Napier, J.A. (1997). Expression of a borage desaturase cDNA containing an N-terminal cytochrome b5 domain results in the accumulation of high levels of a Δ6-desaturated fatty acids in transgenic tobacco. *Proceedings of the National Academy of Sciences, USA*, **94**, 4211–16.

Schmidt, H. & Heinz, E. (1993). Direct desaturation of intact galactolipids by a desaturase solubilised from spinach chloroplast envelopes. *The Biochemical Journal*, **289**, 777–82.

Sevanian, A. & Kim, E. (1985). Phospholipase A2 dependent release of fatty acids from peroxidised membranes. *Free Radical Biology and Medicine*, **1**, 263–71.

Slack, C.R., Roughan, P.G. & Browse, J.A. (1979). Evidence for an oleoyl phosphatidylcholine desaturase in microsomal preparations from cotyledons of safflower seed. *The Biochemical Journal*, **179**, 649–56.

Slack, C.R., Roughan, P.G., Browse, J.A. & Gardiner, S.E. (1985). Some properties of cholinephosphotransferase from developing safflower cotyledons. *Biochimica et Biophysica Acta*, **833**, 438–48.

Smith, M.A. (1992). Electron-transport components of the Δ12 desaturase in oil seeds. PhD thesis, University of Bristol, UK.

Smith, M.A., Cross, A.R., Jones, T.G., Griffiths, T., Stymne, S. & Stobart, A. K. (1990). Electron-transport components of the 1-acyl-2-oleoyl-sn-glycerol-3-phosphocholine Δ12-desaturase in microsomal preparations from developing safflower cotyledons. *The Biochemical Journal*, **272**, 23–9.

Smith, M.A., Jonsson, L., Stymne, S. & Stobart, K. (1992*a*). Evidence for cytochrome b5 as an electron donor in ricinoleic acid biosynthesis in microsomal preparations from developing castor bean. *The Biochemical Journal*, **287**, 141–4.

Smith, M.A., Stobart, K., Bafor, M., Jonsson, L. & Stymne, S. (1992*b*). The biosynthesis of ricinoleic and vernolic acid in microsomal preparations from developing endosperm of castor bean and *Euphorbia lagascae*. In *Proceedings of the 10th International Symposium on the Metabolism, Structure and Utilization of Plant Lipids*, ed. A. Cherif, B. Marzouk, A. Smaoui, D. Ben Miled & M. Zarrouk, pp. 148–51. Tunis: Centre National Pedagogique.

Smith, M.A., Stobart, A.K., Shewry, P.R. & Napier, J. A. (1994*a*). Tobacco cytochrome b5: cDNA isolation, expression analysis and *in-vitro* protein targeting. *Plant Molecular Biology*, **25**, 527–37.

Smith, M.A., Napier, J.A., Stymne, S., Shewry, P.R. & Stobart, A.K. (1994*b*). Expression of a biologically active plant cytochrome b5 in *Escherichia coli*. *The Biochemical Journal*, **303**, 73–9.

Soldatov, K.I. (1976).Chemical mutagenesis in sunflower breeding. In *Proceedings of the 7th International Sunflower Conference*, pp. 352–7. Krasnodar, USSR.

Somerville, C. & Browse, J. (1991). Plant lipids: metabolism, mutants, and membranes. *Science*, **252**, 80–7.

Sperling, P. & Heinz, E. (1993). Isomeric *sn*-1-octadecenyl and *sn*-2-octadecenyl analogues of lysophosphatidylcholine as substrates for acylation and desaturation by plant microsomal membranes. *European Journal of Biochemistry*, **213**, 965–71.

Sperling, P., Schmidt, H. & Heinz, E. (1995). A cytochrome b(5)-containng fusion protein similar to plant acyl lipid desaturases. *European Journal of Biochemistry*, **232,** 798–805.

Sperling, P., Hammer, U., Friedt, W. & Heinz, E. (1990). High oleic sunflower: studies on composition and desaturation of acylgroups in different lipids and organs. *Zeitschrift für Naturforschung*, **45**, 166–72.

Stahl, U., Banas, A. & Stymne, S. (1995). Plant microsomal phospholipid acyl hydrolases have selectivities for uncommon fatty acids. *Plant Physiology*, **107**, 953–62.

Stobart, A.K. & Stymne, S. (1985*a*). The regulation of the fatty acid composition of the triacylglycerols in microsomal preparations from avocado mesocarp and the developing cotyledons of safflower. *Planta*, **163**, 119–25.

Stobart, A.K. & Stymne, S. (1985*b*). Interconversion of diacylglycerol and phosphatidylcholine during triacylglycerol production in microsomal preparations of developing cotyledons of safflower. *The Biochemical Journal*, **232**, 217–21.

Stobart, A.K., Stymne, S. & Hoglund, S. (1986). Safflower microsomes catalyse the accumulation of oil: a model system. *Planta*, **169**, 33–7.

Stobart, A.K., Stymne, S. & Shewry, P.R. (1992). Manipulation of Cereal Protein and Oilseed Quality. In *Control of Plant Gene Expression*, ed. D.P.S. Verma, pp. 499–533. Florida: CRC Press.

Stobart, A.K., Stymne, S. & Mancha, S. (1996). Transacylation reactions in oil seeds In *Oils, Fats and Lipids*, Vol. 1, pp. 88–90. Bridgewater: Barnes and Associates.

Stobart, A.K., Mancha, M., Lenman, M., Dahlqvist, A. & Stymne, S. (1997). Triacylglycerols are synthesised and utilised by transacylation reactions in microsomal preparations of developing safflower seeds. *Planta*, **203**, 58–66.

Stymne, S. & Stobart, A.K. (1984). Evidence for the reversibility of the acyl-CoA : lysophosphatidylcholine acyltransferase in microsomal preparations from developing safflower cotyledons and rat liver. *The Biochemical Journal*, **223**, 305–14.

Stymne, S. & Stobart, A.K. (1987). Triacylglycerol biosynthesis. In *The Biochemistry of Plants: A Comprehensive Treatise*, ed. P.K. Stumpf, vol. 9, pp. 175–214. New York: Academic Press.

Stymne, S. & Stobart, A.K. (1993). Triacylglycerol biosynthesis. In *Seed Storage Compounds: Biosynthesis, Interactions and Manipulation*, eds. P.R. Shewry & A.K. Stobart, pp. 96–114. Oxford: Clarendon Press.

Stymne, S., Griffiths, G. & Stobart, K. (1987). Desaturation of fatty acid on complex lipid substrates. In *The Metabolism, Structure and Function of Plant Lipids*, ed. P.K. Stumpf, J.B. Mudd & W.D. Nes, pp. 405–12. New York: Plenum Press.

Stymne, S., Bafor, M., Jonsson, L., Wiberg, E. & Stobart, A.K. (1990). Triacylglycerol assembly. In *Plant Lipid Biochemistry, Structure and Utilization*, ed. P.J. Quinn & J.L. Harwood, pp. 191–7. London: Portland Press.

Stymne, S., Tonnet, M.L. & Green, A.G. (1992). Biosynthesis of linolenate in developing embryos and cell-free preparations of high-linolenate linseed and low linolenate mutants. *Archives of Biochemistry and Biophysics*, **294**, 557–63.

Taylor, D.C., Weber, N., Barton, D.L., Underhill, E.W., Hogge, L.R., Weselake, R.J. & Pomeroy, M.K. (1991). Triacylglycerol bioassembly in microspore-derived embryos of *Brassica napus*. *American Journal of Plant Physiology*, **97**, 65–79.

Tonnet, M.L. & Green, A.G. (1987). Characterization of the seed and leaf lipids of high and low linolenate flax genotypes. *Archives of Biochemistry and Biophysics*, **252**, 646–54.

van de Loo, F.J., Broun, P., Turner, S. & Somerville, C. (1995). An oleate12-hydroxylase from *Ricinis communis* is a fatty acyl desaturase homolog. *Proceedings of the National Academy of Sciences, USA*, **92**, 6743–7.

van Kuizk, F.J.G.M., Sevanian, A. & Handelman, G.J. (1987). New role for phospholipase A2 : protection of membranes from lipid peroxidation damage. *Trends in Biochemical Science,* **12**, 31–4.

Verma, D.P.S. (ed.) (1992). *Control of Plant Gene Expression.* Boca Raton, Florida: CRC Press.

Vogel, G. & Browse, J. (1996). Cholinephosphotransferase and diacylglycerol acytransferase: substrate specificities at a key branch point in seed lipid metabolism. *American Journal of Plant Physiology*, **110**, 923–31.

White, R.E. & Coon, M.J. (1980). Oxygen activation by cytochrome P-450. *Annual Reviews of Biochemistry*, **49**, 315–56.

Wiberg, E., Tillberg, E. & Stymne, S. (1992). Specificities of diacylglycerol acyltransferases in microsomal fractions from developing oil seeds. In *Proceedings of the 10th International Symposium on the Metabolism, Structure and Utilization of Plant Lipids*, ed. A. Cherif, B. Marzouk, A. Smaoui, D. Ben Miled & M. Zarrouk, pp. 83–6. Tunis: Centre National Pedagogique.

M. FRENTZEN and F.P. WOLTER

10 Molecular biology of acyltransferases involved in glycerolipid synthesis

Introduction

Glycerolipids are the most abundant group of plant lipids including glyco- and phosphodiacylglycerols, the major polar membrane lipids and triacylglycerols (TAG), which serve as storage lipids in most plant species. The biosynthesis of these lipids occurs in different subcellular compartments, namely plastids, mitochondria and endomembranes, mainly the endoplasmic reticulum (ER). In each compartment an *sn*-glycerol-3-phosphate acyltransferase (GPAT) catalyses the first step of *de novo* glycerolipid synthesis, which introduces the first acyl group at the *sn*-1 position of glycerol-3-phosphate (Fig. 1). In the next step, the synthesised *sn*-1-acylglycerol-3-phosphate is utilised by an *sn*-1-acylglycerol-3-phosphate acyltransferase (LPAAT) which catalyses the second acylation reaction at its *sn*-2 position so that 1,2-diacylglycerol-3-phosphate (phosphatidic acid, PA), the central intermediate of glycerolipid synthesis, is formed (Fig. 1). The acyltransferases found in the various compartments are probably all encoded in the nucleus, but differ markedly in their substrate specificities and selectivities (for review see Frentzen, 1993). In this way, they play an important role in establishing the typical fatty acid patterns of the various glycerolipids in the different cell compartments. It has been shown that these fatty acid patterns can be a major determinant of certain plant traits such as chilling tolerance (for review see Gibson *et al.*, 1994; Nishida & Murata, 1996).

During fruit and seed development, a microsomal GPAT and a microsomal LPAAT are also involved in TAG biosynthesis, where the third acylation reaction at the *sn*-3 position of 1,2-diacylglycerol is catalysed by a 1,2-diacylglycerol acyltransferase (DAGAT, Fig. 1). The substrate specificities of these enzymes, especially those of the LPAAT, are decisive for establishing the fatty acid pattern of TAG, and thus its quality and usability for specific markets. Hence, the activities of plant acyltransferases are indispensable for the biosynthesis of both membrane and storage lipids. In addition, their properties can determine

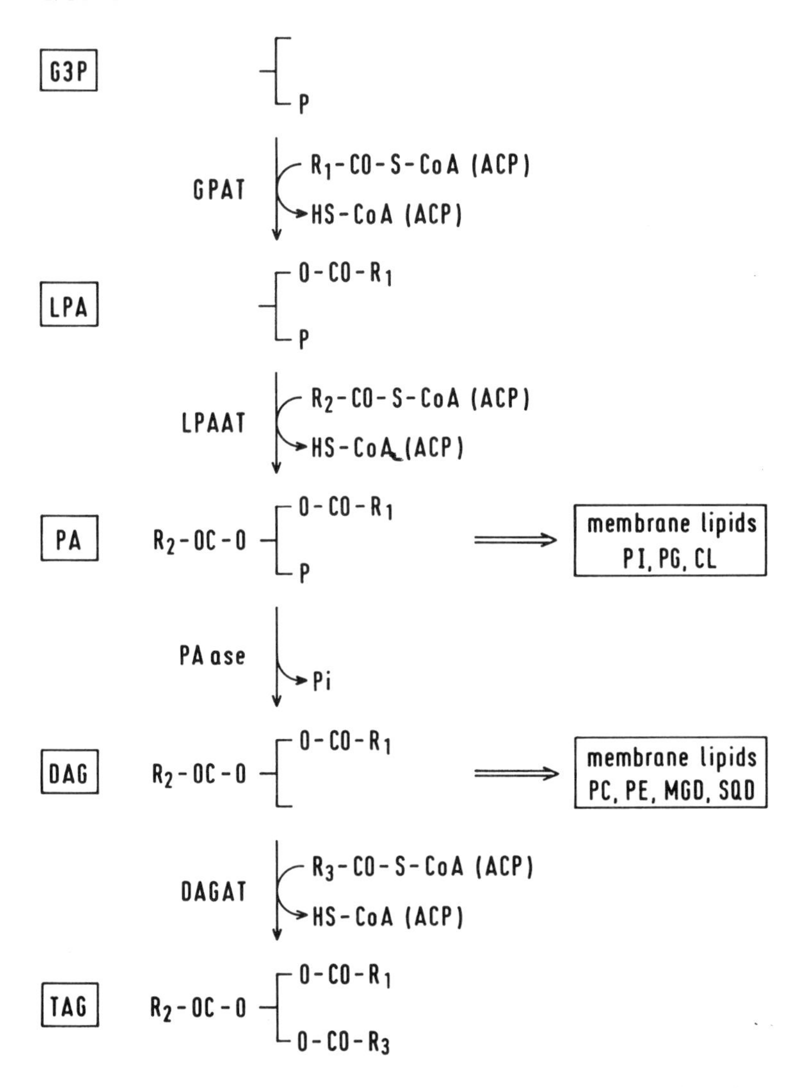

Fig. 1. *De novo* glycerolipid synthesis via GPAT, LPAAT, phosphatidic acid phosphatase (PAase) and DAGAT reaction (G3P, *sn*-glycerol-3-phosphate; LPA, lysophosphatidic acid; DAG, diacylglycerol; PI, phosphatidylinositol; CL, cardiolipin; PC, phosphatidylcholine; PE, phophatidylethanolamine; MGD, monogalactosyldiacylglycerol; SQD, sulphoquinosyldiacylglycerol).

certain plant traits such as chilling tolerance and oil quality. In this regard, the plastidial GPAT and the microsomal LPAAT are of particular importance. Due to agricultural demands for improved chilling tolerance of horticultural crops, as well as for improved oil quality of traditional oilseed crops, and in view of the progress in plant genetic engineering, research on plant acyltransferases has, in recent years, been concentrated on the plastidial and microsomal enzymes. To date, these are the only plant acyltransferases for which the respective genes have been cloned. Therefore, with regard to the molecular biology of plant acyltransferases, GPATs and LPAATs of plastids and microsomes are the main subjects of this article. The progress achieved in this field of research will be described in the following sections, but first a brief overview of the properties of these acyltransferases will be presented in order to lay the framework necessary for discussing the special subjects.

Plastidial and microsomal acyltransferases

Most of the plant acyltransferases are integral membrane proteins. For example, the microsomal enzymes are all membrane bound, and this also applies to the plastidial LPAAT which is located in the envelope. One of the exceptions is the plastidial GPAT of higher plants, which behaves like a soluble stroma protein. It is probably associated with the fatty acid synthesis enzymes, stearoyl-ACP desaturase and acyl-ACP thioesterase, in a multienzyme complex that has the potential to link thylakoid and envelope by spanning the stroma (Roughan & Ohlrogge, 1996). The solubility of the plastidial GPAT explains why this enzyme was the first successfully purified plant acyltransferase. Purification of the GPAT revealed that, depending on the plant species, the enzyme occurs in one to three isomeric forms, but in only one case could different catalytic properties be demonstrated (Frentzen, 1993). Only recently Knutzon *et al.* (1995) succeeded in purifying the first LPAAT, a microsomal form from coconut endosperm. Hence, the properties of the various acyltransferases have been determined predominantly by using subcellular fractions or partially purified acyltransferase fractions as the enzyme source.

The plastidial GPAT and LPAAT preferentially use acyl–acyl carrier protein (ACP) thioesters, the direct products of *de novo* fatty acid synthesis in plastids. Depending on the plant species, the plastidial GPAT either has a pronounced selectivity for oleoyl(18 : 1)-ACP compared to both palmitoyl(16 : 0)- and stearoyl(18 : 0)-ACP or catalyses a largely unselective *sn*-1 acylation (Frentzen, 1993). On the other hand, the plastidial LPAAT invariably directs 16 : 0 to the *sn*-2 position of the glycerol backbone, and discriminates against 18 : 1 as well as 18 : 0. In contrast to the plastidial acyltransferases, the microsomal ones are only active with acyl-CoA thioesters. Furthermore, the microsomal enzymes differ from the plastidial ones with regard to their fatty acid selectivities. The microsomal GPAT shows a slight preference for 16 : 0 and the microsomal LPAAT has a pronounced one for unsaturated C_{18} acyl groups (Frentzen, 1993). Lipid analyses of transgenic tobacco plants provide evidence that the microsomal LPAAT preferentially directs not only unsaturated C_{18}, but also unsaturated C_{16} acyl groups to the *sn*-2 position (Grayburn, Collins & Hildebrand, 1992; Polashock, Chin & Martin, 1992). In leaves of untransformed plants, where unsaturated C_{16} acyl groups are not available to the microsomal LPAAT, PA is formed, which specifically carries unsaturated C_{18} at its *sn*-2 position. This PA subsequently serves as a precursor for the biosynthesis of the various phosphodiacylglycerols via the microsomal pathway.

During fruit and seed development, the microsomal pathway is also used for TAG synthesis (Fig. 1). Unlike membrane lipids, the structures of fatty acids esterified in TAG are remarkably diverse within the plant kingdom. However, only a few fatty acids, which are identical to those found most commonly in membrane lipids, contribute to about 80% of world oil production (Murphy, 1996). The fatty acid distributions of these storage lipids reflect the typical pattern of glycerolipids synthesised via the microsomal pathway, where saturated acyl groups are excluded from the *sn*-2 position. Consistent with the observed fatty acid pattern, the microsomal GPAT and LPAAT involved in TAG biosynthesis of most oil seed crops possess properties very similar to those involved in membrane lipid synthesis (Frentzen, 1993). Therefore, in principle, only one extra enzyme, namely the DAGAT, is required to redirect the microsomal membrane lipid pathway to the production of TAG.

Many plant species from various plant families, however, synthesise unusual fatty acids, which are specifically esterified in TAG, but are largely excluded from the membrane lipids (van de Loo, Fox & Somerville, 1993). In these plant species, the microsomal LPAAT involved in TAG synthesis shows properties different to those of the respective isoenzymes involved in membrane lipid synthesis. For instance, in accordance with the fatty acid pattern of the seed oil of *Limnanthes* species, enriched with erucic acid (22 : 1) at the *sn*-2 position, the microsomal LPAAT involved in TAG synthesis effectively utilises 22 : 1-CoA as substrate (Frentzen, 1993). On the other hand, the respective enzyme from coconut, which has oil rich in 12 : 0, displays a preference for 12 : 0-CoA (Davies, Hawkins & Nelson, 1995). These data suggest seed-specific expression of defined LPAATs, concomitant with the accumulation of seed oil. This supposition was recently confirmed by Northern blot analysis (Brown *et al.*, 1995; Hanke *et al.*, 1995).

In contrast to the ER, glycerolipid synthesis in plastids is used for the formation of only a part of their own membrane lipids. Due to the fatty acid selectivities of the plastidial acyltransferases, *de novo* glycerolipid synthesis in this compartment results in the formation of PA which specifically carries 16 : 0 at its *sn*-2 position. In all plant species, this PA serves as precursor for the biosynthesis of phosphatidylglycerol (PG), which usually constitutes about 8–10% of chloroplast lipids. In certain plants, such as *Arabidopsis thaliana,* it is also used as precursor for the formation of one part of the glycodiacylglycerols. On the other hand, depending on the plant species, part or all of the plastidial glycodiacylglycerols are synthesised from DAG moieties imported from the ER.

Important features of the plastidial *sn*-glycerol-3-phosphate acyltransferase

It has been shown that the activity of the plastidial GPAT can play a decisive role in controlling the acyl flux via the glycerolipid synthesis pathways (Frentzen, 1993). In the stroma, this enzyme competes with the acyl-ACP thioesterase for the products of *de novo* fatty acid biosynthesis and channels the acyl groups into the plastidial pathway, whereas the acyl-ACP thioesterase effects the export of the acyl groups from the plastids and largely channels them into the microsomal pathway. Consequently, the acyl flux through the different pathways can be directly controlled by the relative activities of the GPAT and the acyl-ACP thioesterase.

Moreover, the fatty acid selectivities of the GPAT can be a major determinant for both fatty acid unsaturation of plastidial PG and chilling sensitivity of plants (Nishida & Murata, 1996). As shown by Murata and colleagues (Murata *et al.*, 1982; Murata, 1983), chilling sensitivity in several plant species is correlated with the proportion of non-fluid, plastidial PG species containing only 16 : 0, 18 : 0 or *trans*-3 hexadecenoic acid (16 : 1-*trans*), which has physical properties very similar to those of a saturated fatty acid. Since *cis* double bonds cannot be introduced into saturated fatty acids esterified in PG by plant desaturases (Heinz, 1993), and since the plastidial LPAAT invariably directs 16 : 0 to the *sn*-2 position of the glycerol, the formation of a critical proportion of non-fluid PG species is determined by the fatty acid selectivities of the plastidial GPAT. As mentioned above, fatty acid selectivity of the plastidial GPAT varies depending on the plant species. While the enzyme from chilling tolerant plants preferentially directs 18 : 1 to the *sn*-1 glycerol position, the GPAT from chilling sensitive plants catalyses an unselective *sn*-1 acylation, so that a distinctly higher proportion of non-fluid lipid species is formed in the sensitive than in the resistant plants (Frentzen, 1993). Although a low proportion of non-fluid PG is necessary but not sufficient to confer chilling tolerance in all plant species (Somerville, 1995), for the first time, variation in the properties of a single enzyme was inferred to be responsible for chilling sensitivity, at least in a certain group of plants. Since this raised the possibility to improve the chilling tolerance of horticultural crops by introducing a gene of an 18 : 1-selective GPAT, it appreciably stimulated research on plant acyltransferases and efforts to clone GPAT genes. Recently, the decisive role of GPATs in determining chilling sensitivity was confirmed with transgenic tobacco and *A. thaliana* plants expressing chimeric GPAT genes (Murata *et al.*, 1992; Wolter, Schmidt & Heinz, 1992). Moreover, analysis of the transgenic tobacco plants revealed that critical proportions of non-fluid PG species cause a low-temperature-induced photoinhibition by reducing the repair rate of damaged photosystem II. These data provide the first direct evidence for a causal link between lipid composition and injury of plants induced by low, non-freezing temperatures (Moon *et al.*, 1995; Nishida & Murata, 1996).

Sequences of *sn*-glycerol-3-phosphate acyltransferase

Due to the pioneering work of Murata and colleagues, to date, cDNAs encoding a plastidial GPAT have been cloned from several plant species (Table 1). The cDNAs from pea, spinach and *A. thaliana* encode 18 : 1-selective GPATs, while the cDNAs from squash, cucumber and French bean encode unselective ones. The properties of the GPAT from safflower have not been determined (S.L. MacKenzie, personal communication), but it putatively represents an unselective enzyme. Cloning was achieved by different techniques including immunological screening of expression libraries with antibodies raised against purified GPAT proteins (Ishizaki *et al.*, 1988; Weber *et al.*, 1991), screening with heterologous probes (Nishida *et al.*, 1993) or with homologous ones obtained via polymerase chain reaction with oligonucleotide primers deduced from known GPAT sequences (Johnson, Schneider & Somerville, 1992; Bhella & MacKenzie, 1994; Fritz, Heinz & Wolter, 1995; Ishizaki-Hishizawa *et al.*, 1995; Wolter, 1996). Analyses of these GPAT cDNAs revealed the typical features of a plastidial protein en-

Table 1. *Plastidial GPAT preproteins deduced from the cloned cDNAs of* Curcurbita moschata *(squash),* Cucumis sativus *(cucumber),* Phaseolus vulgaris *(French bean),* Pisum sativum *(pea),* Spinacia oleracea *(spinach),* Carthamus tinctorius *(safflower) and* Arabidopsis thaliana

cDNA source	Pre-protein length	Transit peptide length	Mature protein mass, kDa	References
C. moschata[a]	437	69	40.9	Ishizaki *et al.*, 1988
C. sativus	470	101	41.2	Johnson *et al.*, 1992
C. tinctorius	463	90	41.7	Bhella & MacKenzie, 1994
P. vulgaris	459	90	41.1	Fritz *et al.*, 1995
P. sativum	457	88	41.0	Weber *et al.*, 1991
A. thaliana	459	90	41.2	Nishida *et al.*, 1993
S. oleracea	472	102	41.4	Ishizaki-Nishizawa *et al.*, 1995; Wolter, 1996

Transit peptide lengths have been deduced from the N-terminal amino acid sequence data of Ishizaki *et al.* (1988).
[a] Sequence has been shown to be incomplete (Nishida *et al.*, 1993).

coded by nuclear gene. The enzyme is synthesised as a preprotein in the cytoplasm and is subsequently transported across the plastidial envelope via a cleavable N-terminal transit peptide (Table 1).

The number of genes encoding a plastidial GPAT has so far only been determined in *A. thaliana* where it is encoded by a single copy gene (Kunst, Browse & Somerville, 1988). This gene designated *ATS1* was cloned as a 4008 base pair DNA fragment (Nishida *et al.*, 1993). It contains 12 intron sequences, one of which is present in the 3′ untranslated region. An intron in this 3′ region was also detected in two GPAT cDNA clones from spinach (Wolter, 1996) at a position similar to that of *ATS1* and with very similar border sequences, but with a different length and a distinctly lower AU content. The low AU content presumably causes inefficient splicing. These data indicate that the intron in the 3′ untranslated region may have special functions in mRNA maturation or its translation, but this awaits clarification.

So far, expression of *ATS1* has been studied during growth at different temperatures. Although the flux through the plastidial pathway is strongly altered by temperatures, and as mentioned above this flux can be controlled by the activity of the GPAT, the level of the *ATS1* gene transcript was found to be constant (C. Somerville, personal communication). Based on these data, one might speculate that regulation of the plastidial GPAT takes place at the post-transcriptional level, similar to most of the plant desaturases (Somerville, 1995), but this remains to be determined experimentally. In addition, it is still unclear whether the isomeric forms of the plastidial GPAT detected in various plant species are derived from different transcripts or result from post-translational modifications.

As given in Table 1, the GPAT preproteins encoded by the cDNAs from the various plant species are very similar in length. The processing site of the preproteins and thus the length of the signal peptides has been deduced from the N-terminal amino acid sequence of one of the purified GPAT isoforms from squash (Ishizaki *et al.*, 1988). A further N-terminal amino acid sequence has been obtained from a purified spinach GPAT (Wolter, 1996). According to this sequence, serine 107 is the N-terminal amino acid of the mature GPAT protein. This processing site is located four amino acid residues downstream from that deduced from the squash sequence. Although it cannot be excluded that these differences are merely due to the spinach protein being truncated at its amino terminus during purification, it is remarkable that serine 107 represents the most N-terminal amino acid residue conserved in all GPAT sequences (Wolter, 1996). Based on the observation that the *A. thaliana* GPAT overexpressed in tobacco plants was estimated by SDS–PAGE to be 43 kDa rather than 41.2 kDa (Table 1), Nishida *et al.* (1993) suggested that the processing site was located about 20 amino acid residues upstream from that deduced from the squash sequence. Consequently, the precise localisation of the processing site is still a matter for discussion.

The alignment of the amino acid sequences deduced from the cDNAs of the different plant species, however, strongly suggest that it is located at, or close to, serine 107, since a highly significant and continuous sequence similarity is found for the GPAT proteins from the various plant species beginning at serine 107. Eleven stretches of 4 to 11 amino acid residues are conserved in all sequences and both number and length of these stretches increase if conservative exchanges are included. These stretches are found predominantly within the central third of the mature GPAT proteins, where all sequences exhibit the highest sequence identity (53.3%).

A comparison of the plastidial GPAT sequences with that of the *Escherichia coli* GPAT encoded by the *plsB* gene (Lightner, Bell & Modrich, 1983), do not reveal any extended sequence similarity. The *E. coli* GPAT differs from the plastidial enzyme by its distinctly higher mass of 91 kDa, and by the fact that it is membrane bound (Lightner *et al.*, 1983). Unlike the plastidial GPATs, other amino acid sequences deduced from genes or cDNAs from both prokaryotic and eukaryotic organisms show significant similarities to that of the *E. coli* GPAT (Fig. 2). In addition to the *plsB* gene of *E. coli*, the cDNA from *Mus musculus* has unequivocally been demonstrated to encode a GPAT (Shin *et al.*, 1991; Yet *et al.*, 1993; Yet, Moon & Sul, 1995). This enzyme is located in the outer mitochondrial membrane and, like the *E. coli* GPAT, it has a mass of 90 kDa. In addition, the respective enzyme from rat liver mitochondria has been purified (Vancura & Haldar, 1994), and according to partial cDNA clones (EMBL accession numbers U36771 to U36773), exhibits more than 90% sequence identity with the murine GPAT. These data suggest that the bacterial GPAT sequences are more closely related to the mitochondrial than to the plastidial ones (Fig. 2). In view of these results, the sequences of cyanobacterial GPATs would be of interest. Searches of the genome database for *Synechocystis* sp. (CyanoBase) with different GPAT sequences revealed just one open reading frame, which more likely encodes an LPAAT than a GPAT (see below).

Molecular basis for *sn*-glycerol-3-phosphate acyltransferase substrate specificity

From the phylogram shown in Fig. 2 it also becomes obvious that the similarities between the plastidial GPAT sequences from the different plant species do not reflect similarities in the enzymic properties but rather relationships of the cDNA sources. Highest similarities are found between GPAT sequences from plant species belonging to the same plant family, such as squash and cucumber or pea and French bean. The two unselective GPAT sequences from cucumber and squash are 90% identical. But, even the sequence of the 18 : 1-selective GPAT from pea shows 82% identity to that of the unselective French bean enzyme. In addition, the 18 : 1-selective GPAT of *A. thaliana* is more similar to the unselective ones of cucumber and squash than to the selective one of spinach. The data suggest that only a few amino acid residues might be critical for fatty acid selectivity. This is in line with the observation that a replacement of one to three amino acids can appreciably alter fatty acid specificities and selectivities of enzymes.

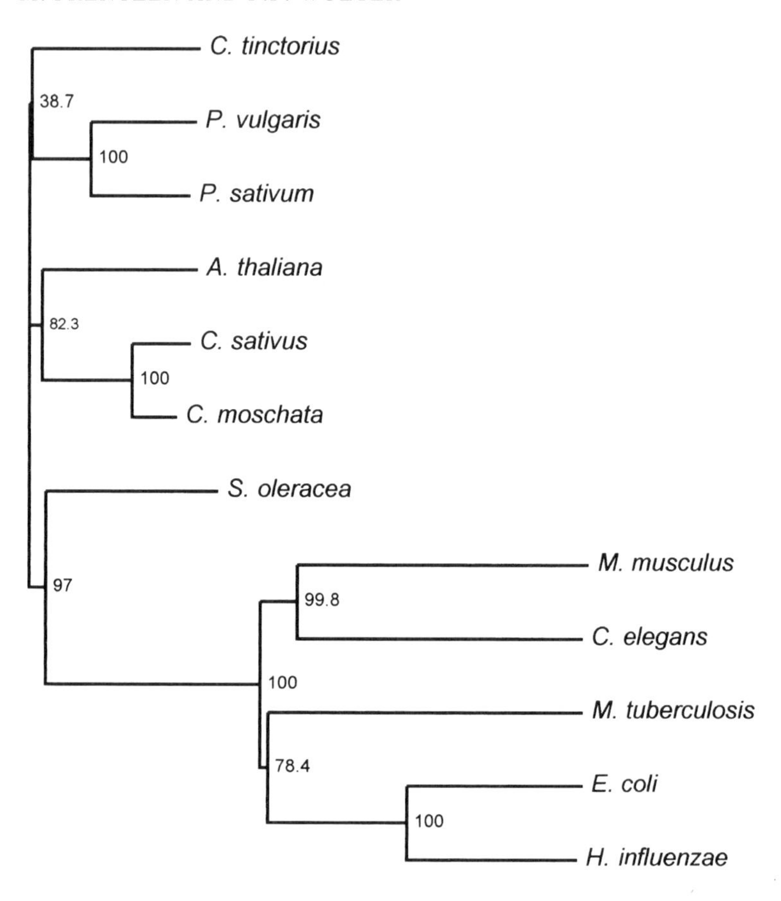

Fig. 2. Phylogram based on the analyses of amino acid sequences of GPATs and putative ones from the given organisms. Genbank/EMBL accession numbers of the sequences are the following: *Mus musculus*, M77007; *Caenorhabditis elegans*, U64847 (gene F08F3.2); *E. coli plsB* gene, K00127; *Mycobacterium tuberculosis*, Z74020 (gene MTCY48.14e), *Haemophilus influenzae*, U32758 (gene HI0748), sequences of the plant species, see Table 1.

This has been shown impressively for the acyl-ACP thioesterase of *Umbellularia californica* (Yuan, Voelker & Hawkins, 1995), the yeast LPAAT (Nagiec *et al.*, 1993) and the human lecithin : cholesterol acyl-transferase (Wang *et al.*, 1997).

Recently, Ferri and Toguri (1997) applied the strategy of using chimeric gene products to define the region of the plastidial GPAT critical for fatty acid specificity. To this end, six different chimeras of the GPAT from *Spinacia oleracea* (S) and *Cucurbita moschata* (C) were created in which the N-terminal, central and C-terminal regions of roughly equal length of the two proteins have been interchanged. Using 16 : 0-CoA and 18 : 1-CoA as acyl donors, the properties of the chimeras were compared with those of the spinach and squash GPAT, which were found to display a specificity for 18 : 1-CoA and 16 : 0-CoA, respectively, at low physiologically relevant glycerol-3-phosphate concentrations of about 0.3 mM (Fig. 3). These analyses show that each

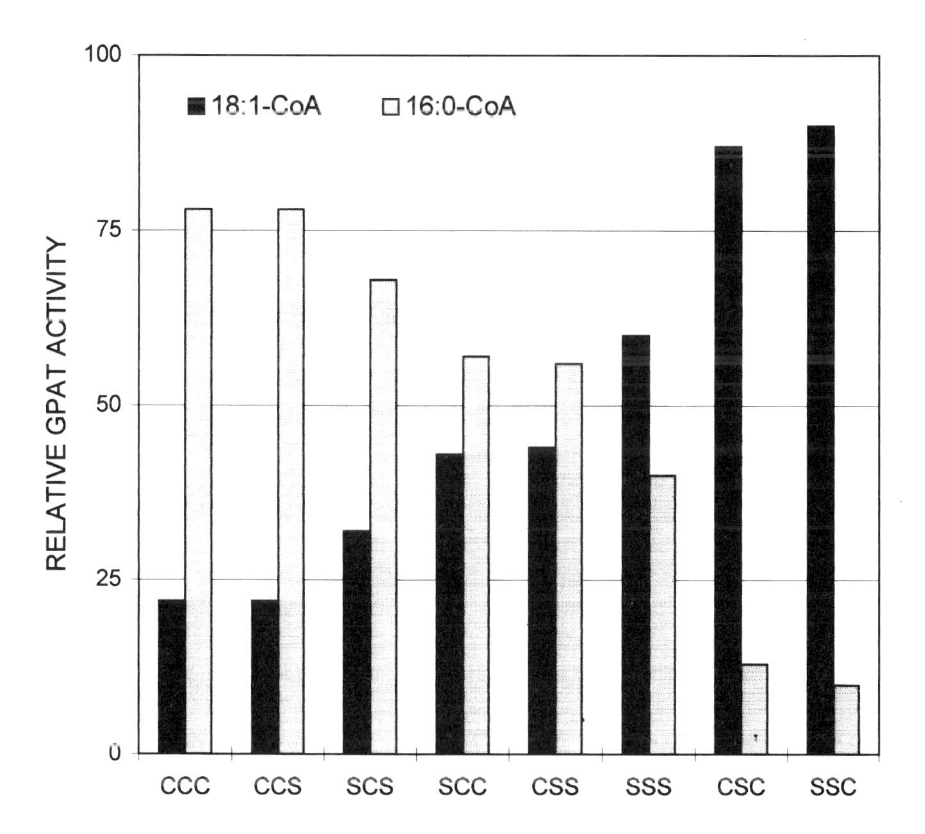

Fig. 3. Fatty acid specificities of the GPAT of *Spinacia oleracea* (SSS) and *Curcurbita moschata* (CCC) as well as the six different chimeras in which the N-terminal, central and C-terminal third of the two plastidial enzymes have been interchanged. GPAT activities determined with either 16 : 0-CoA or 18 : 1-CoA are given as a percentage of the total according to the results of Ferri and Toguri (1997).

chimera differs to a greater or lesser extent in its acyl-CoA specificity and its kinetic constants from the GPAT of squash and spinach, respectively. In comparison with the effects caused by interchanges of the terminal regions, the observed differences were most impressive for the two chimeras CSC and SCS, in which only the central regions have been interchanged. These interchanges caused the CSC chimera of the 16 : 0-CoA specific squash GPAT to become 18 : 1-CoA specific and the SCS chimera of 18 : 1-CoA specific spinach enzyme to become 16 : 0-CoA specific (Fig. 3). Therefore, the central regions of the two GPAT proteins essentially determine fatty acid specificities of the enzymes, but the terminal regions exert additional effects (Ferri & Toguri, 1997). This is clearly illustrated by the greater 18 : 1-CoA specificity of the chimeras CSC and SSC compared with that of the spinach GPAT and in particular that of CSS (Fig. 3). As pointed out by the authors, it remains to be determined whether the chimeras display properties under physiological conditions similar to those shown in Fig. 3. In view of the results so far available, it is, however, likely that the chimeras will display essentially the same 18 : 1 specificities with acyl-ACP as with acyl-CoA and that their selectivities for 18 : 1 will be even more pronounced.

Since the central region of the plastidial GPAT proteins is highly conserved, the data of Ferri and Toguri (1997) reduce the number of amino acid residues which might be decisive for fatty acid specificity to just a few. If their results hold true for all the plastidial GPATs, two amino acid residues, namely serine 243 within box I (Fig. 5) and glutamate 305 between the two boxes of the spinach sequence appear to be the most critical. These residues are the only two conserved in all selective GPATs, but substituted by threonine and aspartate, respectively, in the unselective enzymes of French bean, cucumber and squash as well as safflower, which represents a putative unselective GPAT. Site-directed mutagenesis of these residues could determine the structural features of the central region more precisely.

Moreover, the chimeric proteins having specificities for 18 : 1, even higher than that of the spinach GPAT, appear to be well suited for reducing the level of non-fluid PG species and thus contribute to the improvement of chilling tolerance in horticultural crops by genetic engineering.

Decisive properties of microsomal *sn*-1-acylglycerol-3-phosphate acyltransferases

As mentioned above, plant acyltransferases can also play a decisive role in establishing the fatty acid patterns of storage lipids and in channelling unusual fatty acids into TAG. Since unusual fatty acids are of particular interest for various industrial applications, in recent years efforts have been undertaken to genetically engineer the genomes of oil seed crops in such a way that they accumulate unusual fatty acids in sufficient quantities and in a sufficiently pure form for new economically valuable oils to be produced. This requires manipulation of fatty acid and TAG synthesis so that a defined fatty acid is formed in large quantities and effectively incorporated into each position of TAG. During the last decade great success has been achieved with regard to the synthesis of defined acyl groups (for review see Ohlrogge, 1994; Murphy, 1996) and more recently with regard to the channelling of unusual acyl groups into seed oils (Lassner *et al.*, 1995; Brough *et al.*, 1996; Weier *et al.*, 1997; Zou *et al.*, 1997*a*, *b*). These investigations have provided new insight into the formation and incorporation of unusual fatty acids into TAG, and have appreciably increased our knowledge of acyltransferases, especially of LPAATs involved in storage and membrane lipid synthesis.

With regard to the channelling of unusual acyl groups into TAG catalysed by the three different acyltransferases, the substrate specificities of the LPAAT have been found to be critical. As mentioned above, in most of the traditional oil seed crops such as rapeseed, this enzyme displays substrate specificities typical of the microsomal LPAAT involved in membrane lipid synthesis: it preferentially directs unsaturated C_{18} to the *sn*-2 position, but excludes unusual fatty acids such as 22 : 1 and 12 : 0 from this position (Frentzen, 1993; Voelker *et al.*, 1996). Consequently, due to the properties of the LPAAT, the formation of oil with a homogeneous fatty acid composition such as trierucin or trilaurin is prevented.

In constrast to the enzyme of rapeseed, the seed specifically expressed microsomal LPAATs of certain plant species mentioned above, can effectively incorporate unusual acyl groups at the *sn*-2 position of TAG. Hence, the genes encoding these seed-specific LPAATs appear to be optimally suited for improving the industrial applicability of seed oil from traditional oil seed crops by genetic engineering. To this end, efforts have been undertaken in various research groups to clone LPAAT genes by exploiting different strategies and methods. Knutzon *et al.* (1995) successfully purified the 12 : 0-CoA specific microsomal LPAAT from coconut endosperm. By using partial amino acid sequences of the purified protein, the authors obtained the respective cDNA clones. On the other hand, cDNA clones of the 22 : 1-CoA specific LPAAT from *L. douglasii* (Brown *et al.*, 1995; Hanke *et al.*, 1995) and of another LPAAT from maize (Brown *et al.*, 1994) have been isolated via functional complementation of the *E. coli* mutant

JC201 which is deficient in LPAAT activity (Coleman, 1990). It was also used to clone the first LPAAT gene, namely *plsC* of *E. coli* (Coleman, 1992).

Sequences of *sn*-1-acylglycerol-3-phosphate acyltransferases

To date, cDNA clones are available for five different plant species and, apart from pLAT1 and pRAT1, these cDNAs have unequivocally been demonstrated to encode a microsomal LPAAT (Table 2). Functional expression studies with the cDNA of pAM10, which differs in the 5′ region of its open reading frame from that of pRAT1, showed that it encodes a microsomal LPAAT with substrate specificities typical of the microsomal rapeseed enzyme involved in both membrane and reserve lipid synthesis (Gräfin zu Münster, F.P. Wolter & M. Frentzen, unpublished data). On the other hand, Northern blot analyses and functional expression of the cDNAs from coconut and *L. alba*, as well as the almost identical cDNAs from *L. douglasii* (pCH21 and pLAT2),

Table 2. *cDNA clones encoding microsomal LPAAT and putative ones from* Cocos nucifera *(coconut),* Limnanthes alba *(meadowfoam),* Limnanthes douglasii, Zea mays *(maize) and* Brassica napus *(rapeseed)*

cDNA source	cDNA clone	Encoded protein		References
		length	mass, kDa	
C. nucifera	pCGN5503	308	34.8	Knutzon *et al.*, 1995
L. alba	–	281	31.7	Lassner *et al.*, 1995
L. douglasii 2	pCH21	281	31.7	Hanke *et al.*, 1995
	pLAT2	281	31.7	Brown *et al.*, 1995
L. douglasii 1	pLAT1	377	42.7	Brown *et al.*, 1995
Z. mays	pMAT1	374	42.5	Brown *et al.*, 1994
B. napus 1	pRAT1	311	34.4	A.P. Brown, C.L. Brough, J. Kroon & A.R. Slabas, unpublished data[a]
B. napus 2	pAM10	390	43.7	Gräfin zu Münster, F.P. Wolter & M. Frentzen, unpublished data

[a] EMBL accession number Z49860.

revealed that they encode seed specifically expressed LPAATs which display substrate specificities different from those of the microsomal LPAATs involved in membrane lipid biosynthesis (Brown *et al.*, 1995; Hanke *et al.*, 1995; Knutzon *et al.*, 1995; Lassner *et al.*, 1995).

These seed specifically expressed LPAATs are very similar with regard to their masses and their hydrophobicity profiles, which indicates that they contain at least two predicted membrane spanning domains (Table 2). Moreover, although the sequences encoding LPAATs have appreciably different substrate specificities and are derived from both monocotyledonous and dicotyledonous plants, they exhibit high sequence identity of 57%. Highest sequence identities of approximately 66% were observed within the central region of the proteins.

Unexpectedly, these sequences are more similar to those of the LPAAT from *E. coli* (Coleman, 1992) and yeast (Nagiec *et al.*, 1993, Zou *et al.*, 1997*a*, *b*) than to those deduced from the further group of cDNA clones given in Table 2, where only short stretches of a few amino acid residues are conserved (Brown *et al.*, 1995; Hanke *et al.*, 1995; Knutzon *et al.*, 1995; Lassner *et al.*, 1995). The proteins encoded by these cDNAs have higher molecular masses than the seed-specific LPAATs (Table 2) and show similar hydrophobicity profiles, but lack the hydrophilic stretch of about 30 amino acid residues at the N-terminus of the proteins, typical of the LPAATs specifically involved in TAG synthesis. In contrast to the distinct differences observed when compared to the seed-specific LPAATs, the amino acid sequences deduced from the cDNAs of pLAT1, pMAT1, pRAT1 and pAM10 are highly conserved. They exhibit more than 60% sequence identity and, with regard to the central region of about 150 amino acid residues this value increases to 77%. Consequently, analysis of the different LPAAT cDNA clones provides strong evidence that they represent two distinct gene classes. Within each class, the members are very similar to each other whereas between classes they exhibit distinct differences even if they are derived from the same plant species.

Northern blot analysis revealed that the two LPAATs of *L. douglasii* are differently expressed (Brown *et al.*, 1995). In contrast to the observed seed-specific expression of the gene encoding the cDNA of pLAT2 and pCH21 (Brown *et al.*, 1995; Hanke *et al.*, 1995), the cDNA of pLAT1 hybridised not only to mRNA from developing seeds but also to mRNA from stems and leaves of *L. douglasii* plants (Brown *et al.*, 1995). These data, in conjunction with the high sequence identity between the rapeseed LPAAT encoded by pAM10 and that encoded by pLAT2, strongly suggest that the second class of microsomal LPAATs represents constitutively expressed enzymes involved in membrane lipid synthesis. Differences between members of the two groups of microsomal LPAATs were also observed by Southern blot analysis, which indicate that the seed specifically expressed LPAAT of *L. douglasii* is encoded by a single copy gene whereas the constitutively expressed ones are encoded by a small gene family (Brown *et al.*, 1995).

The plastidial acyl-ACP thioesterase has also been shown to be encoded by two distinct gene classes termed FatA and FatB (Jones, Davies & Voelker, 1995). FatA represents the constitutively expressed 18 : 1-ACP thioesterases, while FatB includes not only the seed-specific enzymes, but also constitutively expressed ubiquitous thioesterases from which the various acyl-ACP thioesterases evolved. To elucidate whether this might apply to the LPAAT genes as well, additional sequences such as those from rapeseed with similarities to the seed-specific LPAATs are required.

Interestingly, database searches with the different plant LPAAT sequences revealed genes and cDNA clones from various organisms which, at the deduced amino acid level, exhibit significant sequence similarities to either the seed-specific or the constitutive LPAAT sequences (Fig. 4). Only the distantly related LPAAT genes of *Neisseria meningitidis* and *N. gonorrhoeae*, which unlike *plsC* of *E. coli* are not essential for bacterial growth (Swartley *et al.*, 1995), and the putative one of *Synechocystis* sp. do not fit in either of the two LPAAT classes (Fig. 4). The number of sequences increases appreciably if expressed sequence tags and further partial cDNA sequences are also considered. However, only one partial cDNA of *A. thaliana* (Genbank accession

---

Fig. 4. Phylogram based on amino acid sequences of LPAATs and putative ones from the given organisms. Sequences which have clearly been demonstrated to encode an LPAAT are marked with an asterisk. Accession numbers of the sequences are the following: *Synechocystis* sp., sll1848; *Neisseria meningitidis*, U21807; *N. gonorrhoeae*, U21806; *Mycoplasma genitalium*, U39698 (g1045898); *M. pneumonia*, AE000052 (g1674233); *Homo sapiens* 1, U89336 (g1841552); *H. sapiens* 2, D86960 (g1503994); *Caenorhabditis elegans* 1, Z72511 (g1301695); *C. elegans* 2, Z73975 (g1403001); *C. elegans* 3, U23526 (g746580); *Saccharomyces cerevisiae* 1, *SLC1* gene, L13282; *S. cerevisiae* 2, Z35911 (P38226); *S. cerevisiae* 3, Z49770 (S54641); *Borrelia burgdorferi*, L32861; *Haemophilus influenzae*, U32756 (g1573755); *E. coli*, *plsC* gene, M63491; *Salmonella typhimurium*, parF gene, M68936; for sequences of the plant species, see Table 2.

number H76089) was found to have significant sequence similarity to LPAAT sequences, namely the constitutively expressed ones.

Several of these sequences are included in the phylogram (Fig. 4) in order to highlight that certain organisms such as yeast, *Caenorhabditis elegans* and *Homo sapiens* appear to possess acyltransferases belonging to both LPAAT classes. It should be pointed out that only the few marked sequences in Fig. 4 have clearly been demonstrated to encode an LPAAT, while most of them have only been identified by their

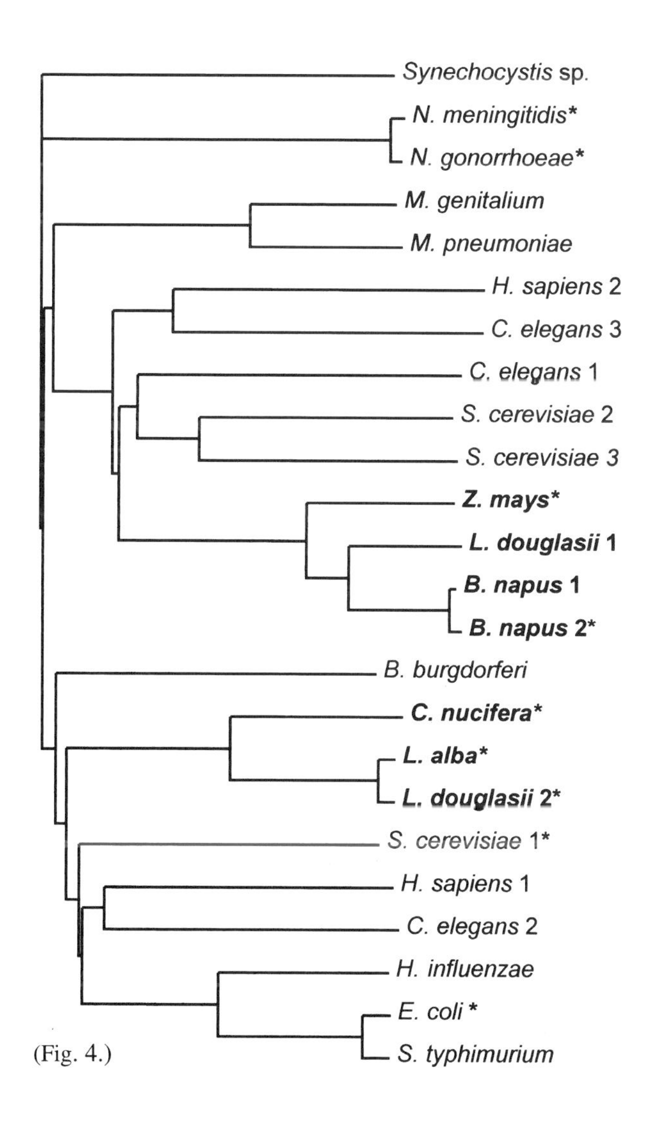

(Fig. 4.)

sequence similarities. In order to further evaluate the relationships depicted in the phylogram, it will be necessary to see whether all these sequences encode LPAATs, in which pathways they are involved and in which compartment the eukaryotic ones are located.

Conserved boxes in acyltransferase sequences

An alignment of the amino acid sequences of different known LPAATs, as well as the putative ones, shows that, apart from the partial sequence of *Borrelia burgdorferi* (Ojaimi *et al.*, 1994) and that encoded by pRAT1 (Table 2), the NHXXXXD box is conserved in all sequences (Fig. 5). In addition, a further box FP/VEGTR, typical of the LPAATs from higher plants and downstream to the first one, is largely conserved in most of the sequences as is the spacing between the two boxes (Fig. 5). Remarkably, the stretches of the *Synechocystis* sp. sequence covering these boxes exhibit a distinctly higher similarity to the sequences of the seed-specific LPAAT than to that of *E. coli* or to those of the other plant LPAATs. Due to its relatively low overall similarity with the plant sequences, this is not reflected in the phylogram (Fig. 4). According to these observations, it is likely that the boxes are conserved not only in microsomal LPAATs but also in the isofunctional enzymes of other compartments.

As shown in Fig. 5, boxes similar to those of the plant LPAATs also exist in the various GPAT sequences as well as in the acyltransferase domain of the 2-acyl-glycero-phospho-ethanolamine acyltransferase of *E. coli* (Jackowski, Jackson & Rock, 1994). The conserved amino acids within these boxes might be critical for acyltransferase activity. In particular, the histidine and aspartate residues of box 1 are likely to be a part of the active site. This is in line with biochemical investigations which showed that the microsomal LPAAT of pea lost its activity when treated with diethylpyrocarbonate and suggest the existence of an active-site histidine (Hares, 1990). As outlined above, the threonine and serine residues, respectively, within box 1 might also be important for the fatty acid specificities of the plastidial GPATs. However, in view of the results obtained with the yeast LPAAT encoded by *SLC1* (Nagiec *et al.*, 1993), it is unlikely that these residues are critical for the LPAAT specificities. A single mutation in *SLC1* changing glutamine 44 to a leucine alters the substrate specificities of the yeast LPAAT and enables it to use very long chain acyl-CoA thioesters as acyl donors.

In order to determine the molecular basis of plant LPAAT substrate specificities, the sequences encoding the microsomal LPAAT of certain lines of *B. oleracea* may be useful. Unlike the typical TAG fatty acid

Sequence source	Box 1	Distance	Box 2
LPAAT			
L. alba	YISNHASPID	← 63 →	LIMFPEGTRS
C. nucifera	YICNHASLVD	← 63 →	LIIFPEGTRS
E. coli	YIANHQNNYD	← 63 →	IWMFPEGTRS
Synechocystis sp.	VVSNHASYFD	← 60 →	VGVFLEGTRT
S. cerevisiae 1	MIANHQSTLD	← 63 →	LWVFPEGTRS
H. sapiens 1	VVSNHQSSLD	← 63 →	VWVFPEGTRN
C. elegans 2	VICNHQSSLD	← 63 →	LWVFPEGTRN
Z. mays	IISNHRSDID	← 69 →	LALFVEGTRF
B. napus 2	VVCNHRSDID	← 69 →	LALFVEGTRF
L. douglasii 1	LICNHRSDID	← 69 →	LALFVEGTRF
M. genitalium	VVANHKSNLD	← 66 →	IAVFAEGTRI
S. cerevisiae 3	IIANHQMYAD	← 89 →	LIMFPEGTNL
C. elegans 1	IVMNHRTRLD	← 71 →	ILLFPEGTDK
C. elegans 3	LLANHLGLLD	← 72 →	VIMYPEGSRL
H. sapiens 2	MLVNHQATGD	← 72 →	IVLFPEGGFL
GPAT			
M. musculus	FLPVHRSHID	← 74 →	LEIFLEGTRS
C. elegans	YLPLHRSHLD	← 74 →	IEFFLEGTRS
E. coli	YVPCHRSHMD	← 68 →	VEYFVEGGRS
H. influenzae	YVPCHRSHID	← 68 →	VEYFIEGGRS
M. tuberculosis	FAFSHRSYLD	← 68 →	LTWSIEGGRT
S. oleracia	LMSNHQSEAD	← 82 →	IWIAPSGGRD
A. thaliana	LISNHQSEAD	← 82 →	IWIAPSGGRD
P. sativum	LMSNHQSEAD	← 82 →	IWIAPSGGRD
P. vulgaris	LMSNHQTEAD	← 82 →	VWIAPSGGRD
C. sativus	LISNHQTEAD	← 82 →	IWIAPSGGRD
C. moschata	LISNHQTEAD	← 82 →	IWIAPSGGRD
C. tinctorius	LISNHQTEAD	← 83 →	IWIAPSGGRD
LPEAT			
E. coli	IPTNHVSFID	← 58 →	VVIFPEGGRI

Fig. 5. Alignment of the conserved sequences stretches of LPAATs (see Fig. 4), GPATs (see Fig. 2) and a 2-acyl-glycero-phosphoethanolamine acyltransferase (LPEAT) from the given organisms. Distance gives the respective numbers of amino acid residues between the two boxes.

pattern of Brassicaceae, in which 22 : 1 is excluded from the *sn*-2 position, certain *B. oleracea* lines have been identified containing appreciable proportions of 22 : 1 at the *sn*-2 position, but no trierucin while, in others, trierucin was also detectable (Taylor *et al.*, 1994; MacKenzie *et al.*, 1997). These data suggest that certain *B. oleracea* lines possess differently mutated LPAAT genes, one which enables the microsomal LPAAT to use 22 : 1-CoA and a further one which enables it to use 22 : 1-CoA even if the acyl acceptor carries a very long chain acyl group.

Impact on agriculture

Quite recently, cloned LPAAT genes, namely those encoding the seed specifically expressed enzymes as well as those of yeast and *E. coli* have been successfully used to alter the fatty acid composition of transgenic rapeseed oil (Lassner *et al.*, 1995; Brough *et al.*, 1996; Zou *et al.*, 1997*a*, *b*; Weier *et al.*, 1997; H.M. Davies, T.A. Voelker & D.S. Knutzon, personal communication; D. Weier, W. Lühs, J. Dettendorfer & M. Frentzen, unpublished results). Expression of very long chain specific LPAATs in developing seeds of transgenic high erucic acid rapeseed plants resulted in the incorporation of significant levels of 22 : 1 into the *sn*-2 position of TAG and in the accumulation of trierucin (Lassner *et al.*, 1995; Brough *et al.*, 1996; Zou *et al.*, 1997*a*, *b*; Weier *et al.*, 1997 and unpublished data). In addition, expression of the coconut LPAAT in transgenic high lauric rapeseed lines randomised the distribution of 12 : 0 into all three glycerol positions so that trilaurin became one of the major TAG species (Voelker *et al.*, 1996; H.M. Davies, T.A. Voelker & D.S. Knutzon, personal communication). These data demonstrate the feasibility of using acyltransferases to achieve the biosynthesis of TAG with homogeneous fatty acid distributions in the highly productive oil seed crops desired for industrial applications. Moreover, expression of the yeast LPAAT gene (*SCL1–1*) in transgenic rapeseed was found to improve oil yield (Zou *et al.*, 1997*a*, *b*).

Consequently, since 1988 when the first plant acyltransferase sequence was published, much progress has been made in the molecular biology of plant acyltransferases, especially with regard to the plastidial GPATs and microsomal LPAATs. These acyltransferase genes have then been successfully used to alter fatty acid compositions of both membrane and storage lipids in transgenic plants. In this way, the decisive role of acyltransferases in determining fatty acid patterns of glycerolipids and therefore, in determining certain agriculturally important traits has been confirmed. In addition, as described for the plastidial

GPATs, the elucidation of the molecular basis of their enzyme specificities will make it possible to design acyltransferases with new or improved properties optimally suited to genetically engineered crops. At present, experiments are well under way to exploit crops as efficient, low polluting chemical factories for the production of new value-added products designed for specific markets.

Acknowledgements

We are grateful to our colleagues for providing us with reprints of their publications and unpublished results and to Dr Jodi A. Scheffler for critically reading the manuscript. Research in the authors' laboratories was supported by Norddeutsche Pflanzenzucht Hans-Georg Lembke KG, Kleinwanzlebener Saatzucht AG, Deutsche Saatveredelung Lippstadt-Bremen GmbH, Hoechst Schering AgrEvo GmbH and the Bundesministerium für Bildung, Wissenschaft, Forschung und Technologie (Grants 0310528F, 0311156 and 316601).

Note added in proof: Since this paper has gone to press, mammalian sequences very similar to that of *Homo sapiens* 1 (Fig. 4) have been demonstrated to encode an LPAAT (e.g. Kume & Shimizu, 1997, *Biochem. Biophys. Res. Commun.*, **237**, 663–6; West *et al.*, 1997, *DNA Cell Biol.* **16**, 691–701). The conserved residues within box 1 (Fig. 5) have been to be essential for the activities of both GPAT and LPEAT of *E. coli* (Heath & Rock, 1998, *J. Bacteriol.*, **180**, 1425–30). In addition, evidence has been provided that the threonine residue 122 of the LPAAT sequence of *E. coli* is critical for the fatty acid specificity of the enzyme (Morand *et al.*, 1998, *Biochem. Biophys. Res. Commun.*, **244**, 79–84).

References

Bhella, R.S. & MacKenzie, S.L. (1994). Nucleotide sequence of a cDNA *from Carthamus tinctorius* encoding a glycerol-3-phosphate acyl transferase. *Plant Physiology*, **106**, 1713–14.

Brough, C.L., Coventry, J.M., Christie, W.W., Kroon, J.T.M., Brown, A.R., Barsby, T.L. & Slabas, A.R. (1996). Towards the genetic engineering of triacylglycerols of defined fatty acid composition: major changes in erucic acid content at the *sn*-2 position affected by the introduction of a 1-acyl-*sn*-glycerol-3-phosphate acyltransferase from *Limnanthes douglasii* into oil seed rape. *Molecular Breeding*, **2**, 133–42.

Brown, A.P., Coleman, J., Tommey, A.M., Watson, M.D. & Slabas, A.R. (1994). Isolation and characterisation of a maize cDNA that complements a 1-acyl *sn*-glycerol-3-phosphate acyltransferase mutant of *Escherichia coli* and encodes a protein which has similarities to other acyltransferases. *Plant Molecular Biology*, **26,** 211–23.

Brown, A.P., Brough, C.L., Kroon, J.T.M. & Slabas, A.R. (1995). Identification of a cDNA that encodes a 1-acyl-*sn*-glycerol-3-phosphate acyltransferase from *Limnanthes douglasii. Plant Molecular Biology*, **29**, 267–78.

Coleman, J. (1990). Characterization of *Escherichia coli* cells deficient in 1-acyl-*sn*-glycerol-3-phosphate acyltransferase activity. *Journal of Biological Chemistry*, **265**, 17215–21.

Coleman, J. (1992). Characterization of the *Escherichia coli* gene for 1-acyl-*sn*-glycerol-3-phosphate acyltransferase (*plsC*). *Molecular and General Genetics*, **232**, 295–303.

Davies, H.M., Hawkins, D.J. & Nelson, J.S. (1995). Lysophosphatidic acid acyltransferase from immature coconut endosperm having medium chain length substrate specificity. *Phytochemistry*, **39**, 989–96.

Ferri, S.R. & Toguri, T. (1997). Substrate specificity modification of the stromal glycerol-3-phosphate acyltransferase. *Archives of Biochemistry and Biophysics*, **337**, 202–8.

Frentzen, M. (1993). Acyltransferases and triacylglycerols. In *Lipid Metabolism in Plants*, ed. T.S. Moore, Jr, pp. 195–220. Boca Raton: CRC Press.

Fritz, M., Heinz, E. & Wolter, F.P. (1995). Cloning and sequencing of a full-length cDNA coding for *sn*-glycerol-3-phosphate acyltransferase from *Phaseolus vulgaris. Plant Physiology*, **107**, 1039–40.

Gibson, S., Falcone, D.L., Browse, J. & Somerville, C. (1994). Use of transgenic plants and mutants to study the regulation and function of lipid composition. *Plant, Cell and Environment*, **17**, 627–37.

Grayburn, W.S., Collins, G.B. & Hildebrand, D.F. (1992). Fatty acid alteration by a Δ9 desaturase in transgenic tobacco tissue. *Bio/Technology*, **10**, 675–8.

Hanke, C., Wolter, F.P., Coleman, J., Peterek, G. & Frentzen, M. (1995). A plant acyltransferase involved in triacylglycerol biosynthesis complements an *Escherichia coli sn*-1-acylglycerol-3-phosphate acyltransferase mutant. *European Journal of Biochemistry*, **232**, 806–10.

Hares, W. (1990). Partielle Reinigung und Charakterisierung der eukaryotenAcyl-CoA : *sn*-1-Acyl-Glycerin-3-Phosphat-Acyltransferase aus Pflanzen. Dissertation, Universität Hamburg.

Heinz, E. (1993). Biosynthesis of polyunsaturated fatty acids. In *Lipid Metabolism in Plants*, ed. T.S. Moore, Jr, pp. 33–89. Boca Raton: CRC Press.

Ishizaki, O., Inishida, I., Agata, K., Eguchi, G. & Murata, N. (1988). Cloning and nucleotide sequence of cDNA for the plastid glycerol-3-phosphate acyltransferase from squash. *FEBS Letters*, **238**, 424–30.

Ishizaki-Nishizawa, O., Azuma, M., Ohtani, T., Murata, N. & Toguri, T. (1995). Nucleotide sequence of cDNA from *Spinacia oleracea* encoding plastid glycerol-3-phosphate acyltransferase (EMBL accession No. X77370). *Plant Physiology*, **108**, 1342.

Jackowski, S., Jackson, P.D. & Rock, C.O. (1994). Sequence and function of the *aas* gene in *Escherichia coli*. *Journal of Biological Chemistry*, **269**, 2921–8.

Johnson, T.C., Schneider, J.C. & Somerville, C. (1992). Nucleotide sequence of acyl–acyl carrier protein: glycerol-3-phosphate acyltransferase from cucumber. *Plant Physiology*, **99**, 771–2.

Jones, A., Davies, H.M. & Voelker, T.A. (1995). Palmitoyl-acyl carrier protein (ACP) thioesterase and the evolutionary origin of plant acyl-ACP thioesterases. *The Plant Cell*, **7**, 359–71.

Knutzon, D.S., Lardizabal, K.D., Nelson, J.S., Bleibaum, J.L., Davies, H.M. & Metz, J.M. (1995). Cloning of a coconut endosperm cDNA encoding a 1-acyl-*sn*-glycerol-3-phosphate acyltransferase that accepts medium-chain-length substrates. *Plant Physiology*, **109**, 999–1006.

Kunst, L., Browse, J. & Somerville, C. (1988). Altered regulation of lipid biosynthesis in a mutant of *Arabidopsis* deficient in chloroplast glycerol-3-phosphate acyltransferase activity. *Proceedings of the National Academy of Sciences, USA*, **85**, 4143–7.

Lassner, M.W., Levering, C.K., Davies, H.M. & Knutzon, D.S. (1995). Lysophosphatidic acid acyltransferase from meadowfoam mediates insertion of erucic acid at the *sn*-2 position of triacylglycerol in transgenic rapeseed oil. *Plant Physiology*, **109**, 1389–94.

Lightner, V.A., Bell, R.M. & Modrich, P. (1983). The DNA sequences encoding *plsB* and *dgk* loci of *Escherichia coli*. *Journal Biological Chemistry*, **258**, 10 856–61.

MacKenzie, S.L., Giblin, E.M., Barton, D.L., McFerson, J.R., Tenashuk, D. & Taylor, D.C. (1997). Erucic acid distribution in *Brassica oleracea* seed oil triglycerides. In *Physiology, Biochemistry and Molecular Biology of Plant Lipids*, ed. J.P. Williams, M.U. Khan & N.W. Lem, pp. 309–21. Dordrecht: Kluwer Academic Publishers.

Moon, B.Y., Higashi, S.I., Gombos, Z. & Murata, N. (1995). Unsaturation of the membrane lipids of chloroplasts stabilizes the photosynthetic machinery against low-temperature photoinhibition in transgenic tobacco plants. *Proceedings of the National Academy of Sciences, USA*, **92**, 6219–23.

Murata, N. (1983). Molecular species composition of phosphatidylglycerols from chilling-sensitive and chilling-resistant plants, *Plant Cell Physiology*, **24**, 81–6.

Murata, N., Ishizaki-Nishizawa, O., Higashi, S., Hayashi, H., Tasaka, Y. & Nishida, I. (1992). Genetically engineered alteration in the chilling sensitivity of plants. *Nature*, **356**, 710–3.

Murata, N., Sato, N., Takahashi, N. & Hamazaki, Y. (1982). Compositions and positional distributions of fatty acids in phospholipids from leaves of chilling-sensitive and chilling-resistant plants. *Plant Cell Physiology*, **23**, 1071–9.

Murphy, D.J. (1996). Engineering oil production in rapeseed and other oil crops. *Trends in Biotechnology*, **14**, 206–13.

Nagiec, M.M., Wells, G.B., Lester, R.L. & Dickson, R.C. (1993). A suppressor gene that enables *Saccharomyces cerevisiae* to grow without making sphingolipids encodes a protein that resembles an *Escherichia coli* fatty acyltransferase. *The Journal of Biological Chemistry*, **268**, 22 156–63.

Nishida, I. & Murata, N. (1996). Chilling sensitivity in plants and cyanobacteria: the crucial contribution of membrane lipids. *Annual Reviews of Plant Physiology and Plant Molecular Biology*, **47**, 541–68.

Nishida, I., Tasaka, Y., Shiraishi, H. & Murata, N. (1993). The gene and the RNA for the precursor to the plastid-located glycerol-3-phosphate acyltransferase of *Arabidopsis thaliana*. *Plant Molecular Biology*, **21**, 267–77.

Ohlrogge, J.B. (1994). Design of new plant products : engineering of fatty acid metabolism. *Plant Physiology*, **104**, 821–6.

Ojaimi, C., Davidson, B.D., Saint Girons, I. & Old, I.G. (1994). Conservation of gene arrangement and an unusual organization of rRNA genes in the linear chromosomes of the Lyme disease spirochaetes *Borrelia burgdorferi*, *B. garinii* and *B. afzelii*. *Microbiology*, **140**, 2931–40.

Polashock, J.J., Chin C.K. & Martin, C.E. (1992). Expression of the yeast Δ-9 fatty acid desaturase in *Nicotiana tabacum*. *Plant Physiology*, **100**, 894–901.

Roughan, P.G. & Ohlrogge, J.B. (1996). Evidence that isolated chloroplasts contain an integrated lipid-synthesizing assembly that channels acetate into long-chain fatty acids. *Plant Physiology*, **110**, 1239–47.

Shin, D.H., Paulauskis, J.D., Moustaid, N. & Sul, H.S. (1991). Transcriptional regulation of p90 with sequence homology to *Escherichia coli* glycerol-3-phosphate acyltransferase. *Journal of Biological Chemistry*, **266**, 23 834–9.

Somerville, C. (1995). Direct tests of the role of membrane lipid composition in low-temperature-induced photoinhibition and chilling sensitivity in plants and cyanobacteria. *Proceedings of the National Academy of Sciences, USA*, **92**, 6215–18.

Swartley, J.S., Balthazar, J.T., Coleman, J., Shafer, W.M. & Stephens, D.S. (1995). Membrane glycerophospholipid biosynthesis in *Neisseria meningitidis* and *Neisseria gonorrhoeae*: Identification, characterization, and mutagenesis of a lysophosphatidic acid acyltransferase. *Molecular Microbiology*, **18**, 401–12.

Taylor, D.C., MacKenzie, S.L., McCurdy, A.R., McVetty, P.B.E.,

Giblin, E.M., Pass, E.W., Stone, S.J., Scarth, R., Rimma, S.R. & Pickard, M.D. (1994). Stereospecific analysis of seed triacylglycerols from high erucic acid Brassicaceae: detection of erucic acid at the *sn*-2 position in *B. oleracea* genotypes. *Journal of the American Oil Chemistry Society*, **71**, 163–7.

Vancura, A. & Haldar, D. (1994). Purification and characterization of glycerophosphate acyltransferase from rat liver mitochondria. *Journal of Biological Chemistry*, **269**, 27 209–15.

van de Loo, F.J., Fox, B.G. & Somerville, C. (1993). Unusual fatty acids. In *Lipid Metabolism in Plants*, ed. T.S. Moore, Jr, pp. 91–126. Boca Raton: CRC Press.

Voelker, T.A., Hayes, T.R., Crammer, A.M., Turner, J.C. & Davies, H.M. (1996). Genetic engineering of a quantitative trait: metabolic and genetic parameters influencing the accumulation of laurate in rapeseed. *The Plant Journal*, **9**, 229–41.

Wang, J., Gebre, A.K., Anderson, R.A. & Parks, J.S. (1997). Amino acid residue 149 of lecithin : cholesterol acyltransferase determines phospholipase A_2 and transacylase fatty acyl specificity. *The Journal of Biological Chemistry*, **272**, 280–6.

Weber, S., Wolter, F.P., Buck, F., Frentzen, M. & Heinz, E. (1991). Purification and cDNA sequencing of an oleate-selective acyl-ACP : *sn*-glycerol-3-phosphate acyltransferase from pea chloroplasts. *Plant Molecular Biology*, **17**, 1067–76.

Weier, D., Hanke, C., Eickelkamp, A., Lühs, W., Dettendorfer, J., Schaffert, E., Möllers, C., Friedt, W., Wolter, F.P. & Frentzen, M. (1997). Trierucoylglycerol biosynthesis in transgenic plants of rapeseed (*Brassica napus* L.). *Fett/Lipid*, **99**, 160–5.

Wolter, F.P. (1996). Conserved intron position in 3′ untranslated region of a cDNA encoding the plastidial *sn*-glycerol-3-phosphate acyltransferase of spinach (Accession No. Z4901) (PGR 96–118). *Plant Physiology*, **112**, 1735.

Wolter, F.P., Schmidt, R. & Heinz, E. (1992). Chilling sensitivity of *Arabidopsis thaliana* with genetically engineered membrane lipids. *The EMBO Journal*, **11**, 4685–92.

Yet, S.F., Lee, S., Hahm, Y.T. & Sul, H.S. (1993). Expression and identification of p90 as the murine mitochondrial glycerol-3-phosphate acyltransferase. *Biochemistry*, **32**, 9486–91.

Yet, S.F., Moon, Y.K. & Sul, H.S. (1995). Purification and reconstitution of murine mitochondrial glycerol-3-phosphate acyltransferase. Functional expression in baculovirus-infected insect cells. *Biochemistry*, **34**, 7303–10.

Yuan, L., Voelker, T.A. & Hawkins, D.J. (1995). Modification of the substrate specificity of an acyl-acyl carrier protein thioesterase by protein engineering. *Proceedings of the National Academy of Sciences, USA*, **92**, 10 639–43.

Zou, J.T., Katavic, V., Giblin, E.M., Barton, D.L., MacKenzie, S.L.,

Keller, W.A. & Taylor, D.C. (1997*a*). Modification of seed oil content and acyl composition in the Brassicaceae utilizing a yeast *sn*-2 acyltransferase (*SLC1–1*) gene. In *Physiology, Biochemistry and Molecular Biology of Plant Lipids*, ed. J.P. Williams, M.U. Khan & N.W. Lem, pp. 407–9. Dordrecht: Kluwer Academic Publishers.

Zou, J.T., Katavic, V., Giblin, E.M., Barton, D.L., MacKenzie, S.L., Keller, W.A., Hu, X. & Taylor, D.C. (1997*b*). Modification of seed oil content and acyl composition in the Brassicaceae by expression of a yeast *sn*-2 acyltransferase gene. *The Plant Cell*, **9**, 909–23.

A.J. KINNEY

11 Production of specialised oils for industry

Introduction

Oilseed plants have been cultivated for their triacylglycerol for over 3000 years. Oil derived from these plants has been used for both food preparation and in certain non-food uses such as candles and soap. In the early days of chemical engineering in the last century, before the advent of cheaper petrochemicals, plant oils also provided a source of aliphatic hydrocarbons for industrial uses (Pattison, 1968). The rise in the consumption of vegetable oils after the invention of hydrogenation coincided with a decline in the utilisation of industrial fatty acids. Today, only about 2% of the 15 billion pounds or so of vegetable oil produced in the United States is used for non-food purposes (Fig. 1).

The bulk of edible oil produced in the US is soybean oil, which is chemically hydrogenated to increase its oxidative stability and, in some cases, also provide functionality for solid fat applications. Hydrogenation results in the formation of *trans* monounsaturated fatty acids,

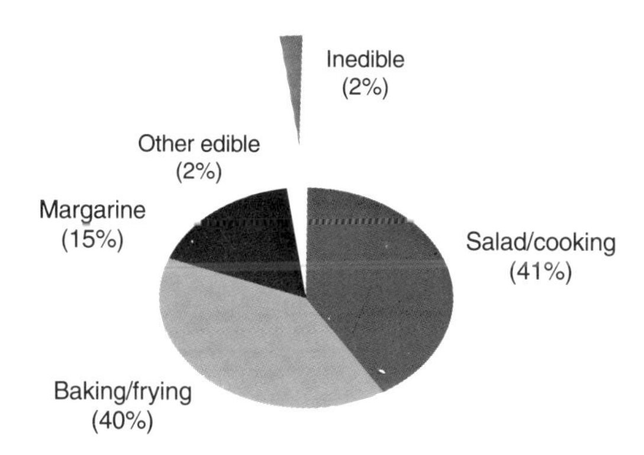

Fig. 1. The market for vegetable oils in the United States. The area of the pie represents about 15 billion pounds of oil per year.

consumption of which has been linked to coronary heart disease in humans (Katan, Mensink & Zock, 1995).

It is now possible to produce, by genetic engineering of soybean plants, edible vegetable oils which are naturally oxidatively stable or are both stable and have desired functionality in solid fat applications (Kinney & Knowlton, 1998). These new oils also have a number of unanticipated industrial applications such as biodegradable lubricants or feedstocks for further chemical modification.

It may also be possible to produce novel industrial fatty acids in soybeans and thus extend some of these non-food uses for vegetable oils (Kinney, 1997). Thus it is likely that there will be a resurgence in the non-food use of plant oils in the next century.

This chapter will describe the development of new edible soybean oils, some non-edible uses for high oleic acid soybean oil and future directions for producing new industrially useful oils from soybean. The principles and techniques described can, and have been, applied to other oilseed plants such as rapeseed and maize oils (Kinney, 1997; also see Knutzon, this volume).

Edible soybean oils with modified fatty acid content

Initial efforts in oilseed modification by genetic engineering, which began in the 1980s, focused on edible oils (Kinney, 1997). The primary target for soybean oil is to produce oils with increased oxidative stability by decreasing the polyunsaturated fatty acid content and increasing the oleic acid content. For solid fat applications, such as margarine, it is also necessary to combine this increased stability with an increased solid fat content (SFC). Thus an oil with increased stearic and oleic acids would have the ideal functionality for margarines and spreads. For salad oils, marketed directly to the consumer, the goal is to increase the perceived health benefits of the oil by decreasing the saturated fatty acid content to close to zero.

Although in developing seeds the pathway for triacylglycerol biosynthesis from acetate to triacylglycerol (see Kinney, 1997, for details), appeared to be somewhat linear (Fig. 2) a number of issues remained unresolved when this work was begun. It was not clear, for example, that simply changing the ratio of different acyl-CoAs in the cytoplasmic pool would be reflected by a similar change in the triacylglycerol. Nor was it known which of the enzymes involved in the synthesis of 16- and 18-carbon fatty acids were regulating the distribution of carbon among the five major fatty acids, or to what extent additional pathways, such as plastid 18 : 1 and 18 : 2 desaturation, might contribute to the final triacylglycerol fatty acid composition.

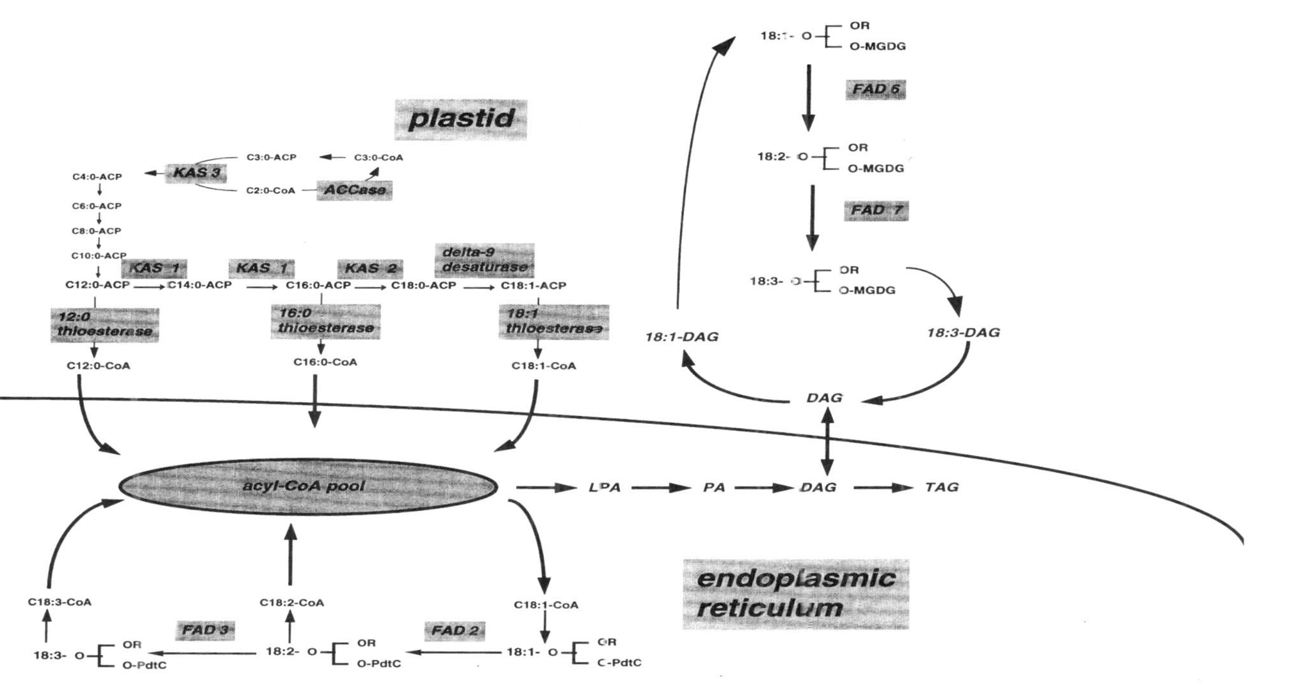

Fig. 2. Triacylglycerol biosynthesis in the developing oilseed. The enzymatic steps for which soybean genes have been cloned are shown and labelled with the gene designation. Abbreviations: ACCase = acetyl CoA carboxylase, KAS = β-ketoacyl synthetase, FAD = fatty acid desaturase, PtdC = phosphatidylcholine, LPA = lysophosphatidic acid, PA = lysophosphatidic acid, DAG = diacylglycerol, TAG = triacylglycerol, MGDG = monogalactosyl diacylglycerol.

Since the genes for most of the important fatty acid biosynthetic enzymes in soybean had been cloned (Kinney 1994, 1995), an initial approach was to over-express or silence each gene in turn and then observe the effect on fatty acid composition of the transgenic seed.

Transforming a gene into soybean cells and then measuring its effect on the triacylglycerol content of transgenic seed can take as long as 24 months (Christou *et al.*, 1990; Finer & McMullen, 1991; Kinney, 1995). However, since the method used to transform soybeans is particle bombardment of somatic embryos (Finer & McMullen, 1991), it was possible to get a preview of the seed fatty acid phenotype of a transgenic line after only a few weeks (Fig. 3).

Initially, the somatic embryo cultures were optimised for frequency of transformation events and for rapid maturation rate rather than for regeneration capability. These cultures could then be used as a model soybean transformation system in which the effect of expressing an embryo-specific transgene on triacylglycerol composition could be observed in less than 3 months. It was then possible to transform regenerable embryo cultures, or other tissue types, with the preferred gene constructs (i.e. those which resulted in the desired phenotype).

It was first important to demonstrate that the modified fatty acid compositions of mature somatic, transgenic embryos were predictive of the final oil composition of seeds from plants regenerated from those embryos. Antisense constructs with cDNAs of two different fatty acid desaturases were used: a seed-specific, microsomal omega-6 oleic acid

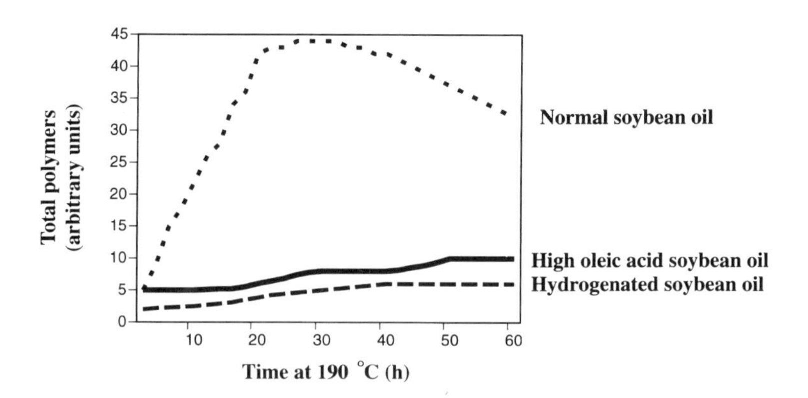

Fig. 3. Polymer formation at high temperature in three different oils: liquid soybean oil, high oleic acid liquid soybean oil and hydrogenated, solid soybean fat. Polymers were measured by high pressure liquid chromatography (data from Susan Knowlton, DuPont).

seed-specific desaturase and a microsomal omega-3 linoleic acid seed-specific desaturase.

In soybean there are at least two genes which encode microsomal oleic acid omega-6 desaturases, Fad 2-1 and Fad 2-2 (Heppard *et al.*, 1996). The Fad 2-2 gene is expressed in all soybean tissues, whereas the Fad 2-1 gene is expressed only in the seed. There are also at least two genes which encode microsomal omega-3 linoleic acid desaturases (Yadav *et al.*, 1993) and one of these is also seed specific (*Fad 3-1*). Embryo-specific suppression of Fad 3-1 leads to a decrease in linolenic acid content of the oil, with a corresponding increase in linoleic acid, suppression of Fad 2-1, a decrease in total polyunsaturates and an increase in oleic acid.

In all cases, when a phenotype was seen in a somatic embryo line, a corresponding phenotype was observed in seeds of plants regenerated from those lines (Kinney, 1996).

Thus the somatic embryo model system was used as a means of selection to determine which genes to manipulate in soybean to produce a desired fatty acid composition. From these types of experiments (Kinney, 1995), it was possible to determine the control points in the synthesis of 16- and 18-carbon fatty acids (Kinney, 1996).

For example, most of the 16-carbon acyl chains and some of the 18 carbon chains which leave the plastid are the result of a 16 : 0-ACP thioesterase (Fat B). By silencing the soybean Fat B gene, it was possible to produce triacylglycerol with a decreased saturated fatty acid content (<4%) and an increased oleate content.

It was also apparent from these somatic embryo experiments that the plastid desaturases contribute only a few per cent of total polyunsaturated fatty acid content of soybean oil. Most of the flux for polyunsaturated fatty acid biosynthesis is through the Fad 2-1 reaction (Kinney, 1995). Finally, it became clear that large changes in the relative content of the acyl-CoA pool was reflected by similar changes in the final triacylglycerol fatty acid content.

Using the information obtained in the somatic embryo model systems, it was possible to produce a number of different elite soybean lines that have oils with modified fatty acid profiles (Table 1).

Suppressing the soybean Fat B gene resulted in seeds with oil having less than 4% total saturates, compared with about 15% in regular soy oil. It is anticipated that commercialisation of this oil will result in healthier salad oils. Experiments with somatic embryos have also led to the conclusion that oils with less than 3% total saturated fats are possible in lines in which the Fat B genes have been completely suppressed.

Table 1. *Fatty acid composition of oil from soybeans of transgenic lines*

Gene (s)	Expression	Relative fatty acid content (%)					Food use
		16 : 0	8 : 0	18 : 1	18 : 2	18 : 3	
Elite line	Wild-type	11	3	22	55	9	Standard soy oil
Fad 2	Silenced	7	3	86	1	3	Frying, baking, etc.
Fad 3	Silenced	12	3	32	50	3	Frying
Fat A	Silenced	2	1	25	64	9	Salad oil
Fat A	Increased	16	10	12	54	9	Margarines, etc.
Fat B	Increased	36	4	7	47	9	Margarines, etc.
Hst1 + *Fad 2*	Silenced	5	30	60	1	4	Margarines, etc.

The numbers represent the relative abundance (%) of major fatty acids in the seed oil in different transgenic soybean lines. The phenotypes were created by silencing or over-expressing phospholipid desaturases (*Fad* genes), acyl-ACP thioesterases (*Fat* genes) or delta-9 desaturase (*Hst1*). The profiles are from soybeans grown in the field.

Oil with a total polyunsaturated content of less than 5% and oleic acid content of 85% has been produced by suppressing the Fad 2-1 gene. The oil from this line also has a palmitic acid content of around 7%, compared with 11% in regular oil. The reason for the reduction in palmitate is not entirely clear but it may be caused by competition for common substrates, palmitoyl-CoA and oleoyl-CoA, by a class of acyltransferase which is able to transfer either of these molecules to the glycerol backbone. If this were the case, a large increase in the oleoyl-CoA pool would be expected to result in a reduced utilisation of palmitoyl-CoA and a reduced palmitate content of the oil. It may be possible to reduce the palmitate content of this line even further by combining it with the Fat B suppression line described above.

The liquid oil from these high oleic acid (Fad 2-1) lines has an oxidative stability comparable to hydrogenated, solid soybean oil (Knowlton, Ellis & Kelly, 1998). For example, one of the standardised methods used to evaluate the oxidative stability of commercial cooking oils is the active oxygen method (AOM). This is an accelerated oxidation test in which an oil is aerated under a constant, elevated temperature

(97.8 °C) and degradation is monitored by measuring peroxide accumulation. The end point, or induction time, is determined by the number of hours required to reach a peroxide value of 100 meq/kg. Thus, the longer the induction time, the more stable the oil. High oleic acid soybean oil has an AOM induction time of about 150 hours compared with 15 hours for normal soybean oil and about 200 hours for heavy duty shortening.

Another standard method now commonly used to evaluate the stability of commercial cooking oils in the US is the oxidative stability index (OSI) measured automatically with a machine manufactured by Ominion, Inc., of Rockland, MA, USA.

In the OSI machine, air is bubbled through oil heated to 110 °C. As the oil oxidises, volatile organic acids, primarily formic acid, are formed which can be collected in distilled water in a cell. The machine constantly measures the conductivity of the distilled water and the induction period is determined as the time it takes for this conductivity to begin a rapid rise. Although the data derived from the two methods do not always have a straight correlation, the OSI induction time values for most oils are generally about half those of the AOM-derived values.

The OSI induction time of high oleic acid soybean oil is about 81 hours. This is over ten times longer than normal soybean oil (7 hours) and similar to the OSI induction time of solid hydrogenated shortening containing antioxidants (85 hours). High oleic acid soybean oil is also more stable in the OSI machine than commercial soybean oils containing the antioxidants TBHQ and silicone (21 hours) as well as liquid, partially hydrogenated soybean frying oil containing the same antioxidants (32 hours).

The high oleic Fad 2-1 line has also been combined with a line that has a mutation in one of the genes encoding delta-9 desaturase (Kinney, 1995) to produce an oxidatively stable oil with a high SFC (Kinney & Knowlton, 1998). The homozygous progeny of this cross-produced oil with a stearic acid content of around 30%, an oleic acid content of 60% and less than 5% polyunsaturated fatty acids. In addition, overexpression of Fat A results in an increase in stearic acid to about 10% of the total fatty acids, from around 3% in regular oil. These oils will be marketed as a replacement for hydrogenated oils in margarine and shortening applications.

Non-edible uses of high oleic soybean oil

Vegetable oils are highly biodegradable when compared with mineral oils. For example, a typical vegetable oil will completely degrade after only about 40 days when exposed (in the ground) to most environmental conditions. In the same amount of time there will have been no significant degradation of a mineral oil. Vegetable oils also have good lubricating properties and thus can make good industrial lubricants (Flider, 1995). This is especially true in applications where biodegradability is important, such as in marine engines and ground drilling applications where the outside of the drill is continuously lubricated with oil.

Most vegetable oils, however, are rich in polyunsaturates and oxidation at high temperatures causes the formation of polymers (see Kinney & Knowlton, 1998). These polymers increase the viscosity of the oil, greatly reducing its lubricating properties and causing excessive machine wear. Although reducing high-temperature oxidation, and hence polymer formation, hydrogenation leads to the formation of saturated fats and high-melting point *trans* mono-unsaturated fatty acids. These fatty acids cause increased viscosity and increased machine-wear at lower temperatures (when the engine or drilling is first started).

High oleic soybean oil is an attractive candidate for an industrial lubricant. It is a low-viscosity, liquid oil with good oxidative stability. When heated to 190 °C over a period of several days, the accumulation of polymer oxidation products in the high oleic acid soybean oil is far less than that of normal soybean oil (Fig. 3). The oil, however, is low in saturates (less than 10%) and contains no *trans* fatty acids. Thus it might be expected that this oil would be an effective lubricant at both high and low temperatures.

Indeed, Susan Knowlton at DuPont, in collaboration with Jim Glancy in the Agricultural Engineering Department at the University of Delaware, has recently demonstrated that high oleic soybean oil, with food-quality antioxidants added, performs almost as well as commercial mineral oil lubricants in machine-wear lubrication tests (S. Knowlton, pers. comm.).

Since the high oleic soybean oil consists of triacylglycerol with a maximum of only three double bonds per molecule, one per fatty acid, it is also an attractive feedstock for other industrial applications. Thus, this oil could be used to provide industrially important molecules with functional properties equivalent to non-renewable, petrochemical-derived molecules.

For example, high oleic acid soybean oil may provide a renewable resource for the production of composites and thermoset resins. Resin transfer moldings (RTMs) are useful composites for bridges, rail cars, car bodies and structural components of cars. Current RTMs are mostly petroleum-derived vinyl ester resins. Richard Wool and his colleagues at the University of Delaware's Center for Composite Materials are able to make RTMs comparable with vinyl ester RTMs from about 60% acetylated, epoxidised soybean oil and 40% vinyl esters. The soybean oil component is limited to 60% of this RTM since epoxidising normal soybean oil results in an average of about 4.5 epoxy groups per triacylglycerol. This is because more than 60% of the fatty acids of normal soybean oil have two or three reactive double bonds. Only two to three functional groups per molecule are needed to cross-link the triacylglycerols in the RTM, the additional unreacted double bonds and epoxy groups act to destabilise the resin. Since epoxidation of high oleic acid soybean oil results in only two to three epoxy groups per molecule, Wool and colleagues predict they will now be able to make RTMs from 100% soybean oil using the high oleic oil as starting material (R. Wool, G. Palmese & S. Kusefoglu, pers. comm.).

New oils for industrial uses

Epoxidised soybean oil has many uses in addition to those described above, including use in paints, coatings, adhesives and cosmetics. It is also currently used as a PVC plasticiser (Eisenhard, 1968) and sells for about $0.75 per pound, which is about three times the price of commodity soybean oil. This oil commands less than 10% of the PVC plasticiser market, however, because of inferior, but cheaper, petrochemical-derived alternatives such as dioctyl phthalate. These alternatives sell for around $0.55 per pound. It would be an economically attractive target to genetically engineer soybeans to directly produce an epoxidised oil. This oil could be sold for about twice the price of commodity soybean oil, and yet still cheaper than petrochemical-derived alternatives, thus expanding the PVC plasticiser market for epoxidised soybean oil.

Furthermore, although soybean oil contains only a handful of different fatty acids, there have been hundreds of different fatty acids identified in the oils of exotic oilseed species (Badami & Patil, 1981). These fatty acids include epoxy, hydroxy, keto, ethylenic, acetylenic, cyclopropanoid and even fluoro fatty acids.

This existing natural diversity not only indicates that there should be no theoretical barriers to producing exotic fatty acids in temperate oilseed crops such as soybeans, but also provides a deep gene pool into which the genetic engineer may dip in order to do so.

One approach to isolating the genes which encode the enzymes responsible for the synthesis of these unusual fatty acids is by high-throughput screening of cDNA libraries, or automated EST sequencing (Fig. 4). The idea would be to construct libraries from the developing

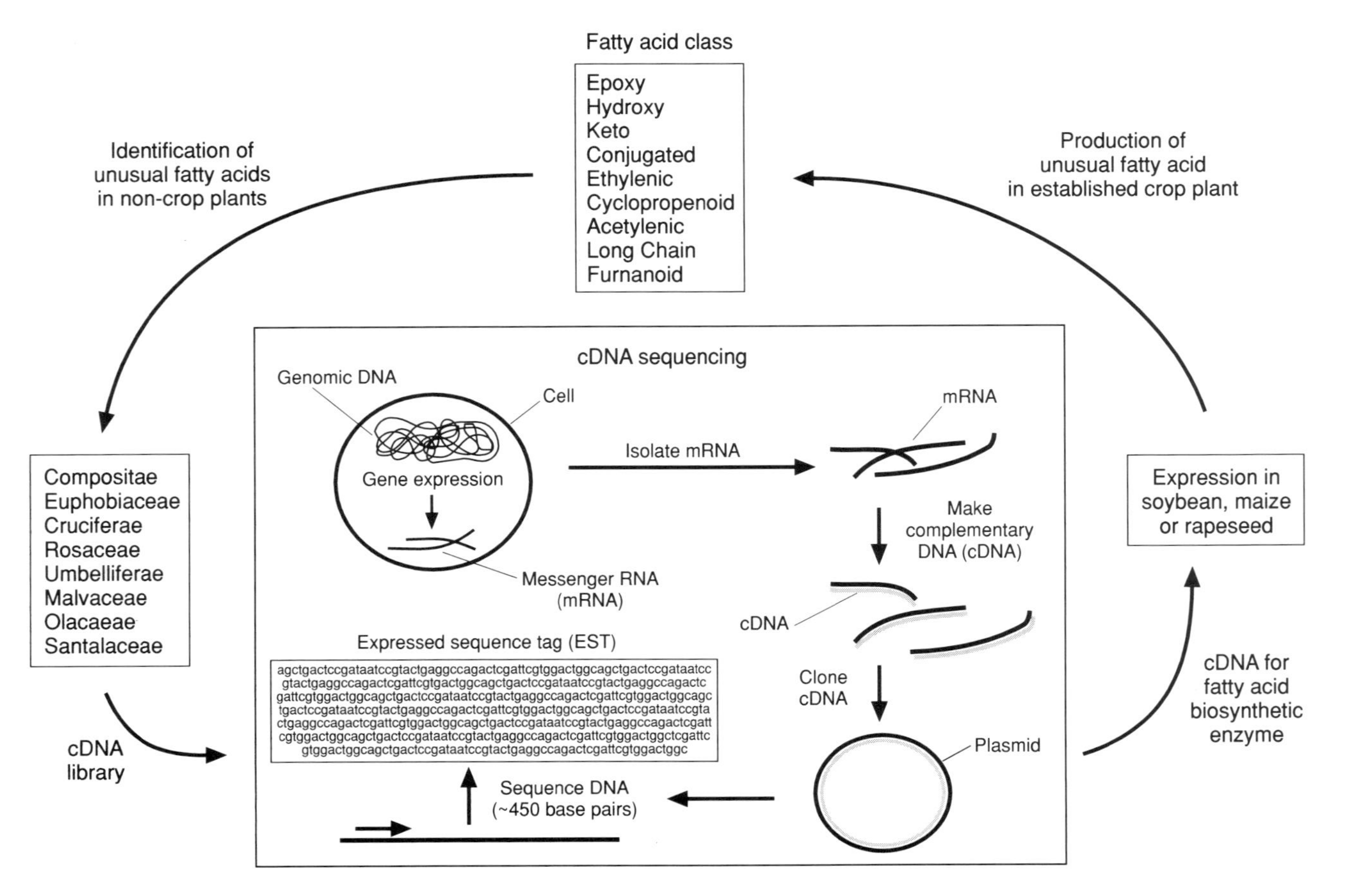

Fig. 4. One possible strategy for producing industrially important fatty acids in temperate oilseed crops. The details are discussed in the text.

seeds of species which produce exotic fatty acids. Automated sequencing of the genes expressed by these seeds would allow the identification of useful genes by their homology to known genes such as fatty acid desaturases. Their identity could be confirmed by expressing them in soybean somatic embryos or other model systems. This method has already proved successful for the identification of a castor bean fatty acid hydroxylase (van der Loo *et al.*, 1995) and many new genes are expected to be identified in the next few years.

Furthermore, structural and functional studies of membrane-bound desaturases and hydroxylases have made it possible to consider designing new enzymes, which will put double bonds or other functional groups at new positions on a fatty acid molecule (Shanklin *et al.*, 1997).

The high oleic acid soybean plants might provide a useful genetic background in which to produce some of these new oils, since oleic acid is the substrate for a number of further enzymatic reactions. A seed with a high substrate content and no competing desaturase activity might then produce higher levels of, for example, hydroxy, epoxy or long-chain fatty acids.

Other traits might require other genetic backgrounds, such as high stearic acid, for optimal expression, and it is clear that the knowledge gained from producing new edible oils in soybean will be useful in optimising the production of new, non-edible fatty acids in the seed.

In summary, it is now becoming a reality to produce industrially important fatty acids by the genetic engineering of oilseed plants such as soybeans and the next century promises to bring a renaissance in the use of plant-derived chemicals in the place of petrochemicals.

Acknowledgements

I thank my colleague Susan Knowlton at DuPont and also Richard Wool at the University of Delaware for providing unpublished data. I also thank my colleagues Bill Hitz and Rich Broglie at DuPont for many helpful discussions.

References

Badami, R.C. & Patil, K.B. (1981). Structure and occurrence of unusual fatty acids in minor seed oils. *Progress in Lipid Research*, **19**, 119–53.

Christou, P., McCabe, D.E., Martinell, B.J. & Swain, W.F. (1990). Soybean genetic engineering – commercial production of transgenic plants. *Trends in Biotechnology*, **8**, 145–51.

Eisenhard, W.C. (1968). Fatty acid-derived plasticizers. In *Fatty Acids and Their Industrial Applications*, ed. E.S. Pattison, pp. 295–303. NY: Marcel Dekker, Inc.

Finer, J.J. & McMullen, M.D. (1991). Transformation of soybean *via* particle bombardment of embryonic suspension culture tissue. *In Vitro Cell and Developmental Biology,* **27**, 175–82.

Flider, J. (1995). Use of rapeseed oil in lubricants. *Inform*, **6**, 1031–5.

Heppard, E.P., Kinney, A.J., Stecca, K.L. & Miao, G-H. (1996). Developmental and growth temperature regulation of two different microsomal omega-6 desaturase genes in soybean. *Plant Physiology*, **110**, 311–19.

Katan, M.B., Mensink, R.P. & Zock, P.L (1995). *Trans* fatty acids and their effects on lipoproteins in humans. *Annual Review of Nutrition,* **15**, 473–96.

Kinney, A.J. (1994). Genetic modification of the storage lipids of plants. *Current Opinion in Biotechnology*, **5**, 144–51.

Kinney, A.J. (1995). Improving soybean seed quality. In *Induced Mutations and Molecular Techniques for Crop Improvement*, pp. 101–13. Vienna, Austria: International Atomic Energy Agency.

Kinney, A.J. (1996). Development of genetically engineered soybean oils for food applications. *Journal of Food Lipids*, **3**, 273–92.

Kinney, A.J. (1997). Genetic engineering of oilseeds for desired traits. In *Genetic Engineering*, ed. J. K. Setlow, vol. 19, pp. 149–66. NY: Plenum Press.

Kinney, A.J. & Knowlton, S. (1998). Designer oils for food applications. In *Genetic Engineering for the Food Industry: A Strategy for Food Quality Improvement*, ed. S. Roller & S. Harlander, pp. 193–213. London: Blackie Academic.

Knowlton, S., Ellis, S.K.B. & Kelly, E.F. (1998). Oxidative stability of high oleic acid soybean oil. *Journal of the American Oil Chemists Society* (*in press*).

Pattison, E.S. (1968). Fatty acids today and tomorrow. In *Fatty Acids and Their Industrial Applications*, ed. E.S. Pattison, pp. 295–303. NY: Marcel Dekker, Inc.

Shanklin, J., Cahoon, E.B., Whittle, E., Lindqvist, Y., Huang, W., Schneider, G. & Schmidt, H. (1997). Structure–function studies on desaturase and related hydrocarbon hydroxylases. In *Physiology, Biochemistry and Molecular Biology of Plant Lipids*, ed. J.P. Williams, M.U. Khan & N.W. Lem, pp. 6–10. Dordrecht: Kluwer Academic Publishers.

van de Loo, F.N., Broun, P., Turner, S. & Somerville, C.R. (1995). An oleate 12-hydroxylase from castor (*Ricinus communis* L.) is a fatty acid desaturase homolog. *Proceedings of the National Academy of Sciences, USA*, **92**, 6743–7.

Yadav, N.S., Wierzbicki, A., Aegerter, M., Caster, C.S., Perez-Grau, L., Kinney, A.J., Hitz, W.D., Booth, J.R., Schweiger, B., Stecca, K.L., Allen, S.M., Blackwell, M., Reiter, R.S., Carlson, T.J., Russell, S.H., Feldmann, K.A., Pierce, J. & Browse, J. (1993). Cloning of higher plant omega-3 fatty acid desaturases. *Plant Physiology*, **103**, 467–76.

D.S. KNUTZON AND V.C. KNAUF

12 Manipulating seed oils for polyunsaturated fatty acid content

The physical properties and functionalities of a vegetable oil are determined by the specific composition of its component fatty acids. Factors such as chain length and degree of unsaturation are important, as is the specific arrangement of the fatty acids on the glycerol backbone. The fatty acid compositions of some vegetable oils in the human diet are shown in Fig. 1. The oils consist almost exclusively of 16- and 18-carbon fatty acids, but the relative degree of unsaturation of the 18-

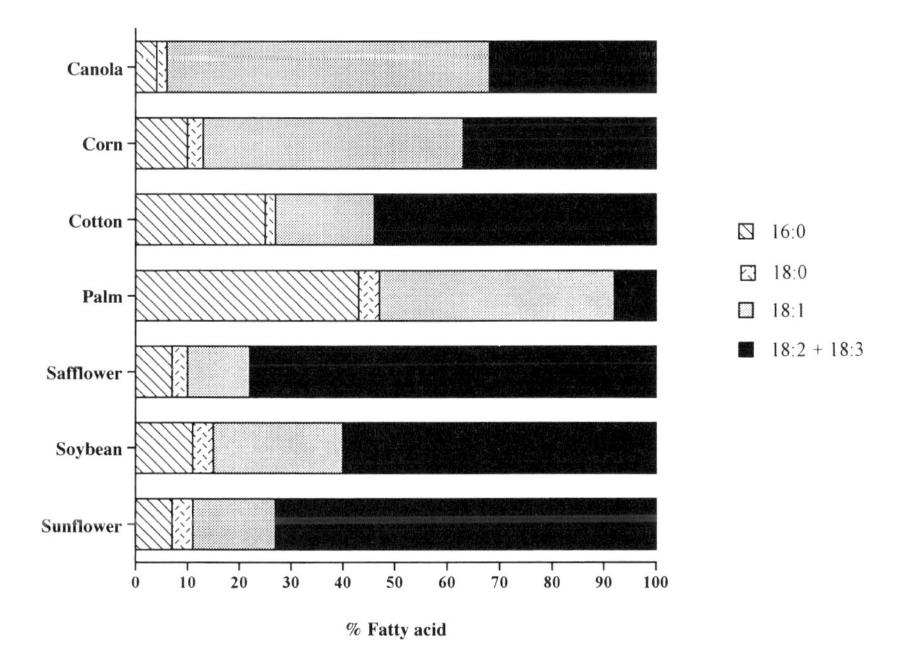

Fig. 1. Fatty acid composition of vegetable oils. Fatty acid abbreviations are: 16 : 0, palmitic acid; 18 : 0, stearic acid; 18 : 1, oleic acid; 18 : 2, linoleic acid; 18 : 3, α-linolenic acid.

carbon fatty acids varies quite a bit. Among these, canola is notable in containing the highest levels of the 18C mono-unsaturated oleic acid.

Besides influences on physical characteristics of the oil such as melting point and oxidative stability, the degree of unsaturation has a large effect on the nutritional attributes of the vegetable oil. The importance of plant oils such as linoleic acid in the diet has long been established. Linoleic acid was identified as an essential fatty acid in 1935 by its ability to reverse the effects of essential fatty acid deficiency : growth retardation, scaliness of the skin, kidney lesions, impaired fertility and increased water consumption and loss by evaporation (Burr & Burr, 1935). Since this early work, both linoleic and α-linolenic acids have come to be known as essential in the human diet since neither can be synthesised *in vivo* and must be derived from the diet. There is currently great interest in such polyunsaturated fatty acids including γ-linolenic acid (GLA), arachidonic acid (ARA), eicosapentaenoic acid (EPA), and docosahexaenoic acid (DHA) for a variety of applications in human health and nutrition. Current sources of these compounds are very limited and genetically engineered plant oils provide an attractive potential source of these valuable fatty acids.

Overview of fatty acid biosynthesis and progress in genetic manipulations

In order to put the prospects of genetic engineering for these highly polyunsaturated fatty acids in perspective, however, it is first necessary to review the foundations that have been laid in the genetic engineering of plant oils. Fig. 2 illustrates the major pathways in plant fatty acid biosynthesis as well as many of the main genetic engineering targets in oil seeds. Representative cDNAs have been cloned from almost every enzymatic step. Most of these classes have been manipulated in transgenic plants to confirm and extend our understanding of the biosynthetic process and control points. The first manipulations involved altering one enzymatic step at a time, but the field is rapidly moving towards gene stacking, or manipulations involving more than one target. This section will briefly go through the fatty acid pathway in oil seeds and give examples of successful genetic manipulations of cDNAs encoding the salient enzymes. The emphasis will be on cloned cDNAs and enzyme activities rather than the pathway *per se*. For a more detailed discussion of fatty acid biosynthesis pathways, the reader is advised to consult other chapters in this volume or one of the recent reviews on the subject (Harwood, 1996; Ohlrogge & Browse, 1995).

De novo fatty acid biosynthesis takes place in the plastid and involves

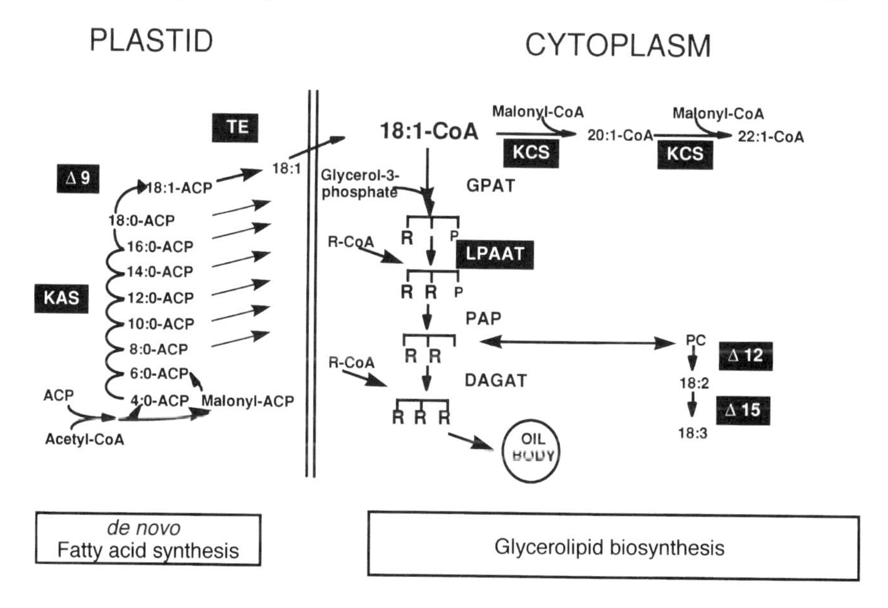

Fig. 2. Outline of seed oil biosynthetic pathway. Shaded boxes indicate enzymatic steps that have been targets for genetic manipulation. Abbreviations are: ACP, acyl carrier protein; DAGAT, diacylglycerol acyltransferase; GPAT, glycerol-3-P acyltransferase; KAS, ketoacyl-ACP synthase; KCS, ketoacyl-CoA synthase; LPAAT, lysophosphatidic acid acyltransferase; PAP, phosphatidic acid phosphatase; PC, phosphatidyl choline; TE, acyl-ACP thioesterase; Δ9,Δ9 desaturase; Δ12,Δ12 desaturase; Δ15,Δ15 desaturase.

the addition of 2-carbon units to a growing acyl chain bound to acyl carrier protein (ACP). A series of ketoacyl-ACP synthases (KAS) with specialized substrate specificities catalyse the elongation of the fatty acyl chain. In *Brassica*, as in most seeds, the major end product of *de novo* fatty acid synthesis is stearate (18 : 0). Stearate is then desaturated in the plastid by the soluble stearoyl-ACP desaturase to produce oleic acid (18 : 1), which is cleaved from the ACP carrier by an oleoyl-ACP thioesterase (TE). Oleate is then transported to the cytoplasm and esterified to CoA to become available for further elongation or incorporation into glycerolipids and storage triacylglycerols (TAG). Seed storage TAG are synthesised via the Kennedy pathway in the ER from acyl-CoA and glycerol-3-P in a series of reactions catalysed by glycerol-3-P acyltransferase (GPAT), lysophosphatidic acid acyl transferase (LPAAT), phosphatidic acid phosphatase (PAP) and diacylglycerol acyltransferase (DAGAT). Other modifications that the oleate may undergo in the cytoplasm include incorporation into the *sn*-2 position of phosphatidyl choline (PC) and further desaturation to linoleic acid (18 : 2) and α-linolenic acid (18 : 3) by the membrane-associated Δ-12 and Δ-15 desaturases. The 18 : 1-CoA molecules can also serve as a substrate for a microsomal fatty acid elongation (FAE) system. Although distinct from the soluble FAS system in plastids, the FAE system is believed to consist of a four-step mechanism analogous to that of FAS. The first step, involving condensation of malonyl-CoA with a long chain acyl-CoA, is catalysed by a β-ketoacyl-CoA synthase (KCS).

Representative cDNAs of the Δ9 desaturase, TE, and KAS enzyme classes were the first to be cloned; the soluble nature of these enzymes made them more amenable to traditional protein biochemistry approaches. Since that time, however, advances in purification of membrane proteins has led to the cloning of genes encoding representatives of the KCS and LPAAT class and advances in genetic approaches have led to the cloning of membrane-bound desaturases.

One of the first oils modifications described was the production of laurate (C12 : 0) in *Arabidopsis* seeds (Voelker *et al.*, 1992). This was accomplished by the purification and cloning of a lauroyl-ACP-specific thioesterase from California bay tree. The idea was to use this enzyme to derail fatty acid biosynthesis at shorter chain lengths than were normally produced, with the hope that the laurate would then be transported out of the plastid, re-esterified to CoA and become available for deposition into triacylglycerols via the acyltransferases in the endoplasmic reticulum. This strategy, in fact, worked and the levels of laurate in *Brassica* seed oil have been increased from essentially none to 56 mol% (Voelker *et al.*, 1996).

Previous studies on the specificities of the various Kennedy pathway enzymes indicated that in *Brassica* the second acyltransferase, LPAAT, has little activity on lauroyl-containing substrates (Oo & Huang, 1989). This was borne out in the transgenic seeds containing bay TE; the laurate accumulates almost exclusively at the *sn*-1 and *sn*-3 positions of the TAG. This commercially produced high laurate canola oil or Laurical™, contains *sn*-1,3 dilauroyltriacylglycerides as its predominant species and this novel TAG structure may impart unique properties to the oil. On the other hand, the inability of *Brassica* LPAAT to incorporate laurate into the *sn*-2 position of the TAG imposes a theoretical limit of 67 mol% for accumulation of laurate in the oil. This is an example of where gene stacking, or the simultaneous expression of two transgenes has been successfully demonstrated. Coconut endosperm contains approximately 50 mol% laurate, which is distributed among all three positions of the TAG and was thus a candidate for containing an LPAAT capable of incorporating laurate into *sn*-2 position. The LPAAT was purified from coconut microsomal membranes and a corresponding cDNA was cloned (Knutzon *et al.*, 1995). Seed from crosses of transgenic canola expressing the bay TE and coconut LPAAT contain significant amounts of laurate in the *sn*-2 position, although the overall levels of laurate are not greatly increased (D.S. Knutzon, H.M. Davies & T.A. Voelker, unpublished observations). Similar results were obtained with the cloning and expression of an erucic acid-specific LPAAT from meadowfoam (Lassner *et al.*, 1995). In this case, the proportion of erucic acid at the *sn*-2 position of a high-erucic acid rapeseed line, Reston, was increased from <0.5% to 15 mol%, but the overall level of erucic acid was not increased. Nevertheless, the demonstration of the ability to alter the fatty acid composition at the *sn*-2 position is a necessary first step to producing trierucin-containing rapeseed oil.

The production of laurate in seed oil confirmed the role of specific acyl-ACP thioesterases in controlling lipid composition. Since the cloning of the bay TE, a number of other TEs have been cloned from plant species that accumulate various chain length fatty acids. These have been split into two classes, based on DNA sequence homologies (Jones, Davies & Voelker, 1995). The FatA class is involved primarily in hydrolysis of 18 : 1-ACP, while the function of the FatB TEs is more variable. Since the prototypical bay TE, FatB TEs involved in hydrolysis of medium chain acyl-ACPs have been cloned from several other species. One such plant is *Cuphea hookeriana,* which accumulates up to 75% caprylate (8 : 0) and caprate (10 : 0) in its seed oil. A cDNA encoding an acyl-ACP TE was identified which, when expressed in *E. coli*, has high activity on 8 : 0- and 10 : 0-ACP substrates. Expression of this cDNA, *Ch FatB2*, in transgenic canola led to the accumulation of 8 : 0 and 10 : 0 in the seed oil (Dehesh *et al.*, 1996). So the idea seemed to bear out, that if you identified a plant accumulating a particular chain length fatty acid, you should be able to clone a TE cDNA capable of driving accumulation of that fatty acid in transgenic plants. In many cases that has turned out to be so, but not always. *In vitro* expression of *Ch FatB2* indicates higher activity on 10 : 0-ACP than 8 : 0-ACP substrates, and the transgenic *Brassica* seeds expressing *Ch FatB2* accumulate more 10 : 0 than 8 : 0. However, the native *C. hookeriana* plants from which the TE was derived accumulate more 8 : 0 than 10 : 0.

Further examples of the discrepancy between substrate specificity of

TE and the oil composition of the tissue from which it was cloned come from nutmeg and elm (Voelker *et al.*, 1997). Nutmeg seeds accumulate 80 mol% myristate (14 : 0); however, the TE cDNA isolated from this species encodes a protein with almost equal activity towards 14 : 0-ACP substrates and longer chain substrates. Expression of this cDNA in transgenic *Brassica* seeds did lead to the accumulation of up to 20% myristate, but even more palmitate (16 : 0) was formed in the seeds. Elm seeds contain 65 mol% 10 : 0 and 10 mol% 8 : 0 and thus seemed a good target tissue from which to obtain a 10 : 0-specific TE. The cDNA obtained, however, encodes a protein with bimodal activity on 10 : 0-ACP and 16 : 0-ACP substrates. Expression of this cDNA in transgenic *Brassica* seeds led to the accumulation predominantly of palmitate, with very little accumulation of 8 : 0 or 10 : 0. These studies all point to the possibility that the interplay of thioesterases and synthases, and perhaps other factors as well, may be the ultimate controlling factor for seed oil composition. This raises yet another potential area for gene stacking – using a KAS to control the flux of fatty acid biosynthesis in combination with TEs of specific substrate specificity. Preliminary indications are that this approach will be successful. The combination of a *C. hookeriana* KAS and a 8 : 0-specific TE from *C. palustris* in transgenic *Brassica* resulted in higher accumulation of 8 : 0 and 10 : 0 in the seed oil than either cDNA did on its own (K. Dehesh, personal communication).

Representatives of the Fat A class of TEs, encoding oleoyl-ACP TE have also been cloned and manipulated. cDNAs encoding oleoyl-ACP thioesterases have been obtained from a wide variety of organisms including safflower (Knutzon *et al.*, 1992*b*), rapeseed (Loader *et al.*, 1993) and recently mangosteen (Kridl & Shewmaker, 1996). With the exception of the mangosteen TE, these clones have encoded proteins with a high degree of specificity for 18 : 1-ACP over other substrates. Modulation of these enzyme activities through genetic engineering has again led to alterations of seed oil composition (Kridl & Shewmaker, 1996; Kinney, 1994).

Another area of manipulation involves the β-ketoacyl-CoA synthases (KCS) present in the cytoplasm that are responsible for the synthesis of very long chain fatty acids. Representative cDNAs encoding this enzyme class have been cloned from several organisms. Using the genetic approach of heterologous transposon tagging, the *Arabidopsis FATTY ACID ELONGATION1* gene was cloned (James *et al.*, 1995). A cDNA clone encoding a KCS of the microsomal fatty acyl-CoA elongation pathway was also isolated from jojoba (Lassner, Lardizabal & Metz, 1996). Expression of this cDNA in a low erucic acid rapeseed line restores KCS activity and drives the accumulation of 20 : 1, 22 : 1, and some 24 : 1 in the seed oil. Based on homology to the jojoba cDNA, a KCS from *Lunaria annua* (commonly called honesty or money plant) was isolated which, when expressed in a high erucic acid rapeseed line, drives the production of nervonic acid (24 : 1) at levels up to 20% of the seed oil (Lassner, 1997). This oil is being investigated for its functionality in making slip agents for the manufacture of polyethylene sheeting.

The last major class of enzymes involved in fatty acid biosynthesis that has been targeted for genetic engineering is the desaturases. The first of these to be manipulated was the soluble, plastid-localised stearoyl-ACP desaturase (Δ9 desaturase). Representative cDNAs have been cloned in a number of labs from a number of organisms including *Brassica*. Modulation of stearoyl-ACP desaturase expression via antisense expression led to a high-stearate phenotype in which 30% of the *Brassica* seed fatty acids were stearate (Knutzon *et al.*, 1992*a*). More recently, genetic approaches have been used to isolate cDNAs encoding the membrane-bound desaturases of the endoplasmic reticulum. The cloning and genetic manipulation of these desaturases is dealt with elsewhere in this volume (see Browse and Kinney contributions) and will not be dealt with in detail here. Manipulation of both Δ12 and Δ15 desaturases have been shown to influence the polyunsaturated fatty acid content in transgenic soybean and *Brassica* plants (Fader, Kinney & Hitz, 1995). An ω-3 desaturase has been cloned from *Caenorhabditis elegans* and shown to function in transgenic *Arabidopsis* plants to introduce an ω-3 double bond into both 18- and 20-carbon fatty acids (Spychalla, Kinney & Browse, 1997).

PUFA terminology and pathways

Any discussion of polyunsaturated fatty acids (PUFAs) needs to begin with a definition of the terminology used to describe the molecules. The most common shorthand notation of a PUFA is similar to that of other fatty acids, and consists of the number of carbons in the fatty acid separated by a colon from the number of double bonds. For example, 18 : 3 indicates an 18-carbon fatty acid with three double bonds. In addition, the position of the double bonds in the fatty acid can be described, since this has significance for both the structure and biological activity of the fatty acid. There are two ways to describe the position of double bonds in unsaturated fatty acids; orientation can be gained from either the methyl- or the carboxyl-terminus of the molecule. Counting carbons from the carboxyl end gives the delta (Δ) designation, and counting from the methyl terminus gives the ω or n-designation. Thus linoleic acid may be described either as Δ9,12–18 : 2 or as 18 : 2n-6 (18 : 2ω-6). Similarly, α-linolenic acid (ALA) can be described as the Δ9,12,15–18 : 3 or 18 : 3n-3. This is differentiated from γ-linolenic acid (GLA), the 18 : 3n-6 species(Δ6,9,12–18 : 3).

There are two main families of PUFAs, the n-6 and n-3 series. Fig. 3 illustrates some of the key PUFAs and possible routes of synthesis. The major endpoint of the n-6 pathway is arachidonic acid (ARA). ARA is produced from linoleic acid by a series of sequential desaturation and elongation steps. The first step is a desaturation at the Δ6 position of linoleic acid to yield GLA. This is followed by a two-carbon chain elongation to produce dihomo-γ-linolenic acid (DGLA, 20 : 3n-6). DGLA is then converted to AA by the action of a Δ5 desaturase. In animals, the Δ6 desaturase is generally believed to be the rate-limiting step in the production of ARA.

In the n-3 pathway, a similar series of reactions is involved in the formation of eicosapentaenoic acid (EPA; 20 : 5n-3) from ALA (Fig. 3). It is believed that the same enzymes catalyse reactions in both the n-6 and n-3 pathways, although this has not been shown definitively. Evidence for this is that the n-6 and n-3 PUFAs compete with one another in the metabolic pathway. The major endpoint of the n-3 pathway in humans is docosahexaenoic acid (DHA, 22 : 6n-3). DHA could be derived from EPA by an additional two-carbon elongation step followed by a Δ4 desaturation; however, there is growing evidence that at least in some systems the actual sequence is two elongations and a Δ6 desaturation to create a 24 : 6 molecule, which is then 'retroconverted' via β-oxidation to create DHA (Δ4,7,10,13,16,19 22 : 6) (Voss *et al.*, 1991; Buzzi, Henderson & Sargent, 1997).

Significance of PUFAs

This section will discuss the significance of PUFAs in the human diet and highlight some of the reasons that PUFAs have been targets for genetic engineering. Four major roles of essential fatty acids, specifically of the n-6 series, have been described (Horrobin, 1992):

(i) Impermeability barrier of the skin. Linoleic acid has been implicated in involvement of maintenance of the epidermal water barrier (Hansen & Jensen, 1985), which is consistent with one of the hallmark symptoms of EFA deficiency: thirst and consumption of large amounts of water yet production of small amounts of urine (Burr & Burr, 1935).

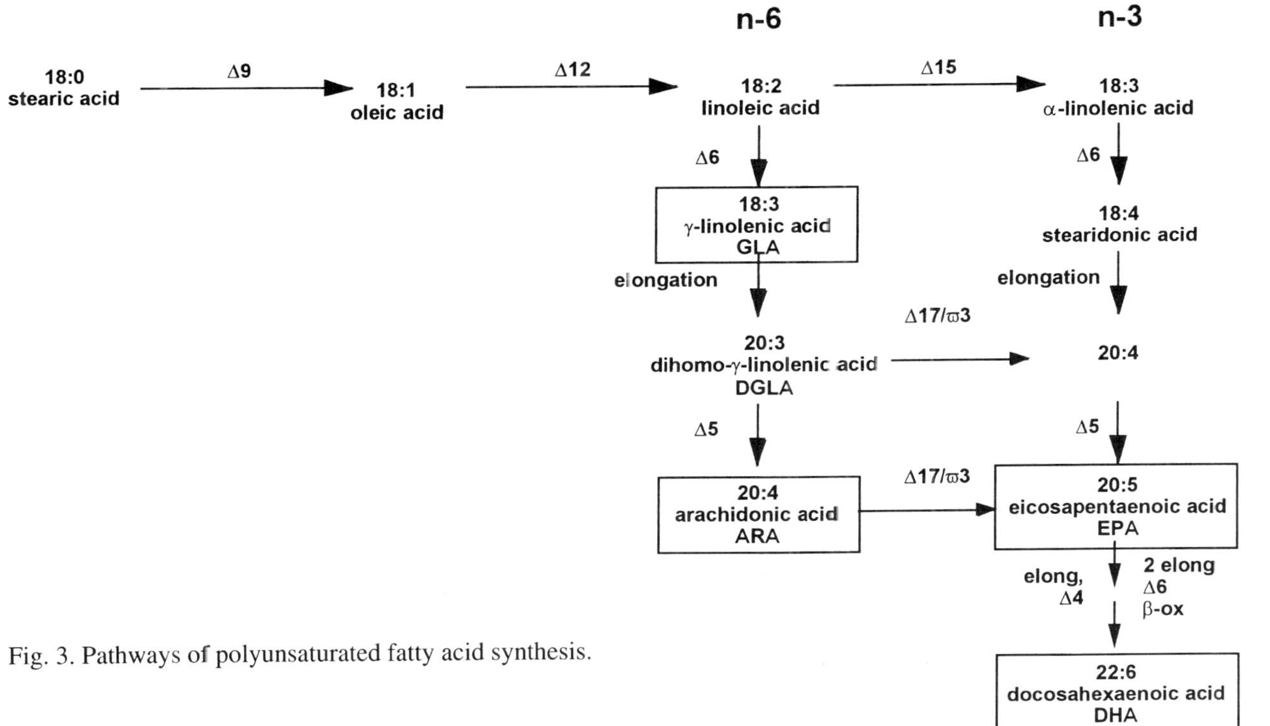

Fig. 3. Pathways of polyunsaturated fatty acid synthesis.

(ii) Modulation of membrane structure. The fluidity and permeability of a membrane is strongly influenced by its lipid components. Membrane fluidity is particularly important in the vascular system. It has been shown that the degree of unsaturation of the membrane environment can affect the behaviour of the proteins embedded in the membrane. ARA and DHA play a role as structural lipids, particularly in nervous system and retinal tissue.

(iii) Involvement in cholesterol transport and metabolism. Cholesterol esters formed with EFA are more soluble and easily cleared than those with saturated or monounsaturated fatty acids (Horrobin, 1992).

(iv) Formation of eicosanoids. Probably the greatest significance of PUFAs in the human diet is their role as precursors of eicosanoids, a group of highly active substances derived from 20C fatty acids with diverse biological actions. Two main metabolic fates of 20C fatty acids are shown in Fig. 4. The cyclooxygenase pathway gives rise to prostaglandins, prostacyclins, and thromboxanes. These compounds play critical roles in the processes of inflammation, vasodilation and vasoconstriction, and platelet aggregation. The number of double bonds in the PUFA precursor determines the specific prostanoid series produced and often the specific biological effect. For example, thromboxane A_2 (TXA_2), formed from ARA is a very potent platelet aggregator, while TXA_1 and TXA_3

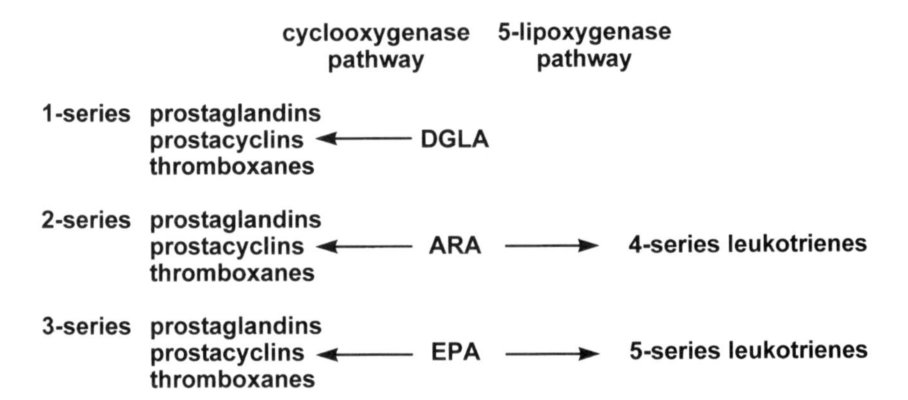

Fig. 4. Main metabolic fates of 20-carbon fatty acids in humans.

have little or no such effect (Sardesai, 1992). The 5-lipoxygenase pathway gives rise to the family of compounds known as leukotrienes, which are also involved in inflammatory processes.

Examination of these major roles for PUFAs leads to an appreciation of the wide range of potential applications of PUFAs to human health and nutrition. As mentioned earlier, the Δ6 desaturation step is believed to be rate limiting for the production of n-6 PUFAs and thus there is much interest in the use of GLA supplementation to overcome this block. Decreases in long-chain n-6 PUFAs have been related to many disease states. Some of the factors associated with decreased Δ6 desaturase activity include: ageing, diabetes, high alcohol intake, nutritional deficiencies of pyridoxine, zinc, magnesium, biotin and calcium. (Horrobin, 1992). Dietary supplementation with GLA bypasses the Δ6 deficiency and can alleviate many clinical symptoms. A shortlist of some of the proposed uses of GLA supplementation includes: cancer, cardiovascular disorders, diabetes mellitus, rheumatoid arthritis, premenstrual breast pain, and atopic eczema.

Currently, there are three major plant sources of GLA: evening primrose (*Oenothera biennis*), borage (*Borago officinalis*) and blackcurrant (*Ribes nigrum*). Their oil compositions are compared in Table 1. In addition to GLA content, the amounts of other fatty acids in the oil may have a bearing on their utility. Evening primrose oil contains the least GLA, but does not contain any n-3 PUFAs such as ALA or SDA as do borage or blackcurrant oils. Borage oil, while a relatively good source of GLA free of n-3 PUFAs, contains significant amounts of longer chain fatty acids. None of these species is well suited for commercial cultivation either. Seed yield, oil content and levels of GLA in evening primrose vary with environmental conditions; seed shattering and indeterminate growth habit limit borage's potential field production (Gunstone, 1992).

Microbial sources of PUFAs include fungi (*Mucor*, *Mortierella*),

Table 1. *Fatty acid composition of major plant sources of GLA*

	Sat	18 : 1	18 : 2	γ-18 : 3	α-18 : 3	18 : 4	22 : 1
Evening primrose	7.3	9.0	74.6	9.1			
Borage	13.5	15.4	37.7	20.0	4.3		3.5
Blackcurrant	7.4	11.6	43.9	18.7	14.5	2.9	

Data taken from Gunstone (1992).

marine bacteria (*Shewanella*, *Vibrio marinus*), algae (*Porphyridium*, *Phaeodactylum*), and cyanobacteria (*Spirulina*, *Synechocystis*) (Ward, 1995). Some of these organisms have been adapted for large-scale growth, but the economies of production are still not attractive. Transgenic oil seeds would provide an attractive vehicle for PUFA production. The production of important PUFAs such as ARA, EPA, and DHA in oil seeds is subject to two main hurdles: the lack of cloned genes with which to work, and the number of enzymatic steps that must be engineered concurrently. In addition, the biosynthetic pathways are not well characterised and it is not even known how many different desaturases or elongase activities are involved.

Engineering canola for GLA production

There are, however, cloned Δ6 desaturases and one can begin to attempt the genetic engineering for GLA accumulation in transgenic oilseeds. The first Δ6 desaturase was cloned from the cyanobacterium, *Synechocystis,* by expression of a cosmid library in *Anabaena*, an organism which contains significant amounts of linoleic acid. Two overlapping *Synechocystis* cosmids were identified that, when introduced into *Anabaena*, were capable of driving production of GLA from linoleic acid (Reddy *et al*., 1993). Subsequently, two groups independently obtained cDNA clones encoding the Δ6 desaturase from borage (Thomas *et al*., 1995; Sayanova *et al*., 1997). Based on amino acid sequence similarities, this enzyme appears to belong to the family of membrane-bound desaturases including the Δ12 and Δ15 desaturases. These enzymes are believed to act on the acyl group at the *sn*-2 position of phosphatidylcholine and utilise cytochrome *b*5 and cytochrome *b*5 reductase for electron transport. Expression of the borage Δ6 desaturase in tobacco under control of the 35S promoter led to the production of GLA in transgenic leaves; approximately 20% of the C18 fatty acids were GLA (Thomas *et al*., 1995). A similar result was seen by Sayanova *et al.* (1997); transgenic tobacco leaves expressing the borage Δ6 desaturase accumulated 13% GLA and 9.6% SDA.

To test the ability of the borage Δ6 desaturase cDNA to drive GLA production in transgenic canola seeds, the coding region of the cDNA was inserted into a seed-specific napin cassette (Kridl *et al*., 1991). This construct was introduced into *Brassica napus* via *Agrobacterium*-mediated transformation and 30 transgenic plants were regenerated. Mature T2 seeds were analysed for fatty acid composition. The seeds from most transgenic plants contained only small amounts of GLA; the levels did not exceed 1% of the total fatty acids in the seed (D.S. Knutzon, unpublished observations). Northern blot analysis of developing seeds from selected events revealed that some borage Δ6-desaturase mRNA was present in the seeds, but the steady-state level was much lower than that typically seen for other cDNAs expressed from the napin cassette. Further experiments will aim to increase the level of Δ6-desaturase message in *Brassica* seeds. Alternatively, one could use analogous constructs for expression in seeds of other crops such as flax or soybean.

Strategies for engineering PUFA production in oil seeds

Production of GLA and/or SDA in an oil seed crop such as *Brassica napus* represents the first step to production of a variety of PUFAs. The production of other n-6 PUFAs such as DGLA or ARA, or ultimately n-3 PUFAs such as EPA or DHA seem to present unique technical challenges. A fundamental challenge will be to identify cDNA clones encoding the necessary desaturases and elongases. To reach ARA, an elongase capable of adding 2 carbons to GLA and a Δ5 desaturase are required. It is possible that the same complement of Δ6 desaturase, elongase, and Δ5 desaturase will be sufficient to make EPA from ALA, but that remains to be shown. To go from EPA to DHA perhaps additional elongases are needed, in addition to a Δ4 desaturase or another Δ6 desaturase. Beyond the technical challenge of identifying these clones comes the issue of expressing the multiple enzyme activities simultaneously in a transgenic tissue. Although, as discussed above, there are already examples of successful metabolic engineering in canola of more than one enzymatic step, PUFA engineering may involve the introduction of 3–6 novel enzyme activities. Issues of promoter regulation and gene silencing may become important. The efficiency with which the fatty acids are pulled through the PUFA pathway may also become important in light of the complex relationships of n-3 and n-6 PUFAs and the sometimes antagonistic effects of their respective metabolites.

An alternative approach to PUFA engineering that may overcome some of these obstacles exists. Certain species of marine bacteria produce PUFAs such as EPA and DHA. A *Shewanella putrefaciens* cosmid clone has been identified, which is capable of driving EPA production in *E. coli* (Yazawa, 1996). Further work led to the identification of five open reading frames from the cosmid that were required for EPA production in *E. coli*. These open reading frames encode components of a polyketide synthase system that produces the EPA (J. Metz, personal communication). Expression of these open reading frames in plants may result in a functional polyketide synthase-type pathway to PUFA synthesis with fewer unwanted side products.

Assuming the successful engineering of PUFAs in seed oils, some interesting additional opportunities exist, beyond merely optimising the amount of certain fatty acids in the seed oil. Some of these relate to triglyceride structure. It has been reported that, although it contains less GLA than borage oil, evening primrose oil is more effective in delivering the GLA in the body and thus is a more biologically effective form of the oil (Horrobin, 1990). This may be due to the composition of other fatty acids, such as linoleic, or to the specific position of GLA in the triglyceride molecule. As mentioned earlier, having shown that cloned LPAATs are able to alter the fatty acid composition at a specific position of the TAG opens the door to creating designer molecules. When the microsomal GPAT and DAGAT cDNAs are found, the opportunities to alter composition at *sn*-1 and *sn*-3 will open up as well.

The demonstration that medium chain specific-acyl-ACP thioesterases are capable of producing C8 and C10 fatty acids in *Brassica* also offers the potential of some novel products. Medium chain fatty acids (MCFA) are an important component of liquid diet formulae for patients with a variety of medical conditions including AIDS, cystic fibrosis, and trauma (Merolli, Lindemann & Del Vecchio, 1997). In addition, patients with these malabsorption conditions may be deficient in essential fatty acids. Current sources of MCFA for these diets are medium chain triacylglycerols containing MCFA at every position of the glycerol backbone. MCTs provide a readily absorbed source of calories, but must be combined with long-chain triacylglycerols to avoid essential fatty acid deficiency. Chemically interesterified triacylglycerols containing MCFAs and essential fatty acids with a random distribution on the glycerol backbone are available as well, but due to positional and substrate specificities of lipases, these may not be the optimal delivery molecule (Christensen *et al.*, 1995). Triacylglycerols containing an essential fatty acid such as 18 : 2 n-6 in the *sn*-2 position with MCFA in the outer positions provide for the highest absorption of 18 : 2 (Jensen, Christensen & Hoy, 1994). This observation has been extended for the absorption of EPA and DHA in the *sn*-2 position as well (Christensen *et al.*, 1995). Transgenic canola seeds would appear to be a good source for these structured triacylglycerols. The specificities of the endogenous acyltransferases of the Kennedy pathway would lead to saturated fatty acids at the outer positions and unsaturated fatty acids in the *sn*-2 position. In addition, the formation of 18 : 2 and perhaps the further desaturated PUFAs takes place at the *sn*-2 position of phosphatidylcholine which may be exchanged directly with the triacylglycerol pool.

In summary, there are currently no good economical, renewable sources of PUFAs. Transgenic plants would provide a good vehicle: production of PUFAs in oil would be economical and easily scaled up to commercial level. Transgenic plants are also attractive from the standpoint that there is a proven track record of modification of the fatty acid biosynthetic pathway and the general plasticity of oil composition obtainable.

References

Burr, G.O. & Burr, M.M. (1935). On the nature and role of the fatty acids essential in nutrition. *Journal of Biological Chemistry*, **86**, 587–621.

Buzzi, M., Henderson, R.J. & Sargent, J.R. (1997). Biosynthesis of docosahexaenoic acid in trout hepatocytes proceeds via 24-carbon intermediates. *Comparative Biochemistry and Physiology*, **116B**, 263–7.

Christensen, M.S., Hoy, C-E., Becker, C.C. & Redgrave, T.G. (1995). Intestinal absorption and lymphatic transport of eicosapentaenoic (EPA), docosahexaenoic (DHA), and decanoic acids: dependence on intramolecular triacylglycerol structure. *American Journal of Clinical Nutrition*, **61**, 56–61.

Dehesh, K., Jones, A., Knutzon, D.S. & Voelker, T.A. (1996). Production of high levels of 8 : 0 and 10 : 0 fatty acids in transgenic canola by over-expression of *Ch FatB2*, a thioesterase cDNA from *Cuphea hookeriana*. *Plant Journal*, **9**, 167–72.

Fader, G.M., Kinney, A.J. & Hitz, W.D. (1995). Using biotechnology to reduce unwanted traits. *Inform*, **6**, 167–9.

Gunstone, F.D. (1992). Gamma linolenic acid-occurrence and physical and chemical properties. *Progress in Lipid Research*, **31**, 145–61.

Hansen, H.S. & Jensen, B. (1985). Essential function of linoleic acid esterified in acylglucosylceramide and acylceramide in maintaining the epidermal water permeability barrier. Evidence from feeding studies with oleate, linoleate, arachidonate, columbinate, and α-linolenate. *Biochimica et Biophysica Acta*, **834**, 357–63.

Harwood, J.L. (1996). Recent advances in the biosynthesis of plant fatty acids. *Biochimica et Biophysica Acta*, **1301**, 7–56.

Horrobin, D.F. (1990). Gamma linolenic acid: an intermediae in essential fatty acid metabolism with potential as an ethical pharmaceutical and as a food. *Reviews in Contemporary Pharmacotherapy*, **1**, 1–45.

Horrobin, D.F. (1992). Nutritional and medical importance of gamma-linolenic acid. *Progress in Lipid Resesarch*, **31**, 163–94.

James, D.W., Jr, Lime, E., Keller, J., Plooy, I., Ralston, E. & Dooner, H.K. (1995). Directed tagging of the *Arabidopsis FATTY ACID ELONGATION1 (FAE1)* gene with the maize transposon *Activator*. *The Plant Cell*, **7**, 309–19.

Jensen, M.M., Christensen, M.S. & Hoy, C-E. (1994). Intestinal absorption of octanoic, decanoic, and linoleic acids: effect of triglyceride structure. *Annals of Nutrition and Metabolism*, **38**, 104–16.

Jones, A., Davies, H.M. & Voelker, T.A. (1995). Palmitoyl-acyl carrier protein (ACP) thioesterase and the evolutionary origin of plant acyl-ACP thioesterases. *The Plant Cell*, **7**, 359–71.

Kinney, A.J. (1994). Genetic modification of the storage lipids of plants. *Current Opinion in Biotechnology*, **5**, 144–51.

Knutzon, D.S., Thompson, G.A., Radke, S.E., Johnson, W.B., Knauf, V.C. & Kridl, J.C. (1992*a*). Modification of *Brassica* seed oil by antisense expression of a stearoyl-acyl carrier protein desaturase gene. *Proceedings of the National Academy of Sciences, USA*, **89**, 2624–8.

Knutzon, D.S., Bleibaum, J.L., Nelsen, J., Kridl, J.C. & Thompson, G.A. (1992*b*). Isolation and characterization of two safflower oleoyl-acyl carrier protein thioesterase cDNA clones. *Plant Physiology*, **100**, 1751–8.

Knutzon, D.S., Lardizabal, K.L., Nelsen, J.S., Bleibaum, J.L., Davies, H.M. & Metz, J.G. (1995). Cloning of a coconut endosperm cDNA encoding a 1-acyl-*sn*-glycerol-3-phosphate acyltransferase that accepts medium-chain-length substrates. *Plant Physiology*, **109**, 999–1006.

Kridl, J.C. & Shewmaker, C.K. (1996). Food for thought: improvement of food quality and composition through genetic engineering. *Annals of the New York Academy of Sciences*, **792**, 1–12.

Kridl, J.C., McCarter, D.W., Rose, R.E., Scherer, D.E., Knutzon, D.S., Radke, S.E. & Knauf, V.C. (1991). Isolation and characterization of an expressed napin gene from *Brassica rapa*. *Seed Science Research*, **1**, 209–18.

Lassner, M. (1997). Transgenic oilseed crops: a transition from basic research to product development. *Lipid Technology*, **9**, 5–9.

Lassner, M.W., Levering, C.K., Davies, H.M. & Knutzon, D.S. (1995). Lysophosphatidic acid acyltransferase from meadowfoam mediates insertion of erucic acid at the *sn*-2 position of triacylglycerol in transgenic rapeseed oil. *Plant Physiology*, **109**, 1389–94.

Lassner, M.W., Lardizabal, K. & Metz, J.G. (1996). A jojoba β-ketoacyl-CoA synthase cDNA complements the canola fatty acid elongation mutation in transgenic plants. *The Plant Cell*, **8**, 281–92.

Loader, N.M., Woolner, E.M., Hellyer, A., Slabas, A.R. & Safford,

R. (1993). Isolation and characterization of two *Brassica napus* embryo acyl-ACP thioesterase cDNA clones. *Plant Molecular Biology*, **23**, 769–78.

Merolli, A., Lindemann, J. & Del Vecchio, A.J. (1997). Medium-chain lipids: new sources, uses. *INFORM*, **8**, 597–603.

Ohlrogge, J. & Browse, J. (1995). Lipid biosynthesis. *The Plant Cell*, **7**, 957–70.

Oo, K-C. & Huang, A.H. C. (1989). Lysophosphatidate acyltransferase activities in the microsomes from palm endosperm, maize scutellum, and rapeseed cotyledon on maturing seeds. *Plant Physiology*, **91**, 1288–95.

Reddy, A.S., Nuccio, M.L., Gross, L.M. & Thomas, T.L. (1993). Isolation of a Δ6-desaturase gene from the cyanobacterium *Synechocystis* sp. strain Pcc6803 by gain-of-function expression in *Anabaena* sp. strain PCC7120. *Plant Molecular Biology*, **27**, 293–300.

Sardesai, V.M. (1992). Biochemical and nutritional aspects of eicosanoids. *Journal of Nutritional Biochemistry*, **3**, 562–79.

Sayanova, O., Smith, M.A., Lapinskas, P., Stobart, A.K., Dobson, G., Christie, W.W., Shewry, P.R. & Napier, J.A. (1997). Expression of a borage desaturase cDNA containing an N-terminal cytochrome *b*5 domain results in the accumulation of high levels of Δ6-desaturated fatty acids in transgenic tobacco. *Proceedings of the National Academy of Sciences, USA*, **94**, 4211–16.

Spychalla, J.P., Kinney, A.J. & Browse, J. (1997). Identification of an animal ω-3 fatty acid desaturase by heterologous expression in *Arabidopsis*. *Proceedings of the National Academy of Sciences, USA*, **94**, 1142–7.

Thomas, T.L., Nunberg, A., Reddy, A.S., Nuccio, M.L. & Beremand, P. (1995). Cloning and expression of Δ6-desaturase genes in transgenic plants. In *Biochemistry and Molecular Biology of Plant Fatty Acids and Glycerolipids Symposium,* ed. National Plant Lipid Cooperative, Abstract no. O-25. South Lake Tahoe, California.

Voelker, T.A., Worrell, A.C., Anderson, L., Bleibaum, J., Fan, C., Hawkins, D., Radke, S. & Davies, H.M. (1992). Fatty acid biosynthesis redirected to medium chains in transgenic oilseed plants. *Science*, **257**, 72–4.

Voelker, T.A., Hayes, T.R., Cranmer, A.M., Turner, J.C. & Davies, H.M. (1996). Genetic engineering of a quantitative trait: metabolic and genetic parameters influencing the accumulation of laurate in rapeseed. *The Plant Journal*, **9**, 229–41.

Voelker, T.A., Jones, A., Cranmer, A.M., Davies, H.M. & Knutzon, D.S. (1997). Broad-range and binary-range acyl-acyl-carrier-protein thioesterases suggest an alternative mechanism for medium-chain production in seeds. *Plant Physiology*, **114**, 669–77.

Voss, A.C., Reinhart, M., Sankarappa, S. & Sprecher, H. (1991). The metabolism of 7, 10, 13, 16, 19-docosapentaenoic acid to 4, 7, 10,

13, 16, 19-docosahexaenoic acid in rat liver is independent of a 4-desaturase. *Journal of Biological Chemistry*, **266**, 1995–2000.

Ward, O.P. (1995). Microbial production of long-chain PUFAs. *INFORM*, **6**, 683–8.

Yazawa, K. (1996). Production of eicosapentaenoic acid from marine bacteria. *Lipids*, **31**, S297–S300.

J.L. HARWOOD

13 Environmental effects on plant lipid biochemistry

Introduction

There is a continuing interest in the effects that environmental factors may have on life on Earth. As our knowledge, particularly of harmful effects, has increased then so has public attention focused on causes and cures. Recent examples of organophosphorus insecticides, acid rain and the 'greenhouse effect' have served to emphasise general awareness, if not knowledge and objective evaluation. Of course, the environment is continuously changing whether or not Man interferes, although it is the latter scenarios which attract our attention and, quite rightly, often give cause for concern.

Plants respond to most environmental changes (Table 1). Where these alterations are capable of inducing potentially injurious effects, then it is said that the plants are stressed (Levitt, 1980). Obviously, if the duration and/or the amount of the external factor is sufficient, then a certain degree of strain can be developed in the organism. Resistance to stress can be inherent (genetically acquired during evolution) or may be developed following exposure to the stress. Surprisingly, it has sometimes been found that exposure to one stress confers subsequent resistance to a different stress (Liljenberg, 1992).

In the following discussion, I will describe various examples of environmental changes which have been noted to give rise to alterations in plant lipids and/or their metabolism. As pointed out before (Harwood *et al.*, 1992), it is not always clear whether such changes are simply a response to changing growth conditions or whether they do, in fact, represent a beneficial adaptation to the alterations. These alternatives should be borne in mind when considering the following cases. Moreover, even in those cases where the modification in lipids is clearly beneficial (e.g. low temperature-induced increases in unsaturation) the mechanism(s) by which metabolism is controlled is often ill defined or even completely unknown.

Table 1. *Environmental changes and typical effects on lipids*

Environmental parameter		Lipid change
Light	– quality	Alterations in chloroplast lipids
	– quantity	High light can lead to oxidative damage Light absolutely required for *trans*-3-hexadecenoate synthesis Light modulates fatty acid synthesis
Temperature	– heat stress	Membrane lipid phase alterations Metabolic balance changed
	– cold stress	Induction of, e.g. desaturase enzymes Phase separation of some membrane lipid species
Water stress (drought)		Peroxidation of fatty acids Decreases in glycosylglycerides Decreases/changes in phosphoglycerides Accumulation of triacylglycerol
Soil constituents	– calcium	Plants divided into calcicole and calcifuge groups. Changes in their lipid metabolism
	– salt stress	Chloroplast lipids particularly degraded
Atmosphere	– ozone	Oxidation of fatty acids. Degradation of glycosylglycerides. Accumulation of triacylglycerol
	– raised CO_2	Changes in seed storage lipid quality and quantity. Balance of 'eukaryotic' and 'prokaryotic' metabolism altered in leaves
Xenobiotics		Several herbicides effective by inhibiting an aspect of fatty acid synthesis

Light

General effects of light on plants are well covered in numerous review articles and books. The ability of plants to adapt to various radiation intensities is seen most noticeably in the so-called 'sun' and 'shade' alterations in photosynthesis and in leaf and chloroplast architecture (Boardman, 1977; Anderson, 1986). In such circumstances, lipid metabolism will be changed as a secondary consequence of the need to modify cellular membranes and organelle distribution or structure.

Light is absolutely required for the synthesis of some lipids in plants. Most noticeable amongst these are the chlorophylls but a particular fatty acid also needs light for its formation. *trans*-Δ3-Hexadecenoate is an unusual fatty acid (*cis* – double bonds are the norm for plant (and other organisms) unsaturated lipids) which is specially located at the *sn*-2 position of phosphatidylglycerol, the only significant phospholipid in chloroplast lamellae (see Chapter 1; Harwood, 1980). The requirement for light in the biosynthesis of *trans*-Δ3-hexadecenoate was first noted during experiments with *Chlorella*, although it was emphasised that light alone is not sufficient because production of chloroplast membranes is also a pre-requisite (see Hitchcock & Nichols, 1971). (In contrast, some green algae such as *Scenedesmus* have been found to form *trans*-Δ3-hexadecanoate in the dark.) The desaturation reaction occurs on phosphatidylglycerol (Bartels, James & Nichols, 1967), a process that has been probed in detail recently by experiments using the salt-tolerant alga, *Dunaliella salina* (Ohnishi & Thompson, 1991).

Because phosphatidylglycerol, the characteristic chloroplast phosphoglyceride (Harwood, 1980) apart from a few mutants, contains significant levels of *trans*-Δ3-hexadecenoate, then it seems obvious to try and find its function in photosynthesis. The first detailed proposal in this regard was a suggestion by Tuquet *et al.* (1997) that the acid was necessary for granal stacking. They based this suggestion on the correlation of *trans*-Δ3-hexadecenoate formation with granal stacking during exposure of aetiolated maize to variable monochromatic wavelengths. However, as discussed more extensively in Gounaris, Barber and Harwood (1986) a lack of correlation has been noted in chlorophyll *b*-less barley mutants (Bolton, Wharfe & Harwood, 1978; Selstam, 1980) and in experiments with different light regimes (Percival *et al.*, 1979) to mention but two experimental systems. Moreover, *trans*-Δ3-hexadecenoate is also found in other organisms that lack the usual higher plant chloroplast morphology. For example, marine brown algae which have triplet lamellar structures or in marine red macroalgae which have phycobilisomes, phosphatidylglycerol is present with 11–20% *trans*-Δ3-hexadecenoate (Jones & Harwood, 1992; Pettitt, Jones & Harwood, 1989).

I suggested that a possible function of *trans*-Δ3-hexadecenoate could be involvement in the State 1/State 2 transitions, which aid efficient light utilisation and minimise stress damage (Harwood, 1984). Such a function would require plants to have the ability to interconvert palmitate and *trans*-Δ3-hexadecenoate *in situ* (see Harwood, 1997, 1998). Indirect evidence for this interconversion was provided by experiments with broad bean (Harwood & James, 1974). Moreover, in redox-induced transitions (Telfer, Hodges & Barber, 1983), we found changes in the ratio of palmitate/*trans*-Δ3-hexadecenoate consistent with the hypothesis (H. Davies & J.L. Harwood, unpublished data). Trémolières, Garnier and Dubertreit (1992) have followed up my suggestion and made a detailed discussion of the possible role of *trans*-Δ3 hexadecenoate in influencing lipid-protein interactions during alterations in the distribution of light energy.

Another potential role for *trans*-Δ3-hexadecenoate is that it may be involved in the oligomeric organisation and stabilisation of light-harvesting chlorophyll–protein complexes. Thus, Trémolières *et al.* (1984) found the acid in isolated complexes and, later (see Trémolières *et al.*, 1992) were able to reconstitute oligomers of LHC-2 *in vitro* by the addition of *trans*-Δ3- hexadecenoate. However, mutants of *Arabidopsis* which lack *trans*-Δ3-hexadecenoate had apparently normal oligomeric LHC-2 (McCourt *et al.*, 1985). Moreover, EPR measurements with phosphoglycerides and phaeophytin (Foley *et al.*, 1988) were also not consistent with the suggestions.

Nevertheless, Huner and associates have provided evidence for an association of *trans*-Δ3-hexadecenoate with functional oligomeric LHC-2. Rye was stressed with low temperatures and its subsequent cold-hardened growth was associated with a specific reduction in *trans*-Δ3-hexadecenoate content and a simultaneous rise in monomeric LHC-2 at the expense of the oligomeric form (Huner *et al.*, 1987). Moreover, phosphatidylglycerol was tightly and specifically associated with purified LHC-2 (Krol *et al.*, 1988). High *trans*-Δ3-hexadecenoate levels in this phospholipid were associated with optimal levels of oligomeric LHC-2 (Krupa *et al.*, 1992). Indeed, differences in oligomeric LHC-2 seen on cold-hardening were thought to be fully explained by changes in the *trans*-Δ3-hexadecenoate content of phosphatidylglycerol (Krupa *et al.*, 1987). Further support came from (phospho) lipase digestion and reconstitution experiments. These are summarised in Table 2. Thus, any reduction of phosphatidylglycerol containing *trans*-Δ3-hexadecenoate caused a decrease in the oligomeric form of LHC-2. Conversely, reconstitution with this specific phospholipid (but not with others) restored oligomer levels (Krupa *et al.*, 1992). Further work in this area is awaited with interest and it has been suggested (Harwood, 1998) that a way forward might be to subject the *trans*-Δ3-hexadecenoate-deficient mutants of *Arabidopsis* (see McCourt *et al.*, 1985) to physiological stresses during growth.

The level of *trans*-Δ3-hexadecenoate was investigated further in winter rye stressed by either temperature or varying irradiance. Decreasing the growth irradiance from 800 to 50 μmol m^{-2} s^{-1} caused a lower-

Table 2. *Demonstration of the importance of* trans-*3-hexadecenoate for oligomerisation of LHC-II in rye leaves*

Treatment of LHC-II	Addition	Ratio LHC oligomer/ monomer
None	None	1.86 ± 0.63
PLC/PLA	None	0.88 ± 0.17
PLC/PLA	18 : 2/16 : 0-PC	0.77 ± 0.29
PLC/PLA	16 : 1-PG	1.87 ± 0.30
PLC/PLA	18 : 1-PG	0.59 ± 0.05

Data re-calculated from Krupa *et al.* (1992) where experimental details may be found. The LHC-II preparation, which contained lipids (including phosphatidylglycerol), was prepared from cold-hardened rye plants. Phospholipase C (PLC) and phospholipase A_2(PLA) digestion destroyed the endogenous phospholipids and reduced the amount of LHC oligomer. Reconstitution with phosphatidylcholine (18 : 2/16 : 0-PC) or with oleoyl-phosphatidylglycerol (18 : 1-PG) failed to restore the content of LHC oligomer. Only phosphatidylglycerol containing *trans*-Δ3-hexadecenoate (16 : 1-PG) was able to do that.

ing of *trans*-Δ3-hexadecenoate in PtdGly from 29 to 16% and a commensurate rise in palmitate. No other fatty acid was changed, nor were other chloroplast lipids affected (Gray *et al.*, 1997). Low temperature growth also lowered *trans*-Δ3-hexadecenoate levels. Both these factors also affect chloroplastic redox poise (Gray *et al.*, 1997).

As might be expected for a photosynthetic organism, light has a major stimulatory action on the primary metabolic pathways of fatty acid and acyl lipid synthesis. Early experiments with leaves showed that the rate of fatty acid biosynthesis (measured by radioincorporation from ^{14}C-acetate) was stimulated up to 20-fold (Hitchcock & Nichols, 1971). This stimulation was thought generally to be due to an increase in the supply of necessary cofactors such as NADPH and ATP (Stumpf, 1984). Using mutants of *Chlamydomonas*, Picaud, Créach and Trémolières (1990) found that much of the stimulation was, indeed, due to photosystem I activity. This activity would, of course, control NADPH (and reduce ferredoxin) supply and could also be used for ATP formation by cyclic electron transport as well as through the Z-scheme.

Of the individual enzymes involved in acyl lipid formation, regulation of carbon flux by a strong control of acetyl-CoA carboxylase has always been an attractive hypothesis. Acetyl-CoA carboxylase catalyses the first, committed, step in fatty acid formation and, furthermore, has been shown to exert strong flux control in mammalian system. In the latter, acetyl-CoA carboxylase is subject to both short-term allosteric control (e.g. tricarboxylic acids, phosphorylation/dephosphorylation) as well as to long-term regulation through gene activation and enhanced enzyme protein levels (see Harwood, 1994*a*). In plants, high acetyl-CoA carboxylase activity was shown to be coincident with high storage oil formation – as befitted a major role in flux control (e.g.Turnam & Northcote, 1983). However, in leaf tissues all evidence for a light stimulation of activity was lacking until Post-Beittenmiller and her colleagues used an indirect approach. She suggested that light could give changes in acyl-thioester pools during active fatty acid formation. The latter was increased either by the use of light versus dark incubations or, additionally, by the inclusion of Triton X-100, a detergent known to accelerate fatty acid biosynthesis (Givan & Stumpf, 1971). Increases in fatty acid formation were accompanied by notable decreases in acetyl-ACP and increases in malonyl-CoA and malonyl-ACP. These changes are consistent with acetyl-CoA carboxylase as a major control of light-driven carbon flux into leaf lipids (Post-Beittenmiller, Jaworski & Ohlrogge, 1991; Post-Beittenmiller, Roughan & Ohlrogge, 1992).

The availability of highly-specific inhibitors of acetyl-CoA carboxylase in the form of the Graminicides, aryloxyphenoxypropionates and cyclohexanadiones, allowed us to measure flux control directly (Page, Okada & Harwood, 1994). We used two sensitive graminaceous species (barley and maize) and challenged them with examples of each of the herbicide classes, fluazifop and sethoxydim. Measurement of the flux control coefficient followed the classical papers of Kacser and Burns (1973) and Heinrich and Rapoport (1974), which proposed that control is shared over the whole pathway and that the sum of all the individual coefficients is one. Measurement of acetyl-CoA carboxylase during light-stimulated leaf lipid labelling in both barley and maize gave flux control coefficients of 0.45–0.61 with a mean of 0.55. Thus, under the experimental conditions used, over half the flux control was at the level of acetyl-CoA carboxylase emphasising, indeed, how important this enzyme was.

The molecular mechanism by which light stimulates acetyl-CoA carboxylase in plants is, at present, unknown. Repeated attempts to show significant stimulation by, for example, tricarboxylic acids have failed (see Harwood, 1988; Herbert *et al.*, 1996), especially at concentrations likely to occur in cellular compartments (Herbert *et al.*, 1996). Supply of ATP, although obviously increased through photosynthesis, in itself is not enough to stimulate acetyl-CoA carboxylase activity sufficiently to account for a 20-fold rise in lipid labelling. However, Nicolau and

Hawke (1984) proposed that changes in ATP, coupled with other changes in stromal ADP, Mg^{2+} and pH, could be enough. Their *in vitro* observations with a partly purified maize acetyl-CoA carboxylase showed that changes for ATP, ADP, Mg^{2+} and pH which were known to occur *in vivo*, were sufficient to cause a 24-fold increase in acetyl-CoA carboxylase activity (Table 3). This paper provides the best current explanation for *in vivo* regulation of chloroplast acetyl-CoA carboxylase.

One feature of light-stimulated lipid synthesis, which is not always appreciated, is the effect on storage oil formation. Although the bulk of carbon for oil formation originates from outside the storage organ, the latter may play some significance in the overall process. Thus, ripening seeds such as oil seed rape are contained within a silique which has significant photosynthetic capacity. In field studies with an analogous crop, mustard (*Brassica campestris* L.), it has been estimated that, in a seed, pods contribute about 60% towards whole canopy photosynthesis in contrast to the leaves which contribute only 35% (Singh, Singh & Sharma, 1986). In agreement, Allen, Morgan and Ridgman (1971) suggested that, in oilseed rape, the leaves had little direct effect on the growth of the pod, which produces the assimilates required for its own growth and for the seeds inside. The efficiency of CO_2 fixation by pods and movement of assimilates into seeds (within 30 seconds) has been examined using detached pods (Sheoran *et al.*, 1991). Moreover, in oil fruits such as avocado and olive, the fruit is green throughout the accumulation period. Recently, the contribution of the olive fruit to its own carbon supply was measured following light stress and defoliation. The results showed that developing olive fruits were able to fix

Table 3. *How changes in the chloroplast's stroma during photosynthesis could increase acetyl-CoA carboxylase activity*

Parameter	*In vivo* change	Increase in ACCase activity *in vitro*
pH	7.1 → 8.0	3.1-fold
Mg^{2+}	2 → 5 mM	2.4-fold
ATP	0.3–0.8 → 8.0–1.4 mM	1.7-fold
ADP	0.6–1.0 → 0.3–0.6 mM	1.9-fold
		24-fold

Data taken from the paper by Nikolau and Hawke (1984) for maize leaf acetyl-CoA carboxylase (ACCase).

atmospheric CO_2 into storage triacylglycerol at significant rates and Sanchez (1994) suggested that this process might contribute appreciably to overall oil deposition in the fruit.

There has been much recent research on the possible harmful effects of UV-B levels on crop production. Because protection from such rays is provided by flavonoids, which can prevent UV-B damaging internal cells and their organelles (Lois & Buchanan, 1994; Landry, Chapple & Last, 1995) and because biosynthesis of these chemicals needs an active cytosolic isoform of acetyl-CoA carboxylase (ACCase) (see Harwood, 1996), the effect of UV-B exposure on ACCase isoforms was examined using fully expanded leaves from 9-day-old pea seedlings. Sasaki and Konishi (1997) showed clearly that UV-B exposure was able to produce a massive rise in the 220 kDa cytoplasmic multifunctional ACCase but negligible effects on the 35 kDa BCCP subunit of the chloroplastic ACCase isoform. Because flavonoid synthesis is concentrated in the epidermis (Beerhues, Robenek & Wiermann, 1988) where the cytosolic ACCase is also located, these results illustrate an important protective function for the latter isoform in providing malonyl-CoA for flavonoid formation.

Temperature

Temperature may affect lipid biosynthesis either through a general action on metabolism (Q_{10} values, etc.) or because it can give rise to stress (either heat or chilling/freezing stress). Changes in plant lipids during responses to low temperature stress (Harwood *et al.*, 1994) and the use of *Arabidopsis* mutants in such studies (Browse *et al.*, 1994) have been reviewed recently. Significant changes have been noted for several aspects of plant membrane lipids in response to low temperatures (Table 4). For example, it has been shown recently that there may be subtle changes in lipid metabolism which give rise to different patterns of lipids and their fatty acid constituents. Thus, following chilling in tomato, there appears to be an inhibition of the 'eukaryotic' pathway for galactolipid synthesis (Yu & Willemot, 1996). In addition, the glycerol 3-phosphate acyl transfer step of the 'prokaryotic' pathway is impaired (Yu & Willemot, 1997).

Molecular species effects

The most subtle of the changes possible for membrane lipids are those relating to molecular species. Either existing fatty acids can be replaced and/or their positions on the glycerol backbone can be swapped. Because the glycerol backbone is not parallel to the surface of the mem-

Table 4. *Alterations in plant lipids in response to lowered environmental temperatures*

Change	Remarks
Molecular species changes and remodelling	Remodelling is a quick ('emergency') response resulting in relatively small changes in membrane fluidity. Transgenic plants created by manipulating acyltransferases
Changes in fatty acid chain length	Commonly observed but functional significance unclear. In theory this could be useful to increase fluidity (as shown in some microbes) but influence complicated because 18C acids are more unsaturated than 16C
Increases in unsaturation	Commonly observed. Insertion of first double bond in an acyl chain has most influence. Induction of expression for desaturase genes has been studied and transgenic plants created
Alterations in proportions of different lipid classes	In theory this could be useful for modifying membrane fluidity but significance of changes *in vivo* is unclear. Some alterations may be secondary to changes in subcellular morphology
Increase in the ratio of membrane lipid to protein	Commonly observed but functional significance unproven

brane, identical fatty acids at the *sn*-1 and *sn*-2 positions project into the hydrophobic interior to different extents. Thus, the acid at the *sn*-1 position is located further into the membrane by the order of about 2 carbons. Since the fluidity of a membrane is dependent on hydrophobic interactions down an acyl chain, then these interactions can be maximised with, for example, palmitate (16 : 0) at the *sn*-1 position and stearate (18 : 0) at the *sn*-2. The opposite positions for acylation of these two acids would result in poorer packing characteristics (Fig. 1) and, less hydrophobic interaction. Thus, the transition temperature for a 1-palmitoyl, 2-myristoyl combination on phosphatidylcholine is about 10 °C lower than for 1-myristoyl, 2-palmitoyl phosphatidylcholine (see Gurr & Harwood, 1991).

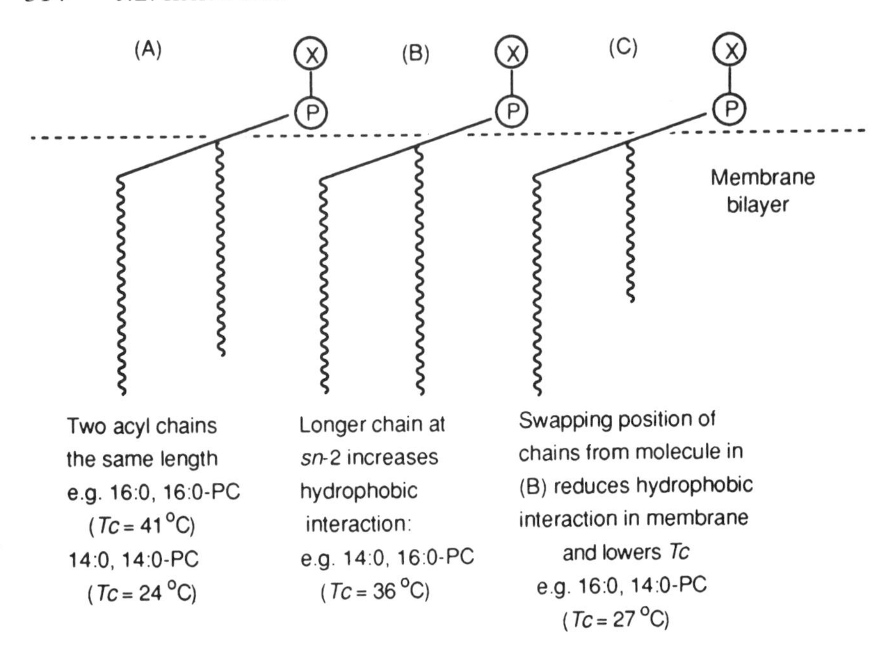

Fig. 1. How the position of acyl attachment can alter the gel to liquid phase transition of a membrane phosphoglyceride. Because the glycerol backbone of membrane lipids is not parallel to the membrane surface, swapping acyl positions can affect fluidity. (Taken from Harwood, 1997 with permission.)

Another possible strategy for changing fluidity properties of membrane lipids is to change the combinations of fatty acids. For example, if only palmitate and oleate were present, in equal quantities, in a membrane's phosphatidylcholine then these could be combined as 50% dipalmitoyl and 50% dioleoyl, 100% 1-palmitoyl, 2-oleoyl, 100% 1-oleoyl, 2-palmitoyl or a mixture of these. For the two palmitoyl/oleoyl species, low temperatures are seen (about 1 °C) but for dipalmitoylphosphatidylcholine a transition of 41 °C is found. Clearly, plants should minimise combinations of two long-chain saturated fatty acids in order to maintain their membrane fluidity (see later).

Re-modelling of molecular species has been shown to occur in some algae or cyanobacteria but has really not been studied systematically in higher plants. For the salt-tolerant alga, *Dunaliella salina*, changes in both plasma membrane as well as chloroplast lipids were demonstrated (Lynch, Norman & Thompson, 1984).

If a homologous series of fatty acids is examined, then increasing chain length is coincident with increasing melting points. Thus, in theory, it should be possible for an organism to adjust to lower environmental temperatures by changing the chain length of its membrane fatty acids. This adaptation has been demonstrated for yeast (*S. cerevisiae*) (Okuyuma *et al.*, 1979) and for (psychrophilic) bacteria like *Micrococcus cryophilis* (Russell, 1984; Sandercock & Russell, 1980). In plants, too, a change in the average chain length parameters has been observed, following exposure to lower environmental temperatures. However, in plants interpretation of these changes is confused because the major C16 fatty acid is saturated (palmitate), whereas the major C18 fatty acids are unsaturated (frequently polyunsaturated). Thus, an increase in C16 fatty acids (rather than C18) may actually *decrease* fluidity because they are more saturated. Therefore, although chain length alterations are often observed (see Harwood *et al.*, 1994), their functional significance in unclear.

Changes in unsaturation

The most usual response by poikilotherms to lowered environmental temperatures is an increase in unsaturation. The strength and rationale for this response is well illustrated in Table 5 where some transitions

Table 5. *Examples of gel to liquid phase transition temperatures for different phosphatidylcholines*

	T_c (°C)
1-14 : 0, 2-14 : 0-PC	24
1-14 : 0, 2-16 : 0-PC	35
1-16 : 0, 2-14 : 0-PC	27
1-16 : 0, 2-16 : 0-PC	41
1-18 : 0, 2-18 : 0-PC	56
1-18 : 0, 2-18 : 1-PC	6
1-18 : 0, 2-18 : 2-PC	−16
1-18 : 0, 2-18 : 3-PC	−13
1-18 : 1, 2-18 : 1-PC	−18

Molecular species are shown with the fatty acids attached at the *sn*-1 and *sn*-2 positions indicated. A shorthand nomenclature for fatty acids is used with the number before the colon giving the carbon number and the figure after showing the number of double bonds. 18 : 1 is oleic acid, 18 : 2 is linoleic acid and 18 : 3 is α-linolenic acid.

for phosphatidylcholines with different degrees of *cis*-unsaturation are illustrated. Insertion of the first *cis*-double bond lowers the melting point of the acyl chain up to 50 °C and a second *cis*-double bond lowers it an additional 15 °C or so. Interestingly, there is no additional advantage (in terms of fluidity) to converting a dienoic to a trienoic fatty acid or adding any further double bonds (Table 5). Therefore, it does not appear to be an advantage, from the point of view of membrane fluidity, for plants to be so enriched in α-linolenic acid (Harwood, 1980) or (marine) algae to be so enriched in arachidonate or eicosapentaenoate (Harwood & Jones, 1989). The evolutionary conservation of such constituents must be due to other considerations – especially in view of the energy expenditure involved in their synthesis and the potential instability (oxidation-prone) of highly polyunsaturated membrane fatty acids.

Nevertheless, in view of the obvious advantages of introducing one or, perhaps, two double bonds into an acyl chain, it is hardly surprising that many organisms (including plants) have adopted this adaptive mechanism. Increasing unsaturation often has an additional advantage in that many desaturases work on complex lipid substrates and, because the fatty acid desaturases themselves are membrane located, they modify membrane lipids *in situ*.

For a plant to be able to increase unsaturation, it follows that it must usually be able to raise the activity of its various fatty acid desaturases. (An alternative strategy (Table 6) of increased turnover of more saturated acyl chains, still relies on significant desaturase activity.) Fatty acid desaturase activity can be increased in four basic ways (Table 6). If substrate availability is limiting them, of course, increased supply of these molecules will give rise to raised desaturation. Of the substrates

Table 6. *Strategies to increase membrane lipid unsaturation*

1. Increase turnover of molecular species, allowing replacement by more unsaturated examples
2. Increase desaturation
   (i) By increasing substrate supply, e.g. oxygen, if limiting
   (ii) By increasing desaturase gene expression
   (iii) By changing desaturase mRNA stability
   (iv) By activating pre-existing desaturases, e.g. by alterations in their membrane environment

for plant desaturases, oxygen is common and, indeed, solubility of this gas is increased at lower temperatures. Thus, James and co-workers in the 1960s, suggested that oxygen solubility would be a natural self-regulating property of the membranes of organisms exposed to lower temperatures (see Harris & James, 1969). Therefore, as the environmental temperature was lowered, the activity of desaturases would increase and, hence, give rise to more unsaturated (i.e. adapted) lipids in the plant's membranes. (Recently, such a scenario has been given more credence by experiments with a soil protozoa, *Acanthamoeba castellanii*, where oxygen has been shown to be not only essential for induced gene expression for a desaturase gene but also to be capable of inducing gene expression *in the absence* of a temperature change (Avery *et al.*, 1996).)

A second method of inducing desaturase activity is to activate a pre-existing enzyme. Thus, lowering environmental temperatures could lead to a change in membrane fluidity which, in turn, would lead either to conformational alterations in the terminal desaturase protein or, alternatively, to a change in its relationship with the other desaturase components (e.g. NADH/cyt b_5 reductase, cytochrome b_5). Evidence for such an adaptive response has been obtained with non-plant organisms such as *Candida lipolytica* (Horvath *et al.*, 1991), where catalytic hydrogenation of membranes led to an immediate increase in Δ12-desaturase activity. Alternatively, in *Acanthamoeba castellanii* incubating membranes at lower temperatures led to significantly raised Δ12-desaturase activity (Jones, Lloyd & Harwood, 1993). Although such experiments have not been carried out with plants, there is no reason to believe that responses by such organisms will be any different.

A third general response is to increase the amount of desaturase protein. Since all proteins turn over, then such a change could be caused by increased synthesis and/or decreased catabolism of desaturase protein. Increased anabolism could itself be caused by raised gene expression or translation. For aerobic desaturases, there are typically three components: an oxidoreductase, a cytochrome (or similar electron carrier) and the terminal oxygen-binding desaturase (Fig. 2). Any one of these could be rate limiting for the overall fatty acid desaturation. However, work with protozoa suggests that it is the terminal component which is rate limiting and, hence, is the one protein that needs to be changed. Indeed, work with *Acanthamoeba castellanii* has shown that low temperature exposure causes an increase in the synthesis of desaturase protein. Moreover, when lowered culture temperatures trigger this change, oxygen is still needed if only for a short period (Avery *et al.*, 1996). Furthermore, it has also been found that oxygen alone can induce

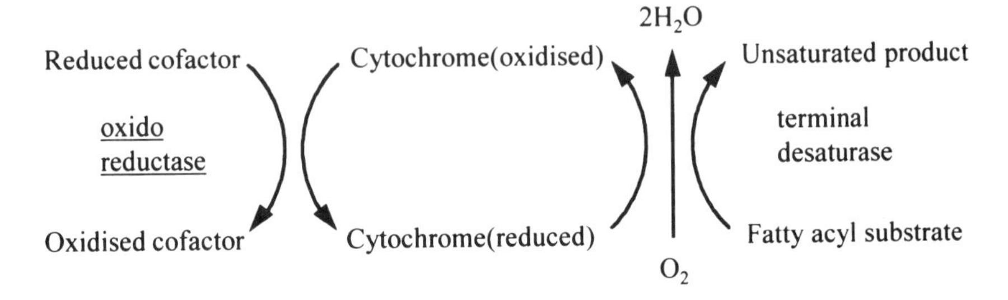

Fig. 2. Typical aerobic fatty acid desaturase reaction.

enhanced desaturase protein synthesis and hence activity (Avery *et al.*, 1996). Since oxygen has enhanced solubility at low temperatures, these data show that one mechanism for inducing desaturase gene expression could be raised oxygen levels – at least in systems, like *Acanthamoeba* where high respiratory rates may make oxygen limiting. As referred to above, the suggestion that oxygen levels could be a trigger for desaturase induction was made originally following experiments with plant tissues (Harris & James, 1969).

A series of elegant experiments have been carried out by Murata using the cyanobacterium *Synechocystis* PCC 6803. A mutant was made which could no longer make linoleic acid and lacked Δ12-desaturase activity. The gene for the latter was then isolated and could be used to restore normal linoleic synthesis in *Synechocystis* and, also, to transform *Anabaena nidulans* (which normally lacks Δ12-desaturase activity) (Wada, Gombos & Murata, 1990). Using the des-12 gene it has been possible to isolate other desaturase genes and to identify concensus sequences (Murata & Wada, 1995). The mechanism for desaturase induction has been probed further. It was found that modifying membrane fluidity using *in situ* catalytic hydrogenation caused an increase in mRNA synthesis (evaluated by Northern blots) commensurate with low temperature induction (Vigh *et al.*, 1993). Thus, it appeared that the cyanobacteria sensed a temperature change through alteration in their cytoplasmic membrane's fluidity (Vigh *et al.*, 1993). Since *Synechocystis* is photosynthetic, it is not known if oxygen is involved in the process – as it appears to be in *Acanthamoeba* (see above). To date the mechanism of desaturase induction has not been studied in higher plants.

Lipid class changes

Some observations indicate that the relative amounts of different lipid classes change following temperature stress. This is, obviously, not a particularly quick response compared to molecular species adaptation or fatty acid modifications (Harwood, 1994*a*). Furthermore, as pointed out before (Harwood *et al.*, 1994), in many experiments there is insufficient data to distinguish between a real change in the composition of a particular membrane or merely an alteration in the balance of subcellular membranes in the tissue being examined.

In the green alga, *Dunaliella salina*, the effect of temperature on lipid metabolism included the isolation of individual organelles and membranes. In chloroplast membranes, for example, there was a clear increase in digalactosyldiacylglycerol (DGDG) at the expense of monogalactosyldiacylglycerol (MGDG) (Lynch & Thompson, 1982). Since MGDG can be converted into DGDG by the action of UDP-galactose galactosyltransferase (Joyard & Douce, 1987), then such a change in their ratios is easily accomplished. Moreover, if one compares the transition temperatures for comparable molecular species of these two galactosylglycerides, then DGDG has a significantly lower value. Thus, the distearoyl-species of MGDG has transition temperatures of 82 °C and 69 °C for the I and II forms, respectively, while distearoyl-DGDG has temperatures of 47 °C and 41 °C (Quinn & Williams, 1983). Therefore, modification of the relative amounts of thylakoid MGDG and DGDG in low temperature-stressed plants would make sense, although such changes are not always consistently seen (Harwood *et al.*, 1994).

Another lipid interconversion easily achieved and which could be of advantage in terms of membrane fluidity is to methylate phosphatidylethanolamine (PtdEtn) to form phosphatidylcholine (PtdCho). For the distearoyl species of PtdEtn and PtdCho, the transitions are 71 °C and 58 °C, respectively (Harwood *et al.*, 1994). Changes in the ratio of these phosphoglycerides have been seen in a number of organisms, including plants, and, although the analysis usually is not detailed enough to tell if a particular membrane has been altered (see above), it seems that there is enough encouragement for further experiments to examine enhanced PtdCho content as an adaptive mechanism.

Lipid to protein ratios in membranes

A general increase in membrane lipid to protein has been seen in a number of studies on low temperature effects (Harwood *et al.*, 1994). This alteration is sometimes seen in the absence of fatty acid composition changes. The physiological basis for such an alteration in membrane composition has been discussed (Harwood, 1997). Nevertheless, increasing lipid to protein ratios does have a useful action in raising fluidity (e.g. Millner *et al.*, 1984) and may, therefore, be a useful adaptation.

Chilling sensitivity

Plants vary considerably in their sensitivity to chilling (lowering temperatures to the 5–10 °C region). If plant membranes show significant phase separation of their lipids after chilling, then that could be a fundamental cause of chilling sensitivity. However, higher plant membranes are usually highly unsaturated and therefore, unlikely to have many molecular species that could undergo lipid to gel transitions in this temperature range. For chloroplast membranes though, there is a reasonably high proportion of 'saturated' acids in the phosphatidylglycerol (PtdGly) and sulphoquinovosyldiacylglycerol (SQDG) components (Harwood, 1980). For PtdGly, Murata (1983) was able to show that over 20% of the PtdGly consisted of dipalmitoyl- or l-palmitoyl, 2-*trans*-Δ3-hexadecenoyl- species (i.e. 'saturated') in chilling-sensitive species but not in chilling-resistant plants. Analyses by others are broadly in agreement with this conclusion (Bishop, 1986; Roughan, 1985; but see Harwood, 1997). Although saturated SQDG could, in theory, also contribute to chloroplast thylakoid chilling-sensitivity (Kenrick & Bishop, 1986), it seems that it is the fatty acid composition of PtdGly which is critical (see Harwood *et al*., 1994).

For agricultural purposes, it would be an advantage to increase chilling resistance and this has been attempted in two ways. During their manipulation of the fatty acid composition of cyanobacteria, Murata and colleagues were able to show that raised unsaturation conferred resistance to lower growth temperatures (Wada *et al*., 1990). Then they specifically addressed the problem of PtdGly in higher plants. During the synthesis of PtdGly, acylases attach fatty acids in succession to the *sn*-1 and *sn*-2 positions of glycerol 3-phosphate (Fig. 3). If the first acylation results in significant amounts of palmitate at the *sn*-1 position, then the 'chilling-sensitive' dipalmitoyl or 1-palmitoyl, 2-*trans*-Δ3-hexadecenoyl-PtdGly can result. Thus, by introducing genes coding for glycerol 3-phosphate acyltransferases of different fatty acyl selectivity into plants, Murata *et al.* were able to increase chilling resistance (Murata *et al*., 1992). Recently, they have transformed plants with a cyanobacterial Δ9-desaturase gene which, unlike higher plant Δ9-desaturases, has high activity towards palmitate. The transformed plants contained elevated levels of palmitoleate and were significantly more chilling resistant (Ishizakini-Nishizawa *et al*., 1996). Conversely, Wolter, Schmidt and Heinz (1992) increased the saturation in PtdGly of *Arabidopsis* by transferring an *E.coli* gene coding for glycerol 3-phosphate acyltransferase. This made *Arabidopsis* chilling sensitive. These experiments show very nicely how it is possible to modify plants geneti-

Pathway A

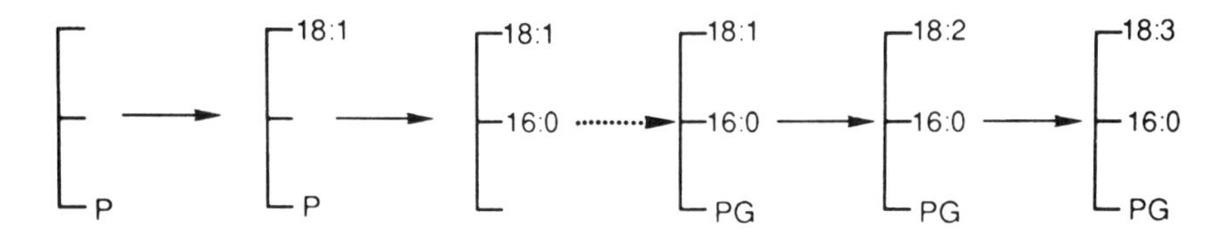

Pathway B

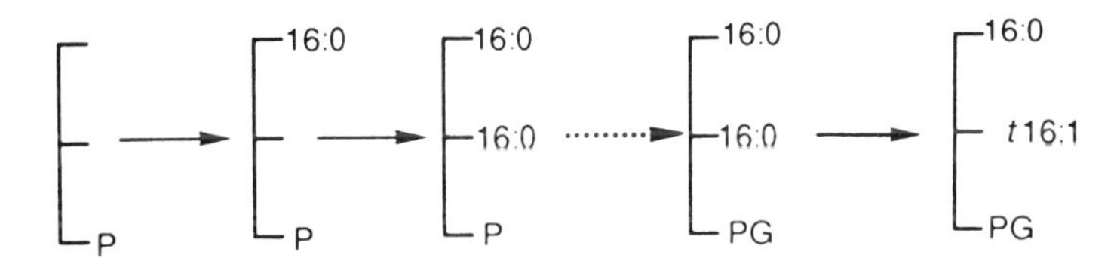

Fig. 3. Biosynthetic pathways for phosphatidylglycerol molecular species in plants. The production of disaturated species of phosphatidylglycerol in temperature-sensitive plants depends on the specificity of their glycerol 3-phosphate acyltransferases. Pathway A is dominant in chilling-resistant plants and produces phosphatidylglycerol molecules of high fluidity (low melting point). Pathways A and B are of comparable activities in chilling-sensitive species and result in an appreciable proportion of the phosphatidylglycerol having a T_c above 32 °C. (Taken from Harwood, 1997 with permission.)

cally following fundamental experiments about their membrane lipid biochemistry.

Part of the reason for chilling sensitivity in chloroplast membranes must relate to the need for fluidity in the lipid bilayer structure for efficient photosynthesis. (Indeed, photosynthesis is frequently used as a test of chilling damage, e.g. Wada *et al.*, 1990.) Ultrastructural changes in thylakoid membranes, following chilling have been examined recently (Quinn, 1977). When fluidity was reduced, PSII oligomeric complexes became aligned and packed more densely. This contrasted to the normal random distribution and would, presumably, affect the interaction of the PSII core complex with associated LHC complexes across the thylakoid membrane.

A thorough review of the role of membrane lipids in chilling sensitivity has recently been made by Nishida and Murata (1996).

The 'greenhouse effect'

In the last decade it has become clear that the global environment is heating up. Release of the so-called greenhouse gasses, particularly carbon dioxide, is causing raised average temperatures as well as instability to global weather patterns. Although estimates are imprecise and subject to considerable discussion, it is thought that an approximate doubling of atmospheric CO_2 (from 350 to 660 ppm) and a 4 °C rise is temperature are likely scenarios. Clearly, vegetation will be exposed to the new conditions but how will it respond? So far as lipids are concerned we have looked at two different tissues, seeds and leaves, mainly in the important cereal crop, wheat.

Although wheat seeds are low in lipid, the lipid content of wheat flour has important nutritional and baking consequences. For a description of the lipids of cereals refer to the review by Morrison (1983), where the unique content of monoacylphosphoglycerides in the starch component is discussed. Because of the technical (Barnes, 1983) and stability properties of flour lipids, then any change in their quality or quantity is of importance to industry. For crops grown under the 'greenhouse conditions', we noted significant changes in both these parameters (Williams, Shewry & Harwood, 1994). Most of the alteration was due to the temperature component rather than to raised CO_2 (Williams *et al.*, 1995). Interestingly, agricultural practice in the pattern of N fertiliser application was able to counteract some of the alterations (Williams *et al.*, 1995). These results show that it may be necessary to change wheat cultivars and/or agricultural practice in the future in order to preserve adequate supplies of suitable wheat flours.

We also considered the effect of raised CO_2 and temperature on the biogenesis of organelles and synthesis of their membrane lipids. For some studies the separate effects of CO_2 and temperature were evaluated. Several surprising findings were made. First, raised CO_2 was capable of accelerating mitochondrial development such that more than a three-fold increase in their number was seen in young wheat leaves (Robertson *et al.*, 1995). This increase in mitochondrial biogenesis was paralleled by a similar rise in diphosphatidylglycerol formation. (Diphosphatidylglycerol is exclusively located in the inner mitochondrial membranes, where it is a major component (Bligny & Douce, 1980; Harwood, 1985).) Moreover, whereas it would be expected that a 4 °C rise in temperature would raise overall biosynthesis appropriately, what was unexpected were the qualitative consequences. It appeared that a change in the apportioning of fatty acids

Table 7. *Raised atmospheric CO_2 alters fatty acid metabolism in wheat leaves*

Labelled lipid	Treatment	Fatty acid labelling pattern %			
		16 : 0	18 : 1	18 : 2	18 : 3
Phosphatidyl-choline	Control	34±5	29±6	31±5	6±3
	Raised CO_2	9±tr ($p<0.01$)	41±1 ($p<0.025$)	48±2 ($p<0.01$)	1±tr ($p<0.01$)
Phosphatidyl-glycerol	Control	22±3	33±4	39±2	3±tr
	Raised CO_2	41±1 ($p<0.01$)	22±2 ($p<0.025$)	27±3 ($p<0.01$)	8±1 ($p<0.01$)

Results are taken from M. Williams and J.L. Harwood (unpublished data, 1995). Leaves from wheat plant (7 days old) grown under atmospheric (350 ppm) or raised (650 ppm) CO_2 were labelled from [1-^{14}C]acetate. Phosphatidylcholine and phosphatidylglycerol accounted for 34% and 17% of the total lipid labelling in control leaves and 40% and 19% in the raised CO_2 samples. Statistical significance by Student's *t*-test.

to the 'prokaryotic' and 'eukaryotic' pathways of lipid metabolism (see Browse & Somerville, 1991) took place (Table 7). Thus, PtdGly, the typical chloroplast phosphatidylglyceride was much better labelled in palmitate while C_{18} acids were channelled into PtdCho (Williams and Harwood, 1995). Because PtdCho is a substrate for the Δ12-desaturase (Harwood, 1996), then the increased flux of oleate through this lipid was thought to account for elevated unsaturation, particularly in linoleate levels, at the raised temperature (Harwood, 1995; Williams & Harwood, 1997). These data emphasise the great difficulties faced by biochemists in trying to predict consequences of environmental changes.

Water stress

Water deficiency, partly caused by rising temperatures and partly by changed weather patterns, is an increasing concern in many parts of the world. It has been estimated that millions of hectares are now cultivated under stressed conditions and were it not for irrigation, then the figure would be much larger. With changes in climate patterns (referred to above), it seems likely that drought will be an escalating problem. Typical physiological responses of plants to drought have been reviewed (Levitt, 1980; Bradford & Hsiao, 1982; Ingram & Bartels, 1996). There is a reduction in stomatal activity, closure of stomata, changes in respiratory rates and alterations in osmotic potential. Some chloroplast enzymes may be inhibited and electron transport reduced (Boyer & Bowan, 1970; Mohanty & Boyer, 1976). The complex signalling pathways invoked by regulators such as abscisic acid in water-stressed plants have been recently reviewed (Bray, 1997).

As with most environmental stresses, the response of plants to drought depends on the severity of water deficiency and also on the plant species (and even variety). Photosynthesis is usually reduced during mild water deficiency because of stromatal closure. Large water deficits can also have a direct inhibitory action on photosynthesis (Smirnoff, 1993). One possibility is that carbon dioxide limitation can give rise to overexcitation of the reaction centres. As a consequence, activated forms of oxygen can be detected and these are now thought to be important generally in the propagation of responses to drought (Price, Atherton & Hendry, 1989; Leprince *et al.*, 1990; Seel *et al.*, 1992; Quartacci & Navari-Izzo, 1992). Large amounts of activated oxygen species may exceed the photoprotective antioxidant defences and oxidative damage will occur (Smirnoff, 1993).

The subject of drought stress on plant lipid metabolism has been reviewed (Liljenberg, 1992). Drought-sensitive plants generally show larger changes in their lipid composition than do drought-resistant species (or cultivars). Common effects observed are decreases in the tissue membrane lipids such as glycosylglycerides or phosphoglycerides. But there are exceptions (see Liljenberg, 1992). Simultaneous to the above decreases, there are often increases in non-polar lipids such as triglycerides (Douglas & Paleg, 1981; Martin, Schoper & Rinne, 1986: Wilson, Burke & Quisenberry, 1987; Navari-Izzo *et al.*, 1990). It has been suggested that the accumulation of triacylglycerols may be a way of removing accumulated non-esterified fatty acids which were formed when membrane lipids were hydrolysed (Douglas & Paleg, 1981; Wilson *et al.*, 1987; Navari-Izzo *et al.*, 1990). These changes in lipid class distribution are not accompanied by any large alterations in fatty acid unsaturation (Liljenberg, 1992), despite the occurrence of significant lipoxygenation (see below).

Membrane lipid hydrolysis leading first to a rise in non-esterified fatty acids and, later, to triacylglycerol accumulation has been frequently recorded (e.g. Douglas & Paleg, 1981; Martin *et al.*, 1986; Navari-Izzo *et al.*, 1990). The availability of polyunsaturated non-esterified fatty acids provides substrates for lipoxygenases (Vick & Zimmerman, 1987) and connects drought stress to a rise in lipid peroxidation (Price & Hendry, 1991). Because some plants seem able to mobilise their defences against radical-mediated lipid peroxidation (Smirnoff & Columbe, 1988), it is not surprising that drought tolerance has been related to the extent of peroxidation for a given species in the field (Seel, Hendry & Lee, 1992). For chloroplasts isolated from drought-treated wheat, increased production of the superoxide radical was observed (Price *et al.*, 1989). In turn, this may take part in the formation of the highly reactive hydroxyl radical. In addition, damage in plants subjected to severe water stress (including that of the photosynthetic machinery) may be associated with a reduction in the activity of typical free radical-processing enzymes. The latter would include glutathione reductase, peroxidase and superoxide dismutase (Leprince *et al.*, 1990).

For an account of the role of active oxygen species in drought responses see Smirnoff (1993) and for detailed aspects of changes in lipid biochemistry see Liljenberg (1992).

Atmospheric constituents

The 'greenhouse effect' and raised carbon dioxide

Aspects of alterations in lipid biochemistry due to the 'greenhouse effect' were discussed in the section on temperature effects. Although many of the manifestations of the 'greenhouse effect' on lipid metabolism are due to the raised temperature component, CO_2 can produce effects in its own right. Notable amongst these are altered patterns of organelle development (Robertson *et al.*, 1995). There are also CO_2-induced changes in lipid metabolism (Williams & Harwood, 1997), some of which may be related to the altered pattern of cellular development.

Ozone pollution

Ozone can affect plants in two ways. First, the reduction in the stratospheric ozone layer permits increased solar radiation including UV-B. These rays are harmful to all organisms and their action on plants has been quite well studied recently (see section on light).

Secondly, local increases in ozone at ground level are caused by photolysis of nitrogen dioxide (particularly from automobile exhaust gases) (Heath, 1984) and occurs in a cyclic manner during the day. Peak levels are known as episodes and levels in excess of 50 ppb (which are thought to be damaging to, at least, some plants) are detected regularly throughout Europe. Moreover, concentrations of ozone at particular world sites, such as urban areas of Japan, have been reported to be as high as 500 ppb on occasion (Sakaki, Kondo & Sugahara, 1983). Because of these very high ozone episodes in places such as Tokyo or the Los Angeles basin, experiments have been performed using short-term exposure of plants to 150–600 ppb ozone. Under these conditions, large decreases in chloroplast galactolipids and their unsaturation levels have been reported (Fong & Heath, 1981; Sakaki *et al.*, 1985; Sakaki, Tanaka & Yamada, 1994; Nouchi & Toyama, 1988). In an interesting series of papers, Sakaki *et al.* (1983, 1985, 1994) have rationalised the changes observed. The initial observation is a rise in non-esterified fatty acids. (Although not mentioned specifically this rise must reflect an increase in acyl hydrolase activity, possible as a result of membrane damage by ozone.) The increase in non-esterified fatty acid levels can activate galactolipid: galactolipid galactosyltransferase (GGGT) but decrease UDP-galactose: diacylglycerol (DAG) galactosyltransferase. This leads to a modification of chloroplast lipid metabolism with an accumulation of DAG which can be acylated to form triacylglycerol (TAG). The accumulation of TAG in ozone-stressed plants is helped by the availability of acyl-CoA because it is formed from non-esterified fatty acids by acyl-CoA synthase. These changes in metabolism account for the decrease in chloroplast thylakoid lipids, rises in non-esterified fatty acids and triacylglycerol and functional impairment of chloroplasts as a result of ozone damage (Wellburn, 1988).

Although the above studies are relevant to some world sites, it is moderate ozone elevation which is generally seen. Thus, the experiments by Sandelius and colleagues using exposure to 65 ppb ozone are more widely applicable. They have shown different responses between young and mature leaves, monocotyledons versus dicotyledons and between 16 : 3 and 18 : 3 (see Browse & Somerville, 1991) plants. For sensitive plants, there seemed to be a general decrease in chloroplast galactolipids but without the change in ratio of MGDG/DGDG reported for high ozone exposure (Sandelius *et al.*, 1995). There was also an increase in non-chloroplast membrane lipid synthesis at the expense of chloroplast lipids. This effect was particularly noticeable for linolenate-containing MGDG (Carlsson *et al.*, 1994). Fatty acid changes in MGDG of Norway spruce exposed to ozone have also been noted. These may then give rise to a decrease in the natural frost resistance (Wellburn, 1988; Wellburn *et al.*, 1994; Wolfenden & Wellburn, 1991).

Other atmospheric constituents

A review of air pollutants capable of injuring plants has been made (Heath, 1980) and many of these compounds have significant effects on lipids (Heath, 1984). Very recently, there has been much interest in nitric oxide, a free radical gas which has many acute effects in animal tissues. It also appears to have growth regulatory action in plants (Lesham & Haramaty, 1996). By using S-nitroso-*N*-acetylpenicillamine, which in aqueous solutions releases NO, Lesham *et al.* (1997) looked for specific effects on lipids. They found some physical effects on monogalactosyldiacylglycerol, possibly due to loose association of NO with π bonds in the double bond systems of the acyl groups. They predict that this might reduce lipoxygenase activity and this was, indeed, found as measured by oxygen consumption (Lesham *et al.*, 1997).

Salt stress and the effect of soil minerals

General discussions of salt effects on plants lipids and their metabolism have been made (Harwood, 1984, 1989; Kuiper, 1980, 1985). Two main aspects have been studied – salt stress and calcium effects.

Salt tolerance

Membranes play a vital role in the ion status of tissues and their organelles through selective permeability. Thus, one might predict that salt tolerance might well be connected with membrane lipids. In a number of experiments, a salt-induced decrease in chloroplast lipids has been found (Harwood, 1984). In other experiments, salt-stress results in a general decrease of all acyl lipids (Harwood, 1983). However, for *Plantago* spp., it has been concluded that salt tolerance is unconnected to alterations in phosphoglyceride content (Erdei, Stuiver & Kuiper, 1980). For sunflower leaves and roots, NaCl stress led to a decrease in fatty acid unsaturation. Moreover, desaturation of exogenous [^{14}C]oleate was also lowered (Ellouze, Gharsalli & Cherif, 1982). Decreases in unsaturation are often particularly marked for chloroplast lipids and, moreover, severe reduction in galactolipid synthesis after salt stress has been observed in both olive (Zarrouk, Seqqat-Dakhma & Cherif, 1995) and in rape (Najine, Marzouk & Cherif, 1995).

In marked contrast to the general decreases in membrane lipid contents (especially for chloroplasts) caused by NaCl stress, calcium (sulphate) exposure gives rise to an increase (Bettaieb, Gharsalli & Cherif, 1980). Moreover, some plants can be divided into calcifuge and calcicole. Thus, plants thriving in soils poor in calcium are called calcifuge (such as *Lupinus luteus*) and those thriving in soils rich in calcium are called calcicole (such as *Vicia faba*). Most experiments have studied root lipid metabolism where calcium inhibits phospholipid synthesis in calcicole plants. Both ethanolaminephosphotransferase and cholinephosphotransferase are sensitive to this inhibition (Oursel, 1979). In addition, calmodulin and calcium antagonists have been noted to stimulate digalactosyldiacylglycerol breakdown in potato leaves (Piazza & Moreau, 1987).

Heavy metals

Heavy metal pollution is known to affect plant growth and metabolism, in addition to affecting other organisms. The topic has been thoroughly covered with regard to effects on algae by Pohl and Zurheide (1979) and by Harwood and Jones (1989). A recent study has investigated copper pollution-induced changes in the lipids of *Selenastrum capricornutum* where fatty acid unsaturation was particularly sensitive (Riches *et al.*, 1997). In addition, a representative study in higher plants (using the oil-yielding sesame) is that of Singh, Bharti and Kumar (1994) and Bharti and Singh (1994) where further references may be found.

Xenobiotics which affect lipid biochemistry

Some examples of xenobiotics which have been reported to affect plant lipids and/or their metabolism are shown in Table 8. Apart from herbicides, there has been relatively little work in this area, so that there are very few individual reports in the literature. The limited number of observations made thus far are sure to be confirmed and extended when further work is done.

Interest in hydrocarbons has principally been applied to aquatic organisms because of concern about chronic exposure or oil spillage disasters. Hydrocarbons are easily taken up by marine organisms (Karydis, 1980) and have been shown, not surprisingly, to produce significant effects (e.g. Karydis & Fogg, 1980) including changes in lipids (Morales-Loo & Goutx, 1990; Petkov, Furnadzieva & Popov, 1992). Experiments have also been carried out on degradative products of petrochemicals including testing the effects of diesel exhaust particles (Liebe & Fock, 1992).

Although herbicides are the obvious particular class which affect plants, crops are treated regularly with fungicides, insecticides, etc. Of course, in an ideal world, such pesticides should have no effect on plants but be specific for their target organisms. Alternatively, if they do alter plant metabolism and growth, they should have only a marginal action – which can be seen as a 'trade-off' against the worse situation of unchecked pest attack.

The insect pest complex threatening olives causes at least a 15% loss

Table 8. *Some reported effects of xenobiotics on lipid biochemistry*

Class of xenobiotic	Example	Effect
Hydrocarbons	Crude oil	Inhibition of synthesis. Alterations in membrane lipid composition
Fungicides	Carbamates	Affect fatty acid patterns in oomycetes. Action on plants negligible or not known
Insecticides	Dimethoate	Inhibition of plant lipid synthesis. Changed patterns of fatty acids made
Herbicides	Substituted pyridazinones	Inhibit fatty acid desaturations, e.g. Δ15-desaturase
	Diflufenican	Inhibits fatty acid synthase
	Cyclohexanediones	Inhibit acetyl-CoA carboxylase
	Aryloxyphenoxy propionates	Inhibit acetyl-CoA carboxylase
	Thiocarbamates	Inhibit fatty acid elongation

in yield and also poorer quality oil in many cases. The olive fruit fly, olive moth and olive scale are the main culprits (Claridge & Walton, 1992). Control often uses organophosphorous insecticides or pyrethum and derivatives. We have examined such compounds for possible effects on lipid accumulation and oil quality in ripening fruits or olive tissue cultures. Commercial preparations of the organophosphorous insecticide dimethoate had significant effects on olive fruit development and oil production (de la Vega, Harwood & Sanchez, 1992) but these were later shown not to be due to dimethoate. On the other hand, dimethoate did have significant effects on lipid metabolism when used at quite high concentrations (>10 μM; Rutter *et al.*, 1994). The question, of course, is whether such concentrations of dimethoate are found on field-grown material and, more particularly, whether they persist long enough to affect lipid accumulation significantly. Since effects on tissue culture lipid metabolism are also seen with 10 μM concentrations of natural and synthetic pyrethroid insecticides (Rutter *et al.*, 1994), these results suggest that pesticides should be evaluated not only for effectiveness against target organisms but also for any side effects on the crops in which they may be used.

Herbicides

So far as herbicides are concerned, all the compounds which have been shown to alter lipid metabolism do so at the level of fatty acid biosynthesis. Three main classes of compounds have been studied: graminicides, substituted pyridazinones and thiocarbamates. These act, respectively, on acetyl-CoA carboxylase, fatty acid (especially Δ15) desaturation and elongation (see Fig. 4). Earlier work on these compounds has been reviewed (Harwood, 1991*a*, *b*), and there has been little more recent work with the pyridazinones apart from their use in elucidating pathways for polyunsaturated fatty acid formation in algae (Cohen, Norman & Heimer, 1993).

Thiocarbamates have been shown to act on fatty acid elongases (Harwood, 1991*a*). While knowledge of these target enzymes has increased considerably in recent years (see Cassagne *et al.*, 1994; Chapter 8), there have also been advances in our understanding of the thiocarbamates. The latter pesticides are believed to be oxidised to their sulphoxides which are the active forms of the herbicides (Hathaway, 1989). Moreover, such a metabolism explains the action of safening chemicals both on plant physiology and on fatty acid elongation (Abdulnaja & Harwood, 1991). Indeed, if thiocarbamates are metabolised by a cytochrome P_{450} mixed-function oxidase (Hathaway, 1989) or by other enzymes such as the peroxygenase (Blée, 1991) to sulphoxide derivatives, then the latter would be expected to be potent against elongase systems. For some thiocarbamates, stable sulphoxide derivatives can be chemically synthesised and, therefore, are available for testing. One such compound is Tillam (Pebulate; s-propylbutylethylthiocarbamate) and we have shown clearly that this is much more effective *in vitro* against pea elongases for stearoyl-CoA and arachidoyl-CoA (Barrett & Harwood, 1993, 1994, 1995). Indeed, when used in whole tissues (germinating pea seeds, young barley leaves), the sulphoxide derivative showed I_{50} values to be about two orders of magnitude less than for the parent herbicide (Barrett & Harwood, 1993).

The graminicides are an important class of herbicide which act by inhibiting acetyl-CoA carboxylase, which catalyses the first committed step in fatty acid and lipid synthesis. (The influence of acetyl-CoA carboxylase in controlling carbon flux to lipids in leaf tissues was pointed out earlier in this chapter and provides a good reason for the effectiveness of the graminicides.) There are two main classes of graminicides: the aryloxyphenoxypropionates (AOPPs) and the cyclohexanediones (CHDs) (Fig. 5), although some other diverse structures are also

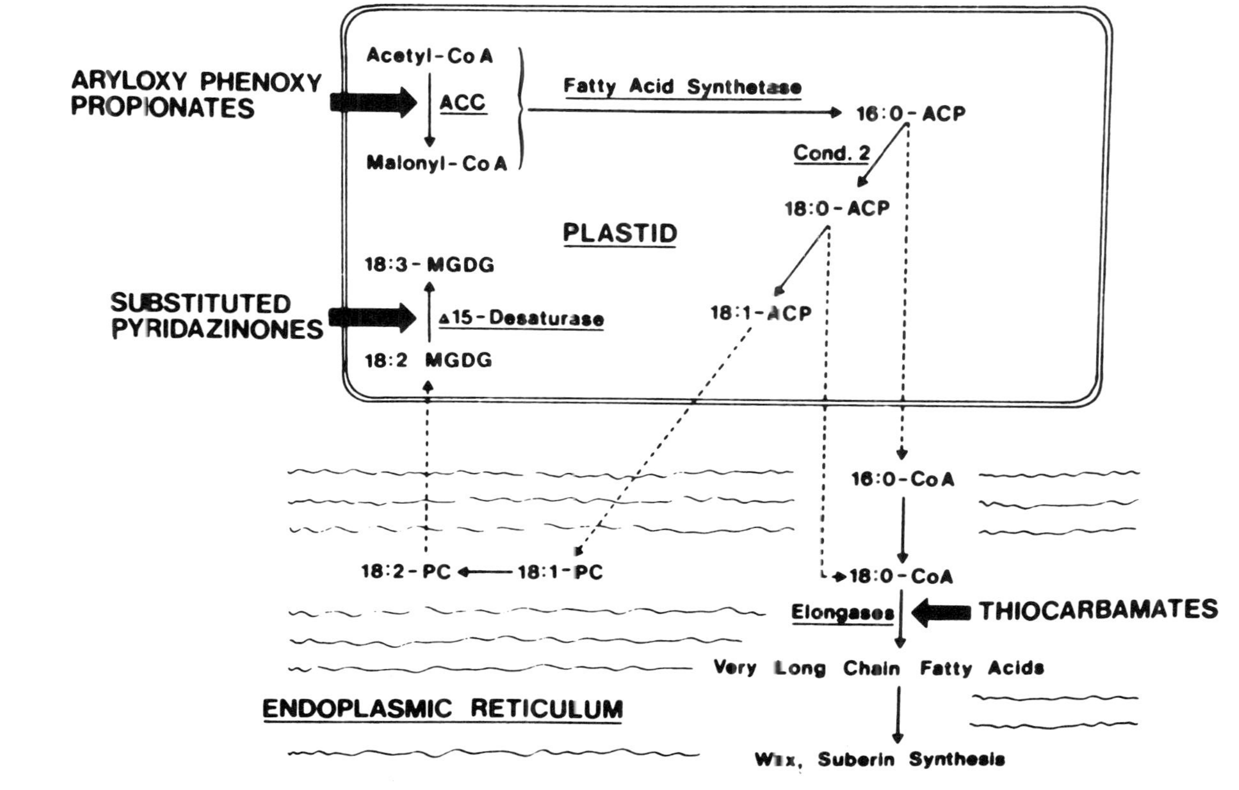

Fig. 4. Site of action of different herbicide classes on plant fatty acid biosynthesis. (Taken from Harwood, 1991*a* with permission.)

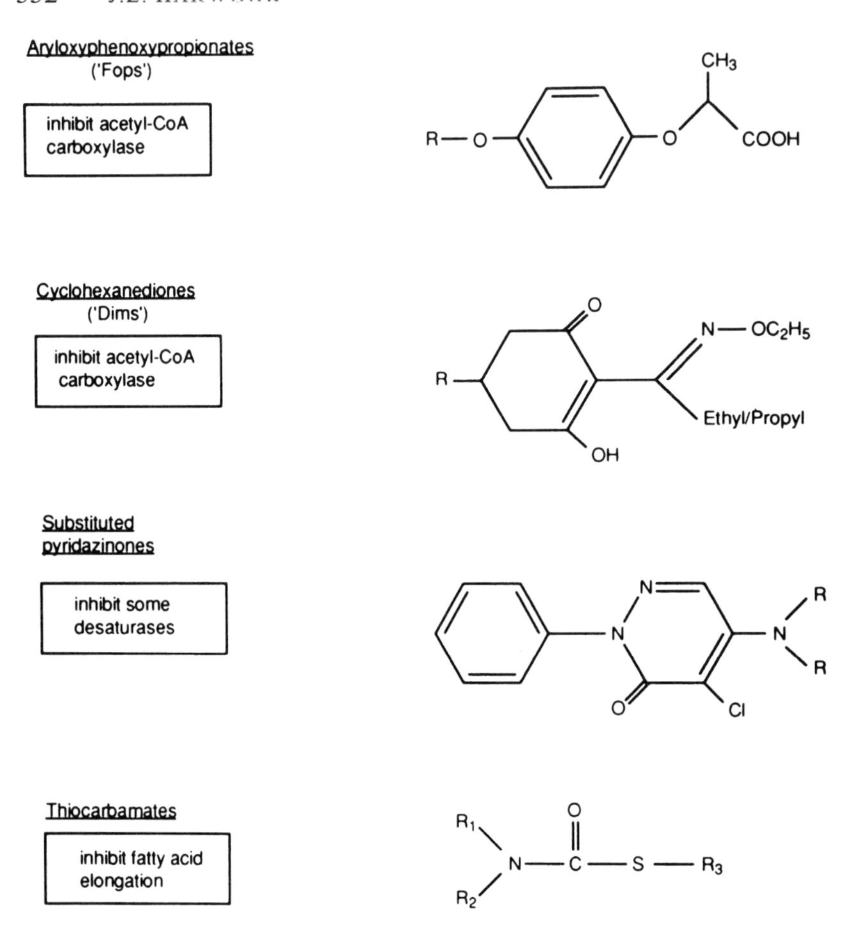

Fig. 5. Some representative examples of different herbicides affecting plant fatty acid (lipid) synthesis. For the aryloxyphenoxypropionates, the R group is usually a benzene or pyridine ring with halogen substitutes. For pyridazones the R group can represent hydrogens or methyl groups and the benzene ring may be substituted. In thiocarbamates R_1 and R_2 are usually alkyl side chains (ethyl, propyl, isopropyl) and R_3 may be ethyl, propyl or a chloride-substituted alkyl group. (Taken from Harwood, 1997 with permission.)

effective (Harwood, 1996). There are different isoforms of acetyl-CoA carboxylase concentrated in the mesophyll chloroplast stroma and the epidermal cytosol, respectively (see Chapter 1). It is only the chloroplast isoform of grasses (not even that of all monocotyledons) which is appreciably sensitive to graminicides. The AOPPs and CHDs show linear, reversible inhibition against different acetyl-CoA carboxylases from susceptible grasses (Burton *et al.*, 1987; Secor, Cseke & Owen, 1989). From various types of studies, it was clear that it was the carboxyltransferase partial reaction which was being inhibited (Harwood, 1996).

In order to understand something about the molecular mechanism by which graminicides are selective in their inhibition of acetyl-CoA carboxylases, several laboratories have conducted research. The first obvious question to answer is why dicotyledons are not sensitive. This was easily solved when it became clear that dicotyledons contain a distinctly different type of acetyl-CoA carboxylase in their chloroplasts when compared with grasses. Thus, the enzyme in chloroplasts from grasses is a multifunctional protein of >200 kDa (Herbert *et al.*, 1997*a*) whereas that in dicotyledons is a multienzyme complex (Alban, Baldet & Douce, 1994; Sasaki *et al.*, 1993) containing four separate proteins.

But what of the multifunctional forms which can differ so much in their sensitivity? In order to try to understand the basis for the varying sensitivity of these enzymes, we have studied the characteristics of acetyl-CoA carboxylase from maize (Herbert *et al.*, 1996) and from resistant grasses such as *Poa annua* (Herbert *et al.*, 1997*b*). In maize the chloroplast isoform is very sensitive to inhibition by graminicides, but the cytosolic isoform is about a thousand times less sensitive. Both isoforms have similar molecular masses, native conformations and Michaelis' constants for the three substrates. Moreover, the reaction characteristics were close to an ordered mechanism in both cases (Herbert *et al.*, 1996). However, there were differences in the binding characteristics of the two isoforms towards aryloxyphenoxypropionates such as quizalofop or fluazifop. The major, chloroplast, isoform showed no co-operativity in the binding of the herbicide to the two subunits of the native enzyme. On the other hand, the minor, cytosolic, isoform (which is relatively insensitive to graminicides) showed strong positive co-operativity for herbicide binding (Herbert *et al.*, 1996). This meant that binding of a graminicide to one half of the native dimer changes its ability to bind to the second site. Interestingly, this co-operativity of binding was also associated with less activity against propionyl-CoA which is an alternative, through poorer, substrate for acetyl-CoA carboxylase (Herbert *et al.*, 1995). Thus we proposed that co-operativity of herbicide binding was an important observation and was indicative of insensitivity of the enzyme towards graminicides.

Supporting evidence for this proposal has come from work with resistant grasses. Thus, *Poa annua* contains an insensitive acetyl-CoA carboxylase (which is the reason for the weed's resistance to graminicides) and partly purified preparations show co-operativity of quizalofop or fluazifop binding (Herbert *et al.*, 1997*b*). Furthermore, when sensitive or resistant biotypes of *Lolium multiflorum* were examined, only the enzyme from the resistant biotype showed significant co-operativity (Evenson, Gronwald & Wyse, 1994).

Although the graminicides have been very successful commercially, acquired resistance is becoming an increasing problem. Because selectivity relies on subtle properties of the target site protein, it is not surprising that most of the acquired resistance seems to be due to altered properties of the acetyl-CoA carboxylase (see Tardiff & Powles, 1993). A typical example of resistance has arisen with blackgrass (*Alopecurus myosuroides*) and we have studied the acetyl-CoA carboxylase activity from susceptible and resistant populations (Price *et al.*, 1997). The enzyme from resistant populations is 100–1000 times less sensitive than acetyl-CoA carboxylase from sensitive populations towards different graminicides. Further studies, which are under way, are clearly needed in order to see whether co-operativity of herbicide binding is significant in the modified enzyme. These experiments may be useful in identifying possible structural modifications of graminicides, which will make them more effective herbicides as well as ones less prone to resistance problems.

The ability to produce effective selective pesticides whose target is such an ubiquitous and important enzyme as acetyl-CoA carboxylase is surprising. It points the way to the possible development of other useful compounds which will act on central metabolic pathways. The episode also emphasises that we still have much to learn about the characteristics and subtleties of plant lipid metabolism and the enzymes responsible!

References

Abdulnaja, K.O. & Harwood, J.L. (1991). Effect of a safener towards thiocarbamates on plant lipid metabolism. *Zeitschrift fur Naturforschung*, **46c**, 931–3.

Alban, C., Baldet, P. & Douce, R. (1994). Localization and characterization of two structurally different forms of acetyl-CoA carboxylase in young pea leaves, of which one is sensitive to aryloxyphenoxypropionates. *The Biochemical Journal*, **300**, 557–65.

Allen, E.J., Morgan, D.G. & Ridgman, W.J. (1971). *Journal of Agricultural Sciences, Cambridge*, **77**, 339–41.

Anderson, J. (1986). Photoregulation of the composition, function and structure of thylakoid membranes. *Annual Review of Plant Physiology*, **37**, 93–136.

Avery, S.V, Rutter, A.J., Harwood, J.L. & Lloyd, D. (1996). Oxygen-dependent low-temperature Δ12-desaturase induction and alteration of fatty acid composition in *Acanthamoeba castellanii. Microbiology*, **142**, 2213–21.

Barnes, P.J. (ed.) (1983). *Lipids in Cereal Technology*. London: Academic Press.

Barrett, P.B. & Harwood, J.L. (1993). The basis of thiocarbamate action on surface lipid synthesis in plants. *Brighton Crop Protection Conference, Weeds 1993*, 183–8.

Barrett, P.B. & Harwood, J.L. (1994). Plant fatty acid elongation: sensitivity to thiocarbamate herbicides and their sulphoxides. *Biochemical Society Transactions*, **22**, 260S.

Barrett, P.B. & Harwood, J.L. (1995). Thiocarbamate action on very long chain fatty acid synthesis in plants. In *Plant Lipid Metabolism*, ed. J-C. Kader & P. Mazliak, pp. 115–17. Dordrecht: Kluwer.

Bartels, C.T., James, A.T. & Nichols, B.W. (1967). Metabolism of *trans*-3-hexadecenoic acid by *Chlorella vulgaris* and lettuce leaf. *European Journal of Biochemistry*, **3**, 7–10.

Beerhues, L.l, Robenek, H. & Wiermann, R. (1988). Chalcone synthesis from spinach. II. Immunofluorescence and immunogold localisation. *Planta*, **173**, 544–53.

Bettaieb, L., Gharsalli, M. & Cherif, A. (1980). The effect of sodium chloride and calcium sulphate on the lipid composition of sunflower lipids. In *Biogenesis and Function of Plant Lipids*, ed. P. Mazliak, P. Benveniste, C. Costes & R. Douce, pp. 243–7. Amsterdam: Elsevier.

Bharti, N. & Singh, R.P. (1994). Antagonistic effect of sodium chloride to differential heavy metal toxicity regarding biomass accumulation and nitrate assimilation in *Sesamum indicum* seedlings. *Phytochemistry*, **33**, 1157–61.

Bishop, D.G. (1986). Chilling sensitivity in higher plants. The role of phosphatidylglycerol. *Plant Cell and Environment*, **9**, 613–16.

Blée, E. (1991). Effect of the safener dichlormid on maize peroxygenase and lipoxygenase. *Zeitschrift fur Naturforschung*, **46c**, 920–5.

Bligny, R. & Douce, R. (1980). A precise localisation of cardiolipin in plant cells. *Biochimica et Biophysica Acta*, **617**, 254–63.

Boardman, N.K. (1977). Comparative photosynthesis of sun and shade plants. *Annual Reviews of Plant Physiology*, **28**, 355–77.

Bolton, P., Wharfe, J. & Harwood, J.L. (1978). The lipid composition of a barley mutant lacking chlorophyll b. *The Biochemical Journal*, **174**, 67–72.

Boyer, J.S. & Bowen, B.I. (1970). Inhibition of oxygen evolution in chloroplasts isolated from leaves with low water potential. *Plant Physiology*, **45**, 612–15.

Bradford, K.J. & Hsiao, T.C. (1982). Physiological responses to water stress. In *Encyclopedia of Plant Physiology*, New series vol. 12B, ed. O.L. Lang, P.S. Nobel, C.B. Osmund & H. Zieger, pp. 263–324. Berlin: Springer-Verlag.

Bray, E.A. (1997). Plant responses to water deficit. *Trends in Plant Sciences*, **2**, 48–54.

Browse, J. & Somerville, C. (1991). Glycerolipid synthesis: biochemistry and regulation. *Annual Review of Plant Physiology and Plant Molecular Biology*, **42**, 467–506.

Browse, J., Miguel, M., McConn, W. & Wu, J. (1994). *Arabidopsis* mutants and genetic approaches to the control of lipid composition. In *Temperature Adaptation of Biological Membranes*, ed. A.R. Cossins, pp. 141–54. London: Portland Press.

Burton, J., Gronwald, J., Somers, D., Connelly, J., Gengenbach, B. & Wyse, D. (1987). Inhibition of plant acetyl-CoA carboxylase by the herbicides sethoxydim and haloxyfop. *Biochemical and Biophysical Research Communications*, **148**, 1039–44.

Carlsson, A.S., Hellgren, L.I., Sellden, G. & Sandelius, A.S. (1994). Effects of moderately enhanced levels of ozone on the acyl lipid composition of leaves of garden pea. *Physiologia Plantarum*, **91**, 754–62.

Cassagne, C., Lessire, R., Bessoule, J.J., Moreau, P., Creach, A., Schneider, F. & Sturbois, B. (1994). Biosynthesis of very long chain fatty acids in higher plants. *Progress in Lipid Research*, **33**, 55–69.

Claridge, M.F. & Walton, M. (1992). The European olive and its pests – management strategies. In *Research Collaboration in European IPM Systems*, ed. P.T. Haskell, pp. 3–12. Farnham: British Crop Protection Council.

Cohen, Z., Norman, H.A. & Heimer, Y.M. (1993). Evaluating the potential of substituted pyridazinones for inducing polyunsaturated fatty acid overproduction in algae. *Phytochemistry*, **32**, 259–63.

Douglas, T.J. & Paleg, L.G. (1981). Lipid composition of *Zea mays* seedlings and water stress-induced changes. *Journal of Experimental Botany*, **32**, 499–508.

Ellouze, M., Gharsalli, M. & Cherif, A. (1982). The effect of sodium chloride on the biosynthesis of unsaturated fatty acids of sunflower plants. In *Biochemistry and Metabolism of Plant Lipids*, ed. J.F.G.M. Wintermans & P.J.C. Kuiper, pp. 419–22. Amsterdam: Elsevier.

Erdei, L., Stuiver, C.E.E. & Kuiper, P.J.C. (1980). The effect of salinity on lipid composition and on activity of Ca^{+} and Mg^{2+} – stimulated ATPase in salt-sensitive and salt-tolerant *Plantago* species. *Physiological Plant Pathology*, **49**, 315–19.

Evenson, K.J., Gronwald, J.W. & Wyse, D.L. (1994). Purification and characterisation of ACCases from diclofop-resistant and -susceptible *Lolium multiflorum*. *Plant Physiology*, **105**, 671–80.

Foley, A.A., Rowlands, C.C., Evans, J.C. & Harwood, J.L. (1988). Electron paramagnetic studies on copper phaeophytin in the presence and absence of phosphoglycerides. *Journal of Inorganic Biochemistry*, **32**, 125–33.

Fong, F. & Heath, R.L. (1981). Lipid content in the primary leaf of bean after ozone fumigation. *Zeitschrift fur Pflanzenphysiologie*, **104**, 109–15.

Givan, C.V. & Stumpf, P.F. (1971). Some factors regulating fatty acid synthesis by isolated spinach chloroplasts. *Plant Physiology*, **57**, 510–15.

Gounaris, K., Barber, J. & Harwood, J.L. (1986). The thylakoid membranes of higher plant chloroplasts. *The Biochemical Journal*, **237**, 313–26.

Gray, G.R., Savitch, L.V., Ivanov, A.G. & Huner, N.P.A. (1996). Photosystem II excitation pressure and development of resistance to photoinhibition. *Plant Physiology*, **110**, 61–71.

Gray, G.R., Krol, M., Khan, M.U., Willimas, J.P. & Huner, N.P.A. (1997). Growth temperature and irradiance modulate *trans*-Δ3-hexadecenoic acid content and photosynthetic light-harvesting complex organisation. In *Physiology, Biochemistry and Molecular Biology of Plant Lipids*, ed. J.P. Williams, M.U. Khan & N.W. Lem, pp. 206–8. Dordrecht: Kluwer.

Gurr, M.I. & Harwood, J.L. (1991). *Lipid Biochemistry*, 4th edn, London: Chapman and Hall.

Harris, P. & James, A.T. (1969). Effect of low temperature on fatty acid biosynthesis in seeds. *Biochimica et Biophysica Acta*, **187**, 13–18.

Harwood, J.L. (1980). Plant acyl lipids: structure, distribution and analysis. In *The Biochemistry of Plants*, ed. P.K. Stumpf, Vol. 4, pp. 1–55. New York: Academic Press.

Harwood, J.L. (1983). Adaptive changes in the lipids of higher plant membranes. *Biochemical Society Transactions*, **11**, 343–6.

Harwood, J.L. (1984). Effects of the environment on the acyl lipids of algae and higher plants. In *Structure, Function and Metabolism of Plant Lipids*, ed. P-A. Siegenthaler & W. Eichenberger, pp. 543–50. Amsterdam: Elsevier.

Harwood, J.L. (1985). Phosphoglycerides in mitochondrial membranes. In *Methods in Enzymology*, ed. L. Packer & R. Douce, vol. 148, pp. 475–85. New York: Academic Press.

Harwood, J.L. (1988). Fatty acid metabolism. *Annual Review of Plant Physiology and Plant Molecular Biology*, **39**, 101–38.

Harwood, J.L. (1989). Lipid metabolism. *CRC Critical Reviews in Plant Science*, **8**, 1–43.

Harwood, J.L. (1991*a*). Lipid synthesis. In *Target Sites for Herbicide Action*, ed. R.C. Kirkwood, pp. 57–94. New York: Plenum.

Harwood, J.L. (1991*b*) Herbicides affecting chloroplast lipid synthesis. In *Herbicides*, ed. N.R. Baker & M.P. Percival, pp. 209–46. Amsterdam: Elsevier.

Harwood, J.L. (1994*a*). Environmental factors which can alter lipid metabolism. *Progress in Lipid Research*, **33**, 193–202.

Harwood, J.L. (1994*b*). Lipid metabolism. In *The Lipid Handbook*, ed. F.D. Gunstone, J.L. Harwood & F.B. Padley, pp. 605–64. London: Chapman and Hall.

Harwood, J.L. (1995). Recent environmental concerns and lipid metabolism. In *Plant Lipid Metabolism*, ed. J-C. Kader & P. Mazliak, pp. 361–8. Dordrecht: Kluwer.

Harwood, J.L. (1996). Recent advances in the biosynthesis of plant fatty acids. *Biochimica et Biophysica Acta*, **1301**, 7–56.

Harwood, J.L. (1997).Plant lipid metabolism. In *Plant Biochemistry*, ed. P.M. Dey & J.B. Harborne, pp. 237–72. London: Academic Press.

Harwood, J.L. (1998). Involvement of chloroplast lipids in the reaction of plants submitted to stress. In *Lipids in Photosynthesis: Structure, Function and Genetics*, ed. P-A. Siegenthaler & N.Murata, in press. Dordrecht: Kluwer.

Harwood, J.L. & James, A.T. (1974). Metabolism of *trans*-3-hexadecenoic acid in broad bean. *European Journal of Biochemistry*, **50**, 325–34.

Harwood, J.L. & Jones, A.L. (1989). Lipid metabolism in algae. *Advances in Botanical Research*, **16**, 1–53.

Harwood J.L., Abdulnaja, A.O., Barrett, P.B., Herbert, D., Rutter, A.J. & Williams, M. (1992). Changes in lipid metabolism caused by environmental factors. In *Metabolism, Structure and Utilization of Plant Lipids*, ed. A. Cherif, D.B. Miled-Daoud, B. Marzouk, A. Smaoui & M. Zarrouk, pp. 319–26. Tunis: Centre National Pedagogique.

Harwood, J.L., Jones, A.L., Perry, H.J., Rutter, A.J., Smith, K.L. & Williams, M. (1994). Changes in plant lipids during temperature adaptation. In *Temperature Adaptation of Biological Membranes*, ed. A.R. Cossins, pp. 107–18. London: Portland Press.

Hathaway, D.E. (1989). *Molecular Mechanism of Herbicide Selectivity*. Oxford: Oxford University Press.

Heath, R.L. (1980). Initial events in injury to plants by air pollution. *Annual Review of Plant Physiology*, **31**, 395–431.

Heath, R.L. (1984). Air pollutant effects on biochemicals derived from metabolism: organic, fatty and amino acids. In *Air Pollutants and Plant Metabolism*, ed. M.J. Koziol & F.R. Whatley, pp. 275–90. London: Butterworth.

Heinrich, R. & Rapoport, T.A. (1974). A linear steady-state treatment of enzymatic chains. *European Journal of Biochemistry*, **42**, 97–105.

Herbert, D., Harwood, J.L., Cole, D.J. & Pallett, K.E. (1995). Characteristics of aryloxyphenoxypropionate herbicide interactions with acetyl-CoA carboxylases of different graminicide sensitivities. *Proceedings of the Brighton Crop Protection Conference*, **1**, 387–92.

Herbert, D., Price, L.J., Alban, C., Dehaye, L., Job, D., Cole, D.J., Pallett, K.E. & Harwood, J.L. (1996). Kinetic studies on two isoforms of acetyl-CoA carboxylase from maize leaves. *The Biochemical Journal*, **318**, 997–1006.

Herbert, D., Walker, K.A., Price, L.J., Cole, D.J., Pallett, K.E., Ridley, S.M. & Harwood, J.L. (1997*a*). Acetyl-CoA carboxylase – a graminicide target site. *Pesticide Science*, **50**, 67–71.

Herbert, D., Cole, D.J., Pallett, K.E. & Harwood, J.L. (1997*b*). Graminicide-binding by acetyl-CoA carboxylase from *Poa annua* leaves. *Phytochemistry*, **44**, 399–405.

Hitchcock, C. & Nichols, B.W. (1971). *Plant Lipid Biochemistry*. London: Academic Press.

Horvath, I., Torok, Z., Vigh, L. & Kates, M. (1991). Lipid hydrogenation induces elevated 18 : 1-CoA desaturase activity in *Candida lipolytica* microsomes. *Biochimica et Biophysica Acta*, **1085**, 126–30.

Huner, N.P.A., Kool, M., Williams, J.P., Maissan, E., Low, P., Roberts, D. & Thompson, J.E. (1987). A low temperature induced decrease in 3-*trans*-hexadecenoic acid content and its influence on LHC II organisation. *Plant Physiology*, **84**, 12–18.

Ingram, J. & Bartels, D. (1996). The molecular basis of dehydration tolerance in plants. *Annual Review of Plant Physiology and Plant Molecular Biology*, **47**, 377–403.

Ishizakini-Nishizawa, D., Fujii, T., Azuma, M., Sekiguchi, K. & Murata, N. (1996). Low-temperature resistance in plants is significantly enhanced by a non-specific cyanobacterial desaturase. *Nature Biotechnology*, **14**, 1003–6.

Jones, A.L. & Harwood, J.L. (1992). Lipid composition of the brown algae *Fucus vesiculosus* and *Ascophyllum nodosum*. *Phytochemistry*, **31**, 3397–403.

Jones, A.L., Lloyd, D. & Harwood, J.L. (1993). Rapid induction of microsomal Δ12 (ω6)-desaturase activity in chilled *Acanthamoeba castellanii*. *The Biochemical Journal*, **296**, 183–8.

Joyard, J. & Douce, R. (1987). Galactolipid synthesis. In *The Biochemistry of Plants*, ed. P.K. Stumpf, vol. 9, pp. 215–74. New York: Academic Press.

Kacser, H.B. & Burns, J.A. (1973). The control of flux. *Symposium of the Society for Experimental Biology*, **27**, 65–104.

Karydis, M. (1980). Uptake of hydrocarbons by the marine diatom *Cyclotella cryptica*. *Microbial Ecology*, **5**, 287–93.

Karydis, M. & Fogg, G.E. (1980). Physiological effects of hydrocarbons on the marine diatom, *Cyclotella cryptica*. *Microbial Ecology*, **6**, 281–90.

Kenrick, J.R. & Bishop, D.G. (1986). The fatty acid composition of phosphatidylglycerol and sulphoquinovosyldiacylglycerol of higher plants in relation to chilling sensitivity. *Plant Physiology*, **81**, 946–9.

Krol, K., Huner, N.P.A., Williams, J.P. & Maissan, E. (1988). Chloroplast biogenesis at cold hardening temperatures. Kinetics of *trans*-3-hexadecenoic acid accumulation and the assembly of LHC II. *Photosynthesis Research*, **15**, 115–32.

Krupa, Z., Huner, N.P.A., Williams, J.P., Maissan, E. & James, D.R. (1987). Development at cold hardening temperatures. The structure and composition of purified rye light harvesting complex II. *Plant Physiology*, **84**, 19–24.

Krupa, Z., Williams, J.P., Khan, M.U. & Huner, N.P.A. (1992). The role of acyl lipids in reconstitution of light-depleted light-harvesting complex II from cold-hardened and non-hardened rye. *Plant Physiology*, **100**, 931–38.

Kuiper, P.J.C. (1980). Lipid metabolism as a factor in environmental adaptation. In *Biogenesis and Function of Plant Lipids*, ed. P. Mazliak, P. Benveniste, C. Costes & R. Douce, pp. 169–76. Amsterdam: Elsevier.

Kuiper, P.J.C. (1985). Environmental changes and lipid metabolism in higher plants. *Plant Physiology*, **64**, 118–22.

Landry, L.G., Chapple, C.C.S. & Last, R.L. (1995). *Arabidopsis* mutants lacking phenolic sunscreens exhibit enhanced UV-B injury and oxidative damage. *Plant Physiology*, **109**, 1159–66.

Leprince, O., Deltour, R., Thorpe, P.C., Atherton, N.M. & Hendry, G.A.F. (1990). The role of free radicals and radical processing systems in loss of desiccation tolerance in germinating maize. *New Phytologist*, **116**, 573–80.

Lesham, Y.Y. & Haramaty, E. (1996). The characterisation of contrasting effects of the nitric oxide free radical in vegetative stress and senescence of *Pisum sativum* Linn. foliage. *Journal of Plant Physiology*, **148**, 258–63.

Lesham, Y.Y., Haramaty, E., Malik, Z. & Sofer, Y. (1997). Chloroplast membrane lipids as possible primary targets for nitric oxide-mediated induction of chlorophyll fluorescence in *Pisum sativum* (*Argentum* mutant). In *Physiology, Biochemistry and Molecular Biology of Plant Lipids*, ed. J.P. Williams, M.U. Khan & N.W. Lem, pp. 157–9. Dordrecht: Kluwer.

Levitt, J. (1980). Water, radiation, salt and other stress. *Responses of Plants to Environmental Stresses*, 2nd edn, vol. II. New York: Academic Press.

Liebe, B. & Fock, H.P. (1992). Growth and adaptation of the green alga *Chlamydomonas reinhardtii* on diesel exhaust particle extracts. *Journal of General Microbiology*, **138**, 973–8.

Liljenberg, C. (1992). The effects of water deficit stress on plant membrane lipids. *Progress in Lipid Research*, **31**, 335–43.

Lois, R. & Buchanan, B.B. (1994). Severe sensitivity to ultraviolet radiation in an *Arabidopsis* mutant deficient in flavonoid accumulation. *Planta*, **194**, 504–9.

Lynch, D.V. & Thompson, G.A. (1982). Low temperature induced alterations in the chloroplast and microsomal membranes of *Dunaliella salina*. *Plant Physiology*, **69**, 1369–75.

Lynch, D., Norman, H.A. & Thompson, G.A. (1984). Changes in membrane lipid molecular species during acclimation to low temperature by *Dunaliella salina*. In *Structure, Function and Metabolism of Plant Lipids*, ed. P-A. Siegenthaler & W. Eichenberger, pp. 567–70. Amsterdam: Elsevier.

McCourt, P., Browse, J., Watson, J., Arntzen, C.J. & Somerville, C.R. (1985). Analysis of photosynthetic antenna function in a mutant of *Arabidopsis thaliana* lacking *trans*-hexadecenoic acid. *Plant Physiology*, **78**, 853–8.

Martin, B.A., Schoper, J.B. & Rinne, R.W. (1986). Changes in soybean glycerolipids in response to water stress. *Plant Physiology*, **81**, 798–801.

Millner, P., Chapman, D.J., Mitchell, R.A.C. & Barber, J. (1984). Modulation of membrane fluidity by chloroplast thylakoid lipid to protein ratio. In *Structure, Function and Metabolism of Plant Lipids*, ed. P-A. Siegenthaler & W. Eichenberger, pp. 433–6. Amsterdam: Elsevier.

Mohanty, P. & Boyer, J.S. (1976). Chloroplast response to low leaf potentials. *Plant Physiology*, **57**, 704–9.

Morales-Loo, M.R. & Goutx, M. (1990). Effects of water-soluble fraction of the Mexican crude oil 'Isthmus Cactus' on growth, cellular contents of chlorophyll a and lipid composition of planktonic microalgae. *Marine Biology*, **104**, 503–9.

Morrison, W.R. (1983). Acyl lipids in cereals. In *Lipids in Cereal Technology*, ed. P.J. Barnes, pp. 11–32. London: Academic Press.

Murata, N. (1983). Molecular species composition of phosphatidylglycerols from chilling-sensitive and chilling resistant plants. *Plant and Cell Physiology*, **24**, 81–6.

Murata, N. & Wada, H. (1995). Acyl lipid desaturases and their importance in the tolerance and acclimatisation to cold of cyanobacteria. *The Biochemical Journal*, **308**, 1–8.

Murata, N., Ishizaki-Nishizawa, O., Higashi, S., Hayashi, H., Tasaka, Y. & Nishida, I. (1992). Genetically engineered alteration in the chilling sensitivity of plants. *Nature*, **356**, 710–13.

Najine, F., Marzouk, B. & Cherif, A. (1995). Sodium chloride effect on the evolution of fatty acid composition in developing rape. In

Plant Lipid Metabolism, ed. J-C. Kader & P. Mazliak, pp. 435–7. Dordrecht: Kluwer.

Navari-Izzo, F., Vangioni, N. & Quartacci, M.F. (1990). Lipids of soybean and sunflower seedlings grown under drought conditions. *Phytochemistry*, **29**, 2119–23.

Nikolau, B.J. & Hawke, C. (1984). Purification and characterisation of maize leaf acetyl-CoA carboxylase. *Archives of Biochemistry and Biophysics*, **228**, 86–96.

Nishida, I. & Murata, N. (1996). Chilling sensitivity in plants and cyanobacteria: the crucial contribution of membrane lipids. *Annual Review of Plant Physiology and Plant Molecular Biology*, **47**, 541–68.

Nouchi, I. & Toyama, S. (1988). Effects of ozone and peroxyacetyl nitrate on polar lipids and fatty acids in leaves of morning glory and kidney bean. *Plant Physiology*, **87**, 638–46.

Ohnishi, M. & Thompson, G.A. (1991). Biosynthesis of the unique *trans*-3-hexadecenoic acid component of chloroplast phosphatidylglycerol: evidence concerning its site and mechanism of formation. *Archives of Biochemistry and Biophysics*, **288**, 591–9.

Okuyuma, H., Saito, M., Joshi, V.C., Gunsberg, S. & Wakil, S.J. (1979). Regulation by temperature of the chain length of fatty acids in yeast. *Journal of Biological Chemistry*, **254**, 12 281–4.

Oursel, A. (1979). Calcium inhibition of phospholipid biosynthesis in a calcicolous plant (*Vicia faba*); comparison with a calcifuge one (*Lupinus luteus*). In *Advances in the Biochemistry and Physiology of Plant Lipids*, ed. L-A. Appelqvist & C. Liljenberg, pp. 421–6. Amsterdam: Elsevier.

Page, R.A., Okada, S. & Harwood, J.L. (1994). Acetyl-CoA carboxylase exerts strong flux control over lipid synthesis in plants. *Biochimica et Biophysica Acta*, **1210**, 369–72.

Percival, M.P., Wharfe, J., Bolton, P., Davies, A.O, Jeffcoat, R., James, A.T. & Harwood, J.L. (1979). Has *trans*-3-hexadecenoic acid a role in granal stacking? In *Advances in the Biochemistry and Physiology of Plant Lipids*, ed. L.A. Appelqvist & C. Liljenberg, pp. 219–24. Amsterdam: Elsevier.

Petkov, G.D., Furnadzieva, S.T. & Popov, S.S. (1992). Petrol-induced changes in the lipid and sterol composition of three microalgae. *Phytochemistry*, **31**, 1165–6.

Pettitt, T.R., Jones, A.L. & Harwood, J.L. (1989). Lipids of the marine red algae *Chondrus crispus* and *Polysiphonia lanosa*. *Phytochemistry*, **28**, 399–405.

Piazza, G.J. & Moreau, R.A. (1987). Lipid metabolism in potato leaf discs: effect of calmodulin antagonists. In *Metabolism, Structure and Function of Plant Lipids*, ed. P.K. Stumpf, J.B. Mudd & W.D. Nes, pp. 321–4. New York: Plenum.

Picaud, A., Créach, A. & Trémoliéres, A. (1990). Light-stimulated

fatty acid synthesis in *Chlamydomonas* whole cells. In *Plant Lipid Biochemistry, Structure and Utilisation*, ed. P.J. Quinn & J.L. Harwood, pp. 393–5. London: Portland Press.
Pohl, P. & Zurheide, F. (1979). Fatty acids and lipids of marine algae and the control of their biosynthesis by environmental factors. In *Marine Algae in Pharmaceutical Science*, ed. H.A. Hoppe, T. Levring & Y. Tanaka, pp. 473–523. Berlin: Walter de Gruyter.
Post-Beittenmiller, D., Jaworski, J.G. & Ohlrogge, J.B. (1991). *In vivo* pools of free and acylated acyl carrier proteins in spinach. *Journal of Biological Chemistry*, **266**, 1858–65.
Post-Beittenmiller, D., Roughan, P.G. & Ohlrogge, J.B. (1992). Regulation of plant fatty acid biosynthesis. Analysis of acyl-CoA and acyl-ACP substrate pools in spinach and pea chloroplasts. *Plant Physiology*, **100**, 923–30.
Price, A.H. & Hendry, G.A.F. (1991). Iron-catalysed oxygen radical formation and its possible contribution to drought damage in nine native grasses and three cereals. *Plant Cell and Environment*, **14**, 477–84.
Price, A.H., Atherton, N. & Hendry, G.A.F. (1989). Plants under drought stress generate active oxygen. *Free Radical Research Communication*, **8**, 61–6.
Price, L.J., Herbert, D., Harwood, J.L., Cole, D.J., Pallett, K.E. & Moss, S.R. (1997). Herbicide-resistant isoforms of acetyl-CoA carboxylases. *Journal of Experimental Botany*, **48**, suppl., 72.
Quartacci, M.F. & Navari-Izzo, F. (1992). Water stress and free radical mediated changes in sunflower seedlings. *Journal of Plant Physiology*, **139**, 621–3.
Quinn, P. (1977). The role of membrane lipids in the arrangement of complexes in photosynthetic membranes. In *Physiology, Biochemistry and Molecular Biology of Plant Lipids*, ed. J.P. Williams, M.U. Khan & N.W. Lem, pp. 148–50. Dordrecht: Kluwer.
Quinn, P.J. & Williams, W.P. (1983). The structural role of lipids in photosynthetic membranes. *Biochimica et Biophysica Acta*, **737**, 223–66.
Riches, C., Rolph, C., Greenway, D. & Robinson, P. (1997). Can lipids from the freshwater alga *Selenastrum capricornutum* serve as an indicator of heavy metal pollution in freshwater environments? In *Physiology, Biochemistry and Molecular Biology of Plant Lipids*, ed. J.P. Williams, M.U. Khan & N.W. Lem, pp. 239–41. Dordrecht: Kluwer.
Robertson, E.J., Williams, M., Harwood, J.L., Lindsay, J.G., Leaver, C.J. & Leech, R.M. (1995). Mitochondria increase three-fold and mitochondrial proteins and lipids change dramatically in post-meristematic cells in young wheat leaves grown in elevated CO_2. *Plant Physiology*, **108**, 469–74.

Roughan, P.G. (1985). Phosphatidylglycerol and chilling sensitivity in plants. *Plant Physiology*, **77**, 740–6.

Russell, N.J. (1984). The regulation of membrane fluidity in bacteria by acyl chain length. In *Membrane Fluidity*, ed. M. Kates & L.A. Manson, pp. 329–47. New York: Plenum.

Rutter, A.J., Sanchez, J. & Harwood, J.L. (1993). Use of olive cultures to evaluate triacylglycerol synthesis. *Grasas y Aceites*, **44**, 130–2.

Rutter, A.J., Sanchez, J. & Harwood, J.L. (1994). The effects of pesticides on lipid synthesis in olive fruits and tissue cultures. *Biochemical Society Transactions*, **22**, 2595.

Sakaki, T., Kondo, N. & Sugahara, K. (1983). Breakdown of photosynthetic pigments and lipids in spinach leaves with ozone fumigation: role of active oxygens. *Physiologia Plantarum*, **59**, 28–34.

Sakaki, T., Ohnishi, J., Kondo, N. & Yamada, M. (1985). Polar and neutral lipid changes in spinach leaves with ozone fumigation. *Plant and Cell Physiology*, **26**, 253–62.

Sakaki, T., Tanaka, K. & Yamada, M. (1994). General metabolic changes in leaf lipids in response to ozone. *Plant and Cell Physiology*, **35**, 53–62.

Sanchez, J. (1994). Lipid photosynthesis in olive fruit. *Progress in Lipid Research*, **33**, 97–104.

Sanchez, J. (1995). Olive oil biogenesis. Contribution of fruit photosynthesis. In *Plant Lipid Metabolism*, ed. J-C.Kader & P. Mazliak, pp. 564–6. Dordrecht: Kluwer.

Sandelius, A.S., Carlsson, A.S., Pleijel, H., Hellgren, L.I., Wallin, G. & Sellden, S. (1995). The leaf acyl lipid composition of plants exposed to moderately enhanced levels of ozone: species, age and dose dependence. In *Plant Lipid Metabolism*, ed. J-C.Kader & P. Mazliak, pp. 459–61. Dordrecht: Kluwer.

Sandercock, S.P. & Russell, N.J. (1980). The elongation of exogenous fatty acids and the control of phospholipid acyl length in *Micrococcus cryophilus*. *The Biochemical Journal*, **188**, 585–92.

Sasaki, Y. & Konishi, T. (1997). Two forms of acetyl-CoA carboxylase in higher plants and effects of UV-B on the enzyme levels. In *Physiology, Biochemistry and Molecular Biology of Plant Lipids*, ed. J.P. Williams, M.U. Khan & N.W. Lem, pp. 32–4. Dordrecht: Kluwer.

Sasaki, Y., Hakamada, K., Nagano, Y., Furusawa, I. & Matsuno, R. (1993). Chloroplast-encoded protein as a subunit of acetyl-CoA carboxylase in pea plant. *Journal of Biological Chemistry*, **268**, 25 118–23.

Secor, J., Cseke, C. & Owen, W.J. (1989). The discovery of the selective inhibition of acetyl-CoA carboxylase by two classes of graminicides. *Proceedings of the Brighton Crop Protection Conference*, **1**, 145–54.

Seel, W.E., Hendry, G.A.F. & Lee, J.A. (1992). The combined effects of desiccation and irradiance on mosses from xeric and hydric habitats. *Journal of Experimental Botany*, **43**, 1023–30.

Selstam, E. (1980). Lipids, pigments, light-harvesting protein complex and structure of a virescent mutant of maize. In *Biogenesis and Function of Plant Lipids*, ed. P. Mazliak, P. Benveniste, C. Costes & R. Douce, pp. 379–83. Amsterdam: Elsevier.

Sheoran, I.S., Sawhney, V., Babbar, S. & Singh, R. (1991). *In vivo* fixation of CO_2 by attached pods of *Brassica campestris* L. *Annals of Botany*, **67**, 425–8.

Singh, D.P., Singh, P. & Sharma, H. (1986). Diurnal patterns of photosynthesis, evapotranspiration and water use efficiency in mustard at different growth phases under field conditions. *Photosynthetica*, **20**, 117–23.

Singh, R.P., Bharti, N. & Kumar, G. (1994). Differential toxicity of heavy metals to growth and nitrate reductase activity of *Sesamum indicum* seedlings. *Phytochemistry*, **35**, 1153–6.

Smirnoff, N. & Columbe, S.V. (1988). Drought influences the activity of enzymes of the chloroplast hydrogen peroxide scavenging system. *Journal of Experimental Botany*, **39**, 1097–108.

Smirnoff, N. (1993). The role of active oxygen in the response of plants to water deficit and desiccation. (Tansley Review no. 52). *New Phytologist*, **125**, 27–58.

Stumpf, P.K. (1984). Fatty acid biosynthesis in higher plants. In *Fatty Acid Metabolism and its Regulation*, ed. S. Numa, pp. 155–79. Amsterdam: Elsevier.

Tardiff, F.J. & Powles, S.B. (1993). Target site-based resistance to herbicides inhibiting acetyl-CoA carboxylase. *Proceedings of the Brighton Crop Protection Conference*, **2**, 533–40.

Telfer, A., Hodges, M. & Barber, J. (1983). Analysis of chlorophyll fluorescence induction curves in the presence of dichlorophenyldimethylurea as a function of magnesium concentration and NADPH-activated light-harvesting chlorophyll a/b protein phosphorylation. *Biochimica et Biophysica Acta*, **724**, 167–75.

Trémolières, A., Remy, R., Ambard-Breteville, F. & Dubacq, J.P. (1984). Role of phosphatidylglycerol containing *trans*-hexadecenoate in the supra-molecular organisation of the dimeric chlorophyll a/b protein complex. In *Structure, Function and Metabolism of Plant Lipids*, ed. P-A. Siegenthaler & W. Eichenberger, pp. 429–32. Amsterdam: Elsevier.

Trémolières, A., Garnier, J. & Dubertreit, G. (1992). Lipid protein interactions in relation to light energy distribution in photosynthetic membranes of eukaryotic organisms. In *Metabolism, Structure and Utilization of Plant Lipids*, ed. A. Cherif, D.B. Miled-Daoud, B. Marzouk, A. Smaoui & M. Zarrouk, pp. 287–92. Tunis: Centre National Pedagogique.

Tuquet, C., Guillot-Salomon, T.D., de Lubac, M.F. & Signol, M.

(1997). Granum formation and the presence of phosphatidylglycerol containing *trans*-3-hexadecenoic acid. *Plant Science Letters*, **8**, 59–64.

Turnham, E. & Northcote, D.N. (1983). Changes in the activity of acetyl-CoA carboxylase during rapeseed formation. *The Biochemical Journal*, **212**, 223–9.

de la Vega, M.G., Harwood, J.L. & Sanchez, J. (1992). Effect of dimethoate on lipid metabolism in olive fruits. In *Metabolism, Structure and Utilization of Plant Lipids*, ed. A. Cherif, D.B. Miled-Daoud, B. Marzouk, A. Smaoui & M. Zarrouk, pp. 368–71. Tunis: Centre National Pedagogique.

Vick, B.A. & Zimmerman, D.C. (1987). Oxidative systems for modification of fatty acids: the lipoxygenase pathway. In *The Biochemistry of Plants*, ed. P.K. Stumpf & E.E. Conn, vol. 9, pp. 53–90. New York: Academic Press.

Vigh, L., Los, D.A., Horvath, I. & Murata, N. (1993). The primary signal in the biological perception of temperature: Pd-catalyzed hydrogenation of membrane lipids stimulated the expression of the *desA* gene in *Synechocystis* PCC6803. *Proceedings of the National Academy of Sciences, USA*, **90**, 9090–4.

Wada, H., Gombos, Z. & Murata, N. (1990). Enhancement of chilling tolerance of a cyanobacterium by genetic manipulation of fatty acid desaturation. *Nature*, **347**, 200–3.

Wellburn, A.R. (1988). *Air Pollution and Acid Rain*. Harlow: Longman Scientific.

Wellburn, A.R., Robinson, D.C., Thompson, A. & Leith, I.D. (1994). Influence of summer episodes of ozone on Δ^5 and Δ^9 fatty acids in autumnal lipids of Norway spruce. *New Phytologist*, **127**, 355–61.

Williams, M. & Harwood, J.L. (1997). Effects of carbon dioxide concentration and temperature on lipid synthesis by young wheat leaves. *Phytochemistry*, **45**, 243–50.

Williams, M., Shewry, P.R. & Harwood J.L. (1994). The influence of the 'greenhouse effect' on wheat grain lipids. *Journal of Experimental Botany*, **45**, 1379–85.

Williams, M., Shewry, P.R., Lawlor, D.W. & Harwood, J.L. (1995). The effects of elevated temperature and atmospheric carbon dioxide concentrations on the quality of grain lipids in wheat grown at two levels of nitrogen. *Plant Cell and Environment*, **18**, 999–1009.

Williams, M., Robertson, E.J., Leech, R.M. & Harwood, J.L. (1998). Lipid metabolism in leaves from young wheat plants grown at two carbon dioxide levels. *Journal of Experimental Botany*, **49**, 511–20.

Wilson, R.F., Burke, J.J. & Quisenberry, J.E. (1987). Plant morphological and biochemical responses to field water deficits. *Plant Physiology*, **84**, 251–4.

Wolfenden, J. & Wellburn, A.R. (1991). Effects of summer ozone on membrane lipid composition during subsequent frost hardening in Norway Spruce. *New Phytologist*, **118**, 323–9.

Wolter, F.P., Schmidt, R. & Heinz, E. (1992). Chilling sensitivity of *Arabidopsis thaliana* with genetically engineered membrane lipids. *EMBO Journal*, **11**, 4685–92.

Yu, H. & Willemot, C. (1996). Inhibition of eukaryotic galactolipid biosynthesis in mature-green tomato fruits at chilling temperature. *Plant Science*, **113**, 33–41.

Yu, H. & Willemot, C. (1997). Chilling injury and lipid biosynthesis in tomato pericarp. In *Physiology, Biochemistry and Molecular Biology of Plant Lipids*, ed. J.P. Williams, M.U. Khan & N.W. Lem, pp. 221–3. Dordrecht: Kluwer.

Zarrouk, M., Seqqat-Dakhma, W. & Cherif, A. (1995). Salt stress effect on plant lipid metabolism in olive leaves. In *Plant Lipid Metabolism*, J-C.Kader & P. Mazliak, pp. 429–31. Dordrecht: Kluwer.

PART IV: Summary

S.W.J. BRIGHT and T.R. HAWKES

14 The future: industry requirements from advances in plant lipid research

Introduction

It is an halcyon time to be working as a plant scientist in industry or in the public sector. The great projects to map and sequence the human genome and model organisms will systematise and broaden the fundamental understanding of biological processes, as well as allow the widespread modification of plants in directed and beneficial ways for use as food, feed, fibre and industrial feedstocks in the twenty-first century. The first genetically modified plant products have now reached the market in the USA, Europe and Australia. It is clear that research now will lead to products in abundance in the future.

The particular area of plant lipid research has already been leading the way in both capturing advances in fundamental understanding and attracting the attention of companies who want to generate and utilise this understanding for making products. The broad range of the contributions in this volume, and their strong industrial content, attests to the timeliness of this synthesis.

A summary of the precise industry requirements for the future is doomed to failure; only hindsight will identify all the gems to be found. However, the requirements for the future can be broadly classified into three areas: products; exploitable knowledge and fundamental science. An attempt is made in this chapter to put flesh on these bones by focusing on specific examples in the area of plant lipid research, although with the hope that the points made may have more generic validity.

Industry needs competitive products

The flow of new products is vital for success in any industry. This is a key competitive factor, especially in technology-intensive areas such as biotechnology. For research scientists outside a particular company it is difficult or impossible to get the whole picture. Decisions are based upon integration of information relating to the technology (feasibility, costs, benefits) with that coming from customers and markets (Fig. 1).

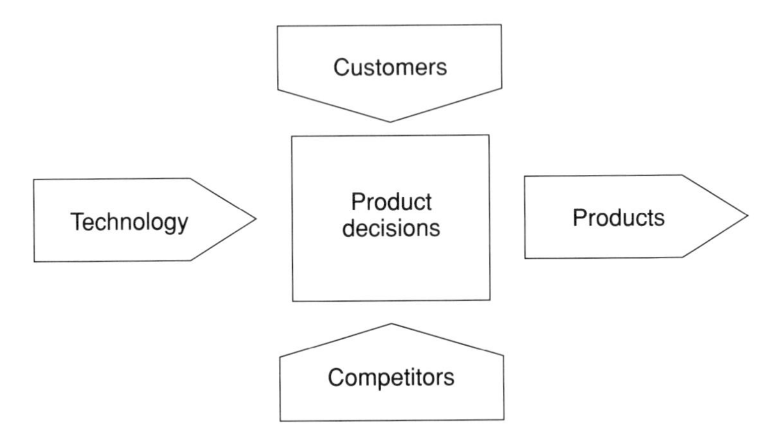

Fig. 1. Product decisions integrate market and technology information.

The customer and market information is used to estimate the size of the business opportunity and its timescales (launch dates, lifetimes). A key consideration is the activity of competitors both now and in the future.

There is a good deal of market information available in published form. Market surveys and analysis of the opportunities specifically in the area of plant lipids have been carried out (e.g. Murphy, 1993) as well as government-sponsored surveys and the formation of coordination activities for specialised aspects such as ACTIN, a network for alternative crops information (ACTIN, 1997). This is, however, an area where at best an imperfect understanding of future opportunities is available; the area of product development is best left to companies. The exception for academic innovators is where the technology developed allows the formation of a specialised start-up company. This is common in the pharmaceutical and health-care industry, but less so in the plant-based industries.

Two other factors need to be understood in trying to make an impact from an academic position. First, product decisions can be high risk; the future is difficult and unpredictable even for the most well-informed companies. They are much happier writing the history of successful products than displaying their future plans. Decisions are subject to sudden changes as particular market or competitor information changes or is re-interpreted. This is difficult to factor into a carefully measured research programme. Secondly, the customer, whilst always being right, may not always appreciate the technology being offered. This is particularly timely to remember at the present time as the USA has been broadly accepting of genetically modified products, whereas the European reaction has been quite heated in political circles, and this could impact dramatically on the future product decisions being made now. Again, the climate can change rapidly and have a big influence on products being developed without this having a logical technological basis. For all these reasons, it is prudent for the longer-term researcher to be wary of too great a product focus; well thought-out areas of research will come into their own, while specific targeted research may come and go on a much shorter timescale.

Industry needs exploitable knowledge

If it is accepted that product decisions are best left to the industry members themselves, then how can the wealth of scientific understanding be best translated into practical application? The plant lipid area provides several examples of how this can occur through the generation of exploitable knowledge. Before moving to specific application areas, it is worth noting in passing two characteristics of exploitable knowledge. The knowledge must be both understandable and accessible to the user.

Being understandable is a challenge for all scientists working in a public context. There are specific challenges also in working with a particular user community where the level of scientific understanding will vary enormously between the different functional groupings. In addition, the familiarity with the latest terminology and results might be sketchy at best. As an example, a clear exposition of the state of the art in lipid synthesis by Ohlrogge and Browse (1995), is provided in a diagram incorporating the genetic and biochemical steps as well as an indication of flux through the pathway. Such clarity can prove enormously helpful. As arcane terminology proliferates, the need for clear interpretation of the state of the art becomes ever more pressing. This is a particular hurdle for smaller companies which have only limited resources for high science, but can offer excellent routes of exploitation.

Accessibility is a further hurdle. Many institutions now have arrangements for technology transfer and the protection of intellectual property. Putting the whole package together can be still quite daunting as well as time-consuming. The lack of precedent in licensing inventions in the plant science area does lead to problems of perception of the value of inventions, given the multi-component nature of any genetically modified product.

Plants as oil factories

Plants are the only renewable resource for the production of oils and other hydrocarbons. The end uses of plant oils can be roughly classified into foods, industrial feedstocks and speciality oils (which might be food or feedstocks). At present, the exploitable understanding of plant lipids has been predominantly gained in the area of changing the structure of the fatty acid components of the harvested plant part. Essentially this addresses the aspect of quality (the value of the product to specific end users). A second area which is yet to be so amenable is the overall yield. Yield of the crop translates very directly into cost of production. Coal and oil are the price yardsticks if in-roads are to be made in the area of industrial feedstocks or energy crops.

Spectacular progress has been made in identifying the genes specifying the enzymes concerned in fatty acid synthesis and desaturation. In spite of its diminutive size, *Arabidopsis* has proved to be a key organism for this work. The identification of mutants and the cloning of the corresponding genes has been groundbreaking in fundamental science as well as allowing exploitation (Ohlrogge & Browse, 1995). Progress in this area is detailed elsewhere in this volume; the general conclusion is that the control of fatty acid chain length and desaturation is now, in principle, open directly to experimental modification. Practical problems remain in getting the specified fatty acids produced and stored in good quantities.

The chain length of fatty acids has important effects on functionality. The ability to specify chain length in a single crop rather than growing a range of crops, which have been identified as having pre-existing storage lipids of the required chain length, is potentially a major agent for change in agriculture. For instance, 12-carbon fatty acids for detergent manufacture could now be sourced from the traditional palm oil in tropical areas or from temperate crops such as soya with the appropriate genetic modification. Clearly, this can have a beneficial effect on price and supply stability, whereas the effects on traditional suppliers might appear less rosy.

The enzymes of fatty acid desaturation have proved difficult to work with at a biochemical level. Now that many of them are available as genes, this limitation is removed; production in heterologous systems allows quantities of pure protein, and the expression in novel combinations allows *in vivo* experimentation to define the limits of what can be made in plants. The range of specialised plant lipids with particular desaturation patterns is being systematised into design rules (Chapter 6). This will allow the production of existing plant lipids in more convenient host plants, for instance oilseed rape instead of jojoba (Chapter 12). More exciting, is the prospect of producing novel lipids in quantities sufficient for testing of the functionalities in a systematic way. In this regard, the isolation of the gene specifying hydroxylation during castor oil biosynthesis has brought forward the prospect of *in vivo* chemical modification (Broun & Somerville, 1997).

The next great target for oil modification is to achieve understanding of the genetics of yield. Historically, breeding for yield has made steady gains by the accumulation of small increments of overall plant yield. Specific quantitative trait loci governing oil yield have been identified through marker assisted breeding programmes (Diers *et al.*, 1992). The molecular basis of these genes is not yet understood. An interesting example of exploitable knowledge is in the area of high oil corn. Classical cyclical selection on the basis purely of high or low oil over multiple generations was undertaken by Alexander and co-workers (Dudley, Lambert & Alexander, 1974). Two points are worthy of mention. First, the direct exploitation of these genetic stocks, available 20 years ago, has only just begun in the last few years with commercial 'TopCross' hybrids from Dupont (Chapter 11). Secondly, the understanding of what the selection has done has indicated that it is easier to change the size of the oil sink in corn (the embryo) than it is to alter the amount of oil storage as a proportion of the sink. This leaves open the question, of very real practical applicability, of how to increase the amount of storage lipid which is accumulated during a plant growing season.

The frontiers at the moment in exploiting the existing knowledge base for the use of plants as oil factories are three-fold. First, the first modified crops with altered oil quality (chain length and desaturation patterns) are just now entering the market-place; their commercial success will stimulate further efforts. Secondly, the challenge is to convert all the available new science already in existence into a steady stream of products. Thirdly, the ability to make new speciality crops will stimulate new product development. The beginnings of this is already seen with the advent of very high oleic soya (Chapter 11). The yield of oils from these modified crops will be crucial for economic success; and if yield can be improved by genetic modification, then perhaps the range of renewable industrial feedstocks can be broadened.

Two contrasting trends will be played out in the coming years. On the one hand, the ability to modify the synthetic apparatus of soya and other temperate-adapted oil seed crops might tend to make them the preferred source of oils. On the other hand, the genetic modification of exotic or niche plant sources of oils could be used to make their cultivation more simple, routine and economic within the agricultural systems currently in place. Both avenues will certainly be explored. In the same vein, the conversion of speciality or niche products which are of high cost to low cost, could turn specialities into commodities. The trick for the commercial producers will be to manage this technology and market interface successfully.

Targets for agrochemical discovery

Agrochemicals have contributed a significant proportion of the improvements in productivity which have been seen over the last 50 years. Commercially successful active ingredients have all been discovered starting from the traditional approach of random screening. However, this will change as improvements in the knowledge base begin to afford new opportunities. Thus, in parallel with trends in the pharmaceutical industry, it is likely that approaches to agrochemical discovery will, in future, become more targeted and 'rational'. Applying the best available scientific understanding will be likely to lead to an improved rate of discovery of useful active ingredients.

Lipid synthesis is a metabolic activity of fundamental importance and it is, perhaps, not surprising that inhibitors should include compounds of potent antibiotic (e.g. the antitubercular 'isoniazid' which inhibits an acyl carrier protein (ACP)-enoyl reductase of mycobacterium, Banerjee *et al.*, 1994) and antifungal (e.g. the polyketide 'soraphen' which inhibits fungal acetylCoA carboxylase, Vahlensieck *et al.*, 1994) as well as herbicidal activity . The major interest in the agrochemical area arose from the discovery of two major classes of grass-killing herbicides (the aryloxyphenoxypropionates and the cyclohexanediones), which were both discovered to act through selective inhibition of the acetylCoA carboxylase enzyme of monocotyledonous but not dicotyledonous plants (*cf* Gronwald, 1991; Harwood, 1991). Current understanding of this area is covered in another chapter of this volume (Chapter 2).

Molecular understanding of the pathways and enzymes involved in fatty acid biosynthesis in plants and fungi is leading to opportunities for new screens, to rational design programmes, is creating options for breeding resistance into crops and for managing pesticide resistance in target organisms. Here, an attempt will be made particularly to highlight the implications for further discovery of agrochemicals arising from this knowledge.

The potential for rational drug design is aided immensely by the availability of protein crystal structures and, in particular, where these structures include bound cofactors, substrates, analogues and inhibitors. As yet, in the plant area, only the structure of the ACP-enoyl reductase component of fatty acid synthase is known (Rafferty *et al.*, 1995) whereas in the microbial area the structures of the *E.coli* malonyl CoA – ACP transacylase (Serre *et al.*, 1995), the structure of the *E. coli* ACP (Holak *et al.*, 1988) as well as those of the ACP-enoyl reductases from *Mycobacterium* (Quemard *et al.*, 1995) and *E. coli* (Baldock *et al.*, 1996) have been solved. In the latter case, the structure revealed the mechanism of action of an experimental antibiotic, diazaborine (Bergler *et al.*, 1994). Consistent with the inhibition studies of Kater *et al.* (1994), the structure revealed that the antibiotic inhibits the enzyme by forming a covalent complex with the oxidised pyridine nucleotide cofactor within the enzyme site. This level of detailed structural information is of obvious potential use for the design of further antibiotics. In the herbicide area, information of similar quality would clearly be useful and industry would thus encourage work to clone, express and solve the structures of other fatty acid synthase components from plants.

A prerequisite for crystallisation studies is the availability of large quantities of pure enzyme from either the native source or produced in recombinant strains. The enzymes produced in this way can also be used in *in vitro* screens for chemical inhibitors. In the pharmaceutical industry such screens are used routinely as the starting point for new chemical synthesis and optimisation for biological activity and specificity. Probably most agrochemical companies carry out some *in vitro* screening (*cf* Ormrod & Hawkes, 1995), although, as yet, no product has been claimed to have originated from this route.

If the pharmaceutical example is a valid parallel, then the rates of screening for new agrochemical compounds and the throughput of exquisitely small sample amounts will increase rapidly over the next few years (cf Leach, 1997). The availability of large numbers of screens based on pure expressed target enzymes or cells will be a key resource for this activity. Systems for validation and rapid evaluation of candidate target sites may then be at a premium.

After the identification and commercialisation of an active ingredient, there are two further avenues where the fundamental science can lead to exploitable knowledge. These are in the areas of resistance and resistance management.

With the identification of the molecular target for a chemical in an organism comes the opportunity to develop resistance. This can be useful if it is directed at alteration of the selectivity of the compound between target pest and crop. Alternatively, if the target organism develops resistance, then the efficacy of the compound is compromised. Several mechanisms are known by which organisms can acquire resistance, such as alteration of the target site, degradation of the chemical compound or alteration of the access of compound to the target. In the case of the ACCase inhibitors, target site changes have been identified in maize (Marshall *et al.*, 1992) leading to the development of resistant hybrids and also in many weed species (cf Evenson, Gronwald & Wyse, 1994). The availability of molecular understanding can help in the design of resistance management strategies for controlling the development and spread of resistant target organisms. Examples of this are in the monitoring of resistant aphid populations (Devonshire & Field, 1991) as well as in profiling the genetic changes responsible for resistance to atrazine and sulphonyl urea herbicides (Gasquez, 1991; Hartnett *et al.*, 1991).

Diet and health

With the realisation of the goal of food security which has been established in the developed world since the last World War, food sufficiency has become a low priority issue. (Alas this is not true in many areas of the developing world as the TV screens show on a regular basis.) Issues that remain and are increasing in importance with the changing demographic profile of developed world economies are ones of diet and health. Fats and oils are in the forefront of this area in both positive and negative associations. Three aspects of the impact of scientific research relating to diet and health are worthy of mention in this context.

First, the current commercial goals of 'light and low' are all focused on total intake of calories, especially fats and oils. Labelling and marketing of such products is ubiquitous if not always elegant. A particularly bad example is the use of terms such as '97% fat free' which begs a number of questions as to the understanding of both the English language and science by customers and marketeers. The approach is scattergun; all sections of the population are targeted; this can lead to problems with particular subgroups with different dietary requirements.

Secondly, the scientific basis for the effects of dietary fats and oils is now being established. Thus the effects of saturated, mono-unsaturated and poly-unsaturated fats are being distinguished in terms of their effects on fat metabolism and cholesterol metabolism in humans (Aro *et al.*, 1997; Harris, 1997). The design of 'healthy oil' products is feasible, indeed the margarine adverts make subtle but significant allusion to this in their battle with butter. However, much remains to be done to be clear on the relative health effects of the different chain lengths and saturation patterns of plant-based oils and fats (as well as the chemical modifications within the food industrial processes).

Thirdly, the science surrounding the analysis of the effects of diet and nutrition within specific subgroups of the population and the analysis of complex interactions amongst food components is proceeding apace. Real opportunities exist for tailoring the diet, for instance for an ageing population, to provide an extension of healthy lifespan and an increased quality of life. Specific current examples are the identification of active ingredients of evening primrose oil (*gamma*-linolenic acid) and of fish oils (n-3 fatty acids) with a metabolic rationale for their effects (Horrobin, 1992; Harris, 1997). Undoubtedly, this is an area of great promise for the future, but one where the scientific basis of health claims will be the subject of considerable scrutiny and debate. The specific effects might be expected to be targeted at subpopulations classified by age, gender, health status or even, more controversially, genetic profile. Such aspects can be anticipated to arise from the 'application fallout' of the human genome programme.

Gene control regions as exploitable knowledge

The advent of relatively routine experimental genetic modification of crop plants has lead to a broad requirement for enabling technology. By this, is meant the technologies and methods to provide a final working plant variety or line with a particular modification. Along the way, this needs efficient methods for gene transfer, identification of transgenic cells and plants and the means to control the activity of inserted genes to give the desired effects. This latter activity encompasses the use of particular promoters, enhancers and other tricks to achieve expression of the inserted gene at the right time and place at the desired level. Current understanding of the fundamental regulation of seed-specific expression of transgenes is fairly rudimentary; some seed-specific transcription factors such as *opaque* and SPA (Albani *et al.*, 1997) are being studied to begin the unravelling of the signal pathways involved. Meanwhile, there is always a need for more and better ways to achieve the targeted expression through the use of the best promoters and other gene regulatory systems. In the future, it is likely that the expression of pathways currently limited to specialised production cells such as glandular hairs, will be amenable to manipulation as the molecular events underlying such expression are understood.

Industry needs an excellent science base

Over the last decade, pressure for applicability and relevance has been felt by the institutes and universities, which provide the bulk of the research drive for any nation. Thus, within the UK, government white papers and initiatives (e.g. 'Realising Our Potential' and the Technology Foresight Programme), as well as the formal assessment of technology interaction in the calculation of institute and university income from government sources are all seen as driving in the same, applied, direction at the expense of pure, curiosity-driven research and researchers. In the USA, the manifestations of this drive are different, for instance, the opportunities and cultural drive to set up new companies are very much more pervasive in the field of biotechnology, and the government programmes to support application are different in structure. But, the feeling from the science base is similar.

However, this debate always has to be about balance. Industry definitely requires a flourishing fundamental science base. First is the requirement for a steady flow of new highly trained recruits to refresh and populate the R&D and other ranks of the technically sophisticated companies of today and tomorrow. It is not possible to specify, except in the broadest way, what training is required, it just has to be excellent. One aspect of this is the capacity to be flexible to be able to seize new developments as they happen.

Secondly, the timescales of product development at 7–12 years in the case of the current products from plant biotechnology, mean that the research agenda is extremely broad; we know that application will happen, but an industrial-style research structure is less likely to generate true novelty on this long time horizon.

Finally, there are some difficult and fundamental questions of direct practical relevance, which need answers by the application of the best fundamental science to their solution. In the area of plant lipids, three such questions might be:

(i) What determines the partitioning of carbon between lipid, starch and protein in the major crops and why do they differ?

(ii) What factors are involved in generating the storage compartment capacity for storage lipids, and what are the constraints on packing in strange and exotic lipids with particular functional properties?

(iii) What are the mechanisms and signals which sense the flux of intermediates to regulate the synthesis and breakdown pathways? Undoubtedly, the solution of such problems will generate new fundamental science and useful applications as well as bringing satisfaction to the curious.

Conclusion

Since the mid-1980s, possibly dating from the seminal publication of Browse, McCourt and Somerville (1985), the area of plant lipid research has turned from an academic specialism on the edge of the limelight to one at the forefront of modern plant science and plant biotechnology. The contributions in this volume attest to the pace of the change and the progress that has been made, both in the area of fundamental understanding and applications. Plant lipid research is now a proving ground for the new technologies.

Much remains to be done; challenges for the new generation of curious scientists exist in abundance, so the area will remain exciting and competitive. It is indeed a timely moment to have taken stock and looked both backwards and forwards in this meeting and in the proceedings which have arisen from it.

References

Albani, D., Hammond-Kosack, M.C.U., Smith, C., Conlan, S., Colot, V., Holdsworth, M. & Bevan, M.W. (1997). *The Plant Cell*, **9**, 171–84.

ACTIN (1997). Alternative Crops Technology Interaction Network website: **http:// www. actin.co.uk**

Aro, A., Jauhianen, M., Partenen, R., Salminen, I. & Mutanen, M. (1997). Stearic acid, *trans* fatty acids and dairy fat: effects on serum and lipoprotein lipids, apolipoproteins, lipoprotein(a) and lipid transfer proteins in healthy subjects. *American Journal of Clinical Nutrition*, **65**, 1419–26.

Baldock, C., Rafferty, J.B., Sedelnikova, S.E., Baker, P.J., Stuitje, A.R., Slabas, A.R., Hawkes, T.R. & Rice, D.W. (1996). A mechanism of drug action revealed by structural studies of enoyl reductase. *Science*, **274**, 2107–10.

Banerjee, A., Dubnau, E., Quemard, A., Balasubramanian, V., Sun Um, K., Wilson, T., Collins, D., de Lisle, G. & Jacobs, Jr, W.R. (1994). *inhA*, a gene encoding a target for isoniazid and ethionamide in *Mycobacterium tuberculosis*. *Science*, **263**, 227–30.

Bergler, H., Wallner, P., Ebeling, A., Leitinger, B., Fuchsbichler, S., Aschauer, H., Kollenz, G., Hogenauer, G. & Turnkowsky, F. (1994). Protein EnvM is the NADH-dependent enoyl ACP reductase (Fab I) of *Escherichia coli*. *Journal of Biological Chemistry*, **269**, 5493–6.

Broun, P. & Somerville, C. (1997). Accumulation of ricinoleic, lesquerolic and densipolic acids in seeds of transgenic *Arabidopsis* plants that express a fatty acyl hydroxylase cDNA from castor bean. *Plant Physiology*, **113**, 933–42.

Browse, J.A., McCourt, P.T. & Somerville, C.R. (1985). A mutant of *Arabidopsis* lacking a chloroplast-specific lipid. *Science*, **227**, 763–5.

Devonshire, A.L. & Field, L.M. (1991). Gene amplification and insecticide resistance. *Annual Reviews of Entomology*, **36**, 1–23.

Diers, B.W., Keim, P., Fehr, W.R. & Shoemaker, R.C. (1992). RFLP analysis of soybean seed protein and oil content. *Theoretical and Applied Genetics*, **83**, 608–12.

Dudley, J.W.R., Lambert, R.J. & Alexander, D.E. (1974). Seventy generations of selection for oil and protein concentration in the maize kernel. In *Seventy Generations of Selection for Oil and Protein Concentration in the Maize Kernel*, ed. J.W. Dudley, pp. 181–212, Crop Science Society of America Special Publication.

Evenson, K.J., Gronwald, J.W. & Wyse, D.L. (1994). Purification and characterisation of acetyl CoA carboxylase from diclofop-resistant and susceptible *Lolium multiflorum*. *Plant Physiology*, **105**, 671–80.

Gasquez, J. (1991). Mutation for triazine resistance within susceptible populations of *Chenopodium album*. In *Herbicide Resistance in Weeds and Crops*, ed. J.C. Caseley, G.W. Cussans & R.K. Atkin, pp. 103–13. Oxford: Butterworth-Heinemann, Ltd.

Gronwald, J.W. (1991). Lipid biosynthesis inhibitors. *Weed Science*, **39**, 435–49.

Harris, W.S. (1997). n-3 fatty acids and serum lipoproteins: human studies. *American Journal of Clinical Nutrition*, **65S**, 1645–54.

Hartnett, M.E., Chui, C-.F., Falco, S.C., Knowlton, C.J., Mauvais, C.J. & Mazur, B.J. (1991). Molecular analysis of sulfonylurea herbicide resistant ALS genes. In *Herbicide Resistance in Weeds and Crops*, ed. J.C. Caseley, G.W. Cussans & R.K. Atkin, pp. 343–53. Oxford: Butterworth-Heinemann, Ltd.

Harwood, J.L. (1991) Herbicides affecting chloroplast lipid synthesis. In *Herbicides*, ed. N.R. Baker & M.P. Percival, pp. 209–46. B.V.: Elsevier Science Publishers.

Holak, T.A., Kearsley, S.K., Kim, Y. & Prestergaard, J.H. (1988). *Biochemistry*, **27**, 6135–42

Horrobin, D.F. (1992). Nutritional and medical importance of *gamma*-linolenic acid. *Progress in Lipid Research*, **31**, 163–94.

Kater, M.M., Koningstein, G.M., Nijkamo, H.J.J. & Stuitje, A.R. (1994). The use of a hybrid genetic system to study the functional relationship between prokaryotic and plant multi-enzyme fatty acid synthase complexes. *Plant Molecular Biology*, **25**, 771–90.

Leach, M. (1997). Discovery on a credit card? *Drug Discovery Today*, **2**, 253–4.

Murphy, D.J. (ed.) (1993). *Designer Oil Crops*. Weinheim: VCH Press.

Ohlrogge, J. & Browse, J. (1995). Lipid biosynthesis. *Plant Cell*, **7**, 957–70.

Ormrod, J.C. & Hawkes, T.R. (1995). Screening practices in the agrochemical industry. *Proceedings of the Brighton Crop Protection Conference*, vol. 1, pp. 97–108.

Quemard, A., Sacchettini, J.C., Dessen, A., Vilcheze, C., Bittman, R., Jacobs, W.R. & Blanchard, J.S. (1995). Enzymatic characterisation of the target for isoniazid in *Mycobacterium tuberculosis*. *Biochemistry*, **34**, 8235–41.

Rafferty, J.B., Simon, J.W., Baldock, C., Artymiuk, P.J., Baker, P.J., Stuitje, A.R., Slabas, A.R. & Rice, D.W. (1995). Common themes in redox chemistry emerge from X-ray structure of oilseed rape (*Brassica napus*) enoyl acyl carrier protein reductase. *Structure*, **3**, 927–38.

Serre, L., Verger, E.C., Dauter, Z., Stuitje, A.R. & Derewanda, Z.S. (1995). The *Escherichia coli* malonyl-CoA: acyl carrier protein transacylase at 1.5 Å resolution. *Journal of Biological Chemistry*, **270**, 12 961–4.

Vahlensieck, H.F., Pridzun, L., Reichenbach, H. & Hinnen, A. (1994). Identification of the yeast ACC1 gene product (acetyl CoA carboxylase) as the target of the polyketide fungicide soraphen A. *Current Genetics*, **25**, 95–100.

Index

Note: page numbers in *italics* refer to figures and tables.